AI-Driven Smart Industrial Technologies

Maki K. Habib
The American University in Cairo, Egypt

IGI Global
Scientific Publishing
Publishing Tomorrow's Research Today

Vice President of Editorial Melissa Wagner
Director of Acquisitions Mikaela Felty
Director of Book Development Jocelynn Hessler
Production Manager Mike Brehm
Cover Design Jose Rosado

Published in the United States of America by
IGI Global Scientific Publishing
701 East Chocolate Avenue
Hershey, PA, 17033, USA
Tel: 717-533-8845
Fax: 717-533-7115
Website: https://www.igi-global.com E-mail: cust@igi-global.com

Library of Congress Cataloging-in-Publication Data

LCCN: 2025008670 (CIP Data Pending)
ISBN13: 9798369379943
Isbn13Softcover: 9798369379950
EISBN13: 9798369379967

British Cataloguing in Publication Data
A Cataloguing in Publication record for this book is available from the British Library.

Table of Contents

Detailed Table of Contents

This chapter highlights the broader implications of adopting 3D CNNs in smart manufacturing. By automating the classification of complex workpieces, industries can achieve higher precision, reduce reliance on manual labor, and enhance production efficiency. Additionally, the integration of voxel-based classification systems into IoT-enabled smart factories is discussed is explored, presenting a pathway toward more intelligent and adaptive manufacturing ecosystems. This chapter also provides a practical guide to implementing 3D CNNs in industrial settings, Addressing both technical and real-world considerations. It contributes to the advancement of AI-driven smart technologies and serves as a valuable resource for researchers and practitioners in smart manufacturing.

This chapter focuses on the use of AI, IoT, and remote sensing technologies in precision agriculture. This explores how the technologies can be used in collaboration to benefit the improvement of yield in agriculture, efficiency in resource use and sustainable utilization of resources. Among them, current practices consist on crop health monitoring, yield prediction, precise irrigation and correct soil management are presented where they are currently being used and the latest applied technologies are showed in the following chapter. Also, it defines the difficulties associated with these solutions: their cost, data compatibility issues, and the necessity of qualified staff. The summarization of the potential of AI and IoT as well as the remote sensing in the farming industry is also provided as well as a focus on the future trends like edge computing and 5G connectivity. It outlines the matters that should be considered by the policy makers and the need for collaboration among various fields as a way of enhancing the uptake of these technologies.

This chapter explores the application of deep learning models, specifically U-Net and YOLOv3, in detecting crack-type defects in industrial materials. The research addresses key challenges, such as the limited availability of defective data samples, the difficulty of distinguishing fine defect details, and the need for precise quantitative evaluations in automated inspection systems. The chapter introduces the significance of defect detection in the context of smart manufacturing and discusses the limitations of traditional manual inspection methods. U-Net, known for its superior pixel-level semantic segmentation capabilities, is utilized for detailed defect localization, while YOLOv3, a real-time object detection model, is optimized for fast detection but has trade-offs in precision, particularly for irregular defect shapes. Both models are trained on a custom dataset provided by a material manufacturer incorporating a novel augmentation method that enhances generalization and improves detection accuracy by expanding defective sample diversity.

Sergei Jatsun, South-West State University, Russia
Jamil Safarov, University of Technology of Azerbaijan, Ganja, Azerbaijan
Egor Tistsov, South-West State University, Kursk, Russia
Dmitrii Efremov, South-West State University, Russia
Gavriil Fursov, South-West State University, Russia

This chapter per the authors reveal the problem of the complexity of selecting the fired charge from the furnace and propose automation of this. They talk about the development of a robot for automatic charge selection, as well as modeling its bucket using a fifth-order polynomial in analog and discrete spaces. The calculation results are presented, as well as graphs of the position, speed, acceleration, torque, and changes in motor current when the bucket is moved. The chapter also contains a description of the sampler job and its main parts, such as a control cabinet, a manipulator arm, and a laser protective barrier. This chapter describes the intended modes of operation of the charge sampling robot, such as timer operation and single sampling.

Alessandro Massaro, Department of Engineering, LUM University, Bari, Italy

The proposed chapter provides a methodology to design electronic circuits and Digital Twin (DT) models in Industry 5.0 environments. Different examples are discussed, by combining the electrical and the electronic behavior with the Artificial Intelligence (AI) algorithms upgrading the DT model about the signal prediction. The analyzed DT models are related to important elements of industrial machines such as filters, amplifiers, sensors, switching and control systems. The goal is to illustrate the setting of open source tools able to implement AI-DT complex models simulating real environment industrial conditions including the disturb effects, noises and interferences of the electronic components.

Nowadays, with the increasing digitalization and widespread use of Internet of Things (IoT) applications, accurate and reliable indoor positioning system applications have become increasingly important. With the increasing need in the areas of use, researchers have proposed various indoor positioning system technologies and methods. In this context, Bluetooth Low Energy (BLE) technology offers a cost-effective, energy-efficient, and easy-to-integrate solution within existing infrastructures. The integration of BLE technology and ANNs for indoor positioning will be the subject of this section, which will cover the theoretical basis and practical applications. A comprehensive review of BLE-based positioning algorithms, performance evaluations, and the design and training of ANN models will be presented. In addition, the limitations of existing systems will be highlighted through the examination of the challenges. The aim of this study is to make a meaningful contribution to the field of indoor positioning and to provide applicable solutions in both academic and industrial environments.

Recently, digital twin has been discussed with the integration of information, communication, intelligence, and robot technologies. In order to conduct multiscale and multiphysics simulations on a real-world phenomenon in the cyber world based on digital twin, topological features and structures should be extracted from given or measured big data. Therefore, the concept of topological twin was proposed to complementarily use digital twin. The physical dynamics is dealt with at the microscopic level, while the spatiotemporal qualitative relationships are used between objects, people, culture, and knowledge at the macroscopic level. Furthermore, a mesoscopic integration method is required to connect microscopic and macroscopic topological features. This chapter explains the multiscopic concept of topological twin and shows an application example to path planning problems and navigation tasks of mobility support robots. Finally, this chapter discusses the applicability and future direction of topological intelligence to Industry 5.0 and Society 5.0.

The chapter explores the various types of equipment failure, techniques to discover the root cause, and different approaches to find the solution to predict the event of the failure. Moreover, this work addresses the equipment failure prediction on the assembly line of truck wheel manufacturing plant. The step-by-step process of the wheel production includes one after another usage of the machines. Any workstation failure leads to the idle time of the subsequent machines and reduction in the working efficiency of the labor. The sensor-based data of the machine is collected using the process automation. Historical data is used to predict the remaining useful life. K-Nearest Neighborhood technique has given the accuracy of 96.40%. The Decision Tree with Gini Index and Entropy have has achieved the 96% accuracy in the machine state prediction. The multi-regression based model developed to calculate the values of coefficient and intercept as learning parameters to be useful in the further deployment.

This chapter introduces an innovative methodology for error and attenuation detection in network services by unifying the REFII time series model with Bayesian networks (BNs), and complementary data science techniques traditionally applied outside temporal analytics. The proposed framework addresses the critical challenge of identifying rare, high-impact events—such as signal degradation in mobile or fixed networks—that compromise service quality but are often obscured by complex temporal dependencies and imbalanced data distributions. This transformation facilitates non-temporal expansion, a novel interpolation process that enriches time series with contextual events and operational parameters lacking inherent temporal markers. Proposed solution unites advanced analytical techniques with decision making process for pattern mining like Bayesian networks, decision trees, FTP, K-Means clustering.

The convergence of Digital Twins and Industry 4.0/5.0, driven by advancements in artificial intelligence, offers transformative potential for smart industrial technologies. This chapter explores how Digital Twins, virtual representations of physical assets, enhance real-time monitoring, predictive maintenance, and operational optimization. The objective is to bridge interdisciplinary research gaps by fostering collaboration across AI, IoT, and sustainability domains. The methodological approach involves a comprehensive literature review, analysis of existing implementations, and case studies demonstrating best practices. Key findings highlight the role of Digital Twins in improving efficiency, reducing environmental impact, and supporting decision-making. The chapter concludes with recommendations for adopting Digital Twins in next-generation industrial practices and outlines future research directions in AI-driven sustainable technologies.

Preface

We are witnessing a pivotal transformation in how industries operate, propelled by the integration of artificial intelligence (AI) with cutting-edge smart technologies. *AI-Driven Smart Industrial Technologies* emerges from this context—where innovation is not only a catalyst for enhanced performance and productivity but also a path toward achieving sustainability and environmental responsibility in industrial domains.

The conception of this volume was inspired by a growing global imperative: to align technological progress with sustainable development. From the advent of Industry 4.0, which ushered in digitalization, to the evolving paradigm of Industry 5.0, which emphasizes human-centric and resilient systems, it is clear that AI and intelligent technologies are not merely tools—they are enablers of a smarter, greener industrial future.

This book brings together an interdisciplinary collection of research and applied knowledge that reflects the fusion of mechatronics, robotics, AI, IoT, and advanced manufacturing technologies. Each chapter showcases how these technologies are being deployed to address critical industrial challenges, optimize resource consumption, reduce waste, and enhance energy efficiency. From predictive maintenance to autonomous systems, from additive manufacturing to digital twins, the contributions highlight how innovation can harmonize with sustainability.

In assembling *AI-Driven Smart Industrial Technologies*, we bring together diverse and cutting-edge research that showcases the transformative potential of AI and smart systems across a range of industrial applications. The chapters in this volume collectively reveal the vital role of artificial intelligence in reshaping industrial operations—enhancing efficiency, sustainability, precision, and adaptability.

CHAPTER OVERVIEW

Chapter 1: 3D CNNs for Smart Manufacturing: Voxel-Based Classification of Industrial Workpieces

This opening chapter sets the stage by demonstrating how 3D Convolutional Neural Networks (CNNs), integrated with voxel-based modeling, are revolutionizing smart manufacturing. The authors present a compelling case for the automation of complex workpiece classification, offering both technical guidance and practical insights. Their work reveals how integrating AI with IoT infrastructure can lead to adaptive and intelligent production lines with minimized manual intervention.

Chapter 2: AI-Based Solutions for Future Trends in IoT and Remote Sensing Integration for Precision Agriculture

Expanding the industrial conversation into the agricultural domain, this chapter explores the confluence of AI, IoT, and remote sensing in optimizing precision farming. From crop health monitoring to intelligent irrigation, the authors highlight real-world implementations and future-focused technologies like edge computing and 5G. Importantly, the chapter also acknowledges the challenges—such as costs and data integration—offering a balanced roadmap for policymakers and technologists alike.

Chapter 3: AI-Powered Defect Detection Using Deep Learning: Enhancing Industrial Automation with U-Net and YOLOv3

This chapter delves into the integration of deep learning for quality assurance, showcasing how models like U-Net and YOLOv3 can effectively detect material defects. The authors offer innovative solutions to overcome limited training datasets and demonstrate how customized augmentation techniques can enhance model generalizability—making this a significant contribution to the field of AI-powered industrial inspection.

Chapter 4: Automation of the Process of Charge Sampling at Metallurgical Enterprises Using Robots

Focusing on robotic automation in metallurgical environments, this chapter details the design and operation of a robotic system for charge sampling. Through mechanical modeling and control system integration, the authors present a functional prototype capable of operating in both timed and on-demand modes. Their work addresses critical safety and efficiency issues, pushing forward automation in high-risk industrial settings.

Chapter 5: Electronic Digital Twin in Industry 5.0: AI-Based Electronic Industry 5.0 Models

Bringing electronic engineering into the realm of Industry 5.0, this chapter provides a methodology for integrating AI with Digital Twin models in electronic circuit design. Through practical examples, the authors demonstrate how signal behavior can be predicted and simulated under real-world industrial conditions, accounting for noise, disturbances, and system interferences—thereby enhancing predictive accuracy and operational readiness.

Chapter 6: Integration of Bluetooth Technology and Artificial Neural Networks in High Accuracy Indoor Positioning

Indoor positioning remains a pressing challenge in smart environments, and this chapter offers a robust solution by combining Bluetooth Low Energy (BLE) with Artificial Neural Networks (ANNs). The authors provide a theoretical foundation along with performance evaluations of BLE-based positioning systems. Their work lays a foundation for scalable, cost-effective, and energy-efficient positioning systems adaptable to both academic and industrial contexts.

Chapter 7: Multiscopic Topological Twin for Cyber-Physical Systems

Introducing the concept of the topological twin, this chapter transcends traditional Digital Twin paradigms. By integrating data across microscopic, mesoscopic, and macroscopic layers, the authors provide a framework for multiscale simulation and spatial reasoning. Their application to mobility support robotics illustrates the relevance of this model to both industrial and societal applications under Industry 5.0 and Society 5.0 paradigms.

Chapter 8: Predictive Maintenance Using Sensor Data Processing

This chapter advances the discussion on predictive maintenance by applying AI models to sensor data from a truck wheel assembly line. Utilizing K-Nearest Neighbors, Decision Trees, and regression models, the authors demonstrate how predictive analytics can minimize downtime, optimize labor productivity, and forecast machine health. The chapter is an excellent example of how historical data, when properly processed, can significantly improve operational foresight.

Chapter 9: Solution for Error and Attenuation Detection in Network Services Based on Time Series Decomposition Model

Focusing on data integrity in industrial networks, this chapter introduces an innovative model combining time series decomposition (REFII), Bayesian networks, and advanced pattern mining techniques. The approach excels at uncovering rare yet impactful anomalies that traditional methods might overlook. This work is a valuable addition for those managing network reliability in smart industrial systems.

Chapter 10: The Role of Digital Twins in AI-Driven Industry 4.0 and 5.0: Enhancing Sustainability and Innovation

Concluding the volume, this chapter explores the intersection of Digital Twins, sustainability, and innovation. Drawing from case studies and literature reviews, the authors illustrate how AI-enhanced digital twins can optimize resources, support real-time monitoring, and drive sustainable decision-making. The chapter culminates in strategic recommendations for future research and adoption, encapsulating the volume's overarching vision.

As the editor, I am pleased to present this collection of scholarly and practical insights that will serve as a resource for a diverse audience—researchers, practitioners, policymakers, and students. Whether one is exploring new research frontiers or seeking solutions to practical industrial challenges, the content of this volume offers valuable guidance and inspiration.

Importantly, this book underscores the necessity of collaboration across domains. The challenges we face—climate change, resource scarcity, and rapid technological change—cannot be addressed in silos. By fostering dialogue between engineering, environmental science, computer science, and industrial practice, we aim to bridge gaps and spark innovation that is both responsible and forward-looking.

I am grateful to the contributors whose expertise and vision shape the core of this book. Their work exemplifies the transformative potential of AI-driven smart technologies and demonstrates how, together, we can chart a course toward sustainable and intelligent industrial progress.

Chapter 1
3D CNNs for Smart Manufacturing Voxel-Based Classification of Industrial Workpieces

Masakazu Hirono
Sanyo-Onoda City University, Japan

Fusaomi Nagata
https://orcid.org/0000-0002-7128-2151
Sanyo-Onoda City University, Japan

Maki K. Habib
https://orcid.org/0000-0001-8088-8043
The American University in Cairo, Japan

Keigo Watanabe
Okayama University, Japan

Hitoshi Nakamura
Research and Development Center, Gunma, Mitsubishi Pencil Co., Ltd., Japan

Hisami Tamano
Research and Development Center, Gunma, Mitsubishi Pencil Co., Ltd., Japan

ABSTRACT

This chapter highlights the broader implications of adopting 3D CNNs in smart manufacturing. By automating the classification of complex workpieces, industries can achieve higher precision, reduce reliance on manual labor, and enhance production efficiency. Additionally, the integration of voxel-based classification systems into IoT-enabled smart factories is discussed is explored, presenting a pathway toward more intelligent and adaptive manufacturing ecosystems. This chapter also provides a practical guide to implementing 3D CNNs in industrial settings, Addressing both technical and real-world considerations. It contributes to the advancement of AI-driven smart technologies and serves as a valuable resource for researchers and

DOI: 10.4018/979-8-3693-7994-3.ch001

practitioners in smart manufacturing.

INTRODUCTION

In actual production lines, many types of defect detection or identification systems using CNN-based image processing techniques are being applied. However, in such cases, it is often heard that detection accuracy is undesirably disturbed by lighting conditions. On the other hand, since any RGB camera is not used, the designed voxel-based 3D CNN for object detection and shape detection is not disturbed by such lighting conditions. Up to now, 3D CNN models have been applied to feature extraction tasks included in movies, however, in this chapter, the authors have applied to the surface shape identification task. The proposed 3D CNN model for the surface shape identification is based on the concept of such a 3D CNN model dealing with movies composed of frames. First of all, it may look like no direct relation with the surface shape detection, a 3D CNN model dealing with movie files for steel spark test is described to introduce one of applications of conventional 3D CNN models.

In contacting a grinding tool to a test steel, scattering sparks can be observed visually. The spark test is a widely used method to identify the composition and content of steel based on the appearance and color of sparks generated during grinding (Mundar et al., 2024). The spark test is standardized in JIS G0566-1980 (JIS G0566, 1980). Up to now, several researchers have explored the automation of the spark test by applying image analysis methods. For example, Nakata (2012) developed an automated spark testing technique by image processing to measure carbon content in steel materials. The system consisted of an automatic grinder, a high-speed camera of 200 images per second, and high-speed image processing computer. The result showed that the carbon content measurement accuracy is about ±0.05% in carbon steel (Nakata, 2012). Yoshioka et al. (2019) developed a classification method for carbon content rate [C%], which is defined in JIS G0566-1980 (JIS G0566, 1980), in unknown steels by analyzing images of sparks captured by a high-speed camera (Yoshioka et al, 2019). Yamashita and Kobayashi (2014) developed a spark test system that can generate stable sparks patterns under controlled conditions and capture video for analysis (Yamashita and Kobayashi, 2014). Building upon these developments, researchers have proposed automated evaluation systems that incorporate deep learning techniques such as convolutional neural networks (CNNs) for spark analysis. Recent advances in computer vision have demonstrated the effectiveness of deep learning in feature extraction for material classification and industrial quality inspection (Choudhary et al., 2022, Prunella et al., 2023). Tran et al. (2015) introduced 3D CNN models to automate the evaluation system for carbon content rate by extracting features of spark scattering from the videos (Tran et al., 2015). Further studies examined the impact of the number of frames used in 3D volume construction and the depth of 3D filters on classification performance (Matsuyama et al., 2021, Nagata et al., 2022). Figure 1 shows an example of frames forming a movie of steel spark test.

Figure 1. Frames extracted from a movie of steel spark test by an inspector (Matsuyama et al., 2021, Nagata et al., 2022).

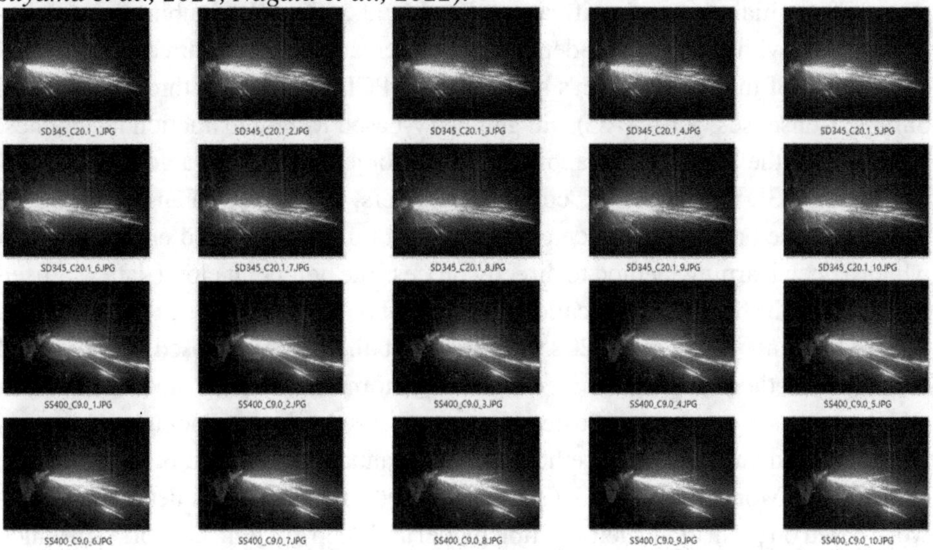

The objective of this chapter is to build a system using the concept of 3D CNN that can accurately classify workpieces with nearly identical contour shapes but different surface features. While conventional CNN-based defect detection systems have been widely used in production lines, these methods rely on capturing 2D images of target products, which introduces challenges in cases where the workpieces have uniform color and nearly identical contours. Up to now, many defect detection systems, such as those proposed by Li et al. (2020) and Usamentiaga et al. (2022), have required extensive images datasets of target products flowing on production lines for training and still face issues related to lighting conditions and camera variability, which complicate achieving stable classification performance. (Li et al., 2020, Usamentiaga et al., 2022). Also, when the contours of workpieces with a single color have the same shape, it is not easy for conventional CNN models to classify them only based on images captured by a camera. Although a large amount of training images have to be accumulated to build a CNN model with high-performance classification ability, it is not also easy for lighting conditions and camera conditions to be kept equal and constant on a long-term basis. Adjusting such conditions will be a complicated process.

Recent studies have shown that point cloud-based models outperform traditional image-based approaches in object classification and industrial defect detection (Qi et al., 2017; Bello et al., 2020). As for the application of point cloud data (PCD), for example, the solution for autonomous robotic grasping proposed by Patel et al. (2025) utilizes an Intel RealSense stereo camera in conjunction with YOLO segmentation models for initial object identification (Patel et al., 2025). It combines segmentation meshes with the camera's depth values to create a detailed three-dimensional point cloud of the target object's surface. This PCD is analyzed through Random Sample Consensus (RANSAC) and geometry-based feature extraction techniques, which allow the algorithm to approximate the object's shape as a combination of fundamental 3D forms, such as cuboids, cylinders, and spheres. Fang et al. (2025) worked on the bolt loosening detection tasks in rail defects based on PointNet, in which a deep learning method to directly process the bolt point cloud data was used and realized the binary classification detection of the bolt loose state through global feature learning, (Fang et al., 2025). Also, Ye et al. (2025) proposed a point cloud clustering method for diagnosing defects on automobile and flat steel surfaces, in which the issue of manual clustering parameter selection, reducing usability, is addressed and an automatic method without manual intervention is proposed (Ye et al., 2025). Moreover, Xu et al. (2025) proposed a new landmark detection framework based on point cloud deformation model and deep learning networks, in which 3D point cloud CNN semantic segmentation models are applied for landmarks in defect areas (defect landmarks). The effectiveness of the proposed framework is demonstrated through the presentation of experimental results (Xu et al., 2025).

As introduced above, 3D cameras, such as Intel RealSense and SiNGRAY Stereo PRO, can measure object depth by emitting infrared light, making them less sensitive to external lighting conditions and more effective at capturing fine surface details. In this study, PCD captured by a 3D camera is converted into voxel data, allowing a more robust feature representation that overcomes the limitations of conventional 2D imaging techniques. A compact 3D CNN model is then designed, trained, and evaluated using voxel data to classify workpieces. The proposed approach is tested on a dataset of 72 workpieces spanning eight different categories, and its classification performance is analyzed. By leveraging recent advancements in 3D CNN architectures, this study aims to enhance the accuracy and robustness in automated workpiece classification task for future smart manufacturing.

VoxNet

In this study, since we are accumulating PCD measured by a 3D camera for training, it is necessary to construct a 3D CNN that can deal with the 3D data given by PCD. Therefore, the VoxNet model as shown in Fig. 2 proposed by Maturana et al. is introduced (Maturana and Scherer, 2015). VoxNet has an architecture to be able to integrate a volumetric occupancy grid representation with a supervised 3D CNN. 3D PCD files are converted into voxel files that can be used as training data for the 3D CNN. When points exist in a voxel or not, the voxel has information of 1 or 0, respectively. Therefore, the information of objects in 3D space is maintained only by binary codes of 1 and 0, and the shape and structure of the target workpiece can be dealt with more accurately. Due to the voxel data, VoxNet has an advantage in terms of computational and memory costs compared to conventional other 3D CNNs. Also, VoxNet, which has the structure of 3D CNN, takes a voxelized shape data as input, and performs convolution and pooling based on the voxel quantity in the 3D direction.

In this study, we designed and 3D-printed a total of 72 (8 categories 9 models) actual 3D models by 3D CAD Creo as shown in Fig. 3 so that each model can have different surface shape. In the next section, we measure the point cloud data (PCD) of these 72 models using a RealSense, and then voxelize them at a resolution of 32×32×32 in 3D space. RealSense is a widely-spread product with depth and tracking technologies designed to give robots and mechatronics depth perception capabilities. After training the VoxNet using the prepared training voxel data, the generalization performance is evaluated.

Figure 2. Overview of VoxNet architecture designed for shape prediction (Maturana and Scherer, 2015).

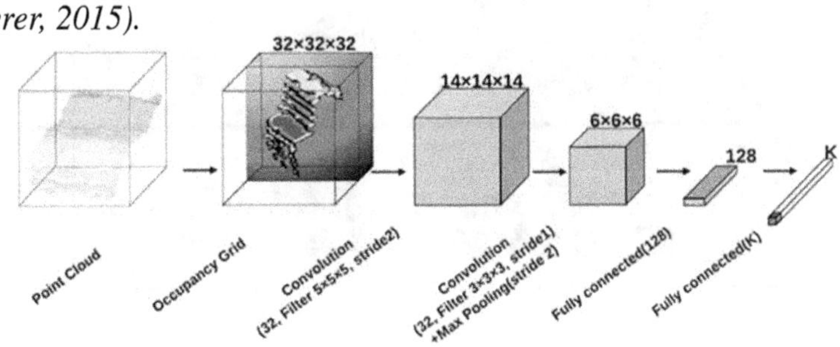

Figure 3. 8 categories × 9 models = 72 models are 3D-printed for accumulating voxel data to train and test 3D CNNs. After the original 72 PCD files were measured from the 72 3D-printed models, the 72 PCD files were augmented to 72×360=25920 by rotating them 1 by 1 degree around z-axis. The 25920 PCD files were used, half (12960) for training and half (12960) for testing.

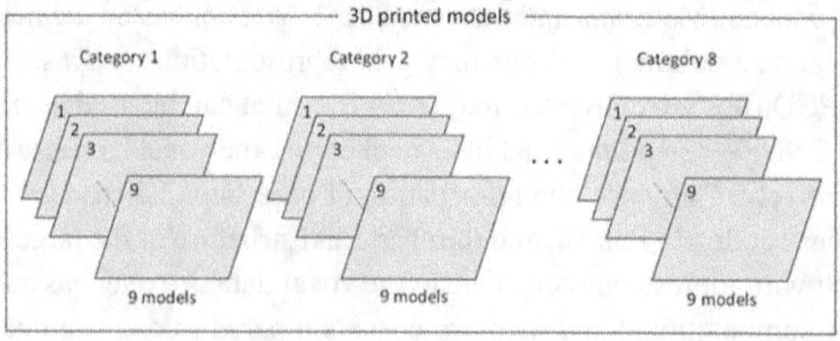

Figure 4. Eight categories of 3D models for classification, i.e., upper figures show examples of 3D CAD models, lower figures show their actual 3D-printed models.

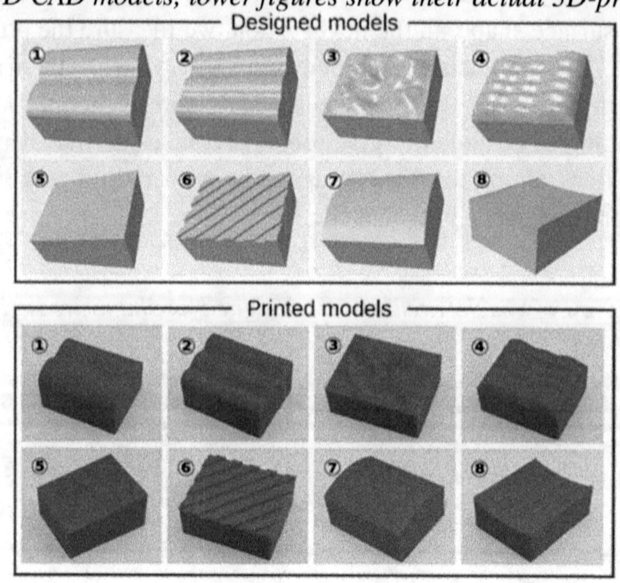

Figure 5. Measurement system of 3D-printed models (left side) and an example of measured PCD (right side).

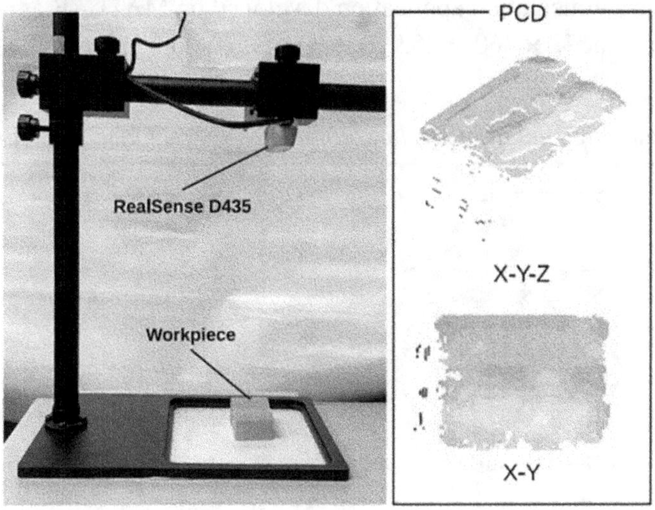

Generation of Voxel Data for Training

A. Fabrication of Work Models and Accumulation of Point Cloud Data (PCD) by 3D Camera

When a model is shot from directly above direction with a 2D camera, it is difficult to reproducibly and stably capture features such as gentle inclines and small undulations in the image, depending on lighting and other shooting conditions. Therefore, we decided to use point cloud data (PCD) captured by a 3D camera. Intel RealSense D435 was used for the 3D camera. Figure 4 shows 3D models of the eight categories with the same color and the same contour viewed from top, which are the targets of classification in this study. Note that each counter shape is the same rectangle, but the top surface of each model has a different surface shape. Furthermore, by using 3D printers we fabricated nine different physical models with a distinct feature as for each category, so that the number of total physical models is 72.

In this experiment, a 3D camera was placed directly above each model as shown in Fig. 5. 72 models were measured by the 3D camera one by one, so that 72 PCD files (.pcd) were obtained. The scan range in the Z-direction is adjusted to remove information about the floor on which the model is placed. We considered a data augmentation method that increases the number of PCD files by rotating the acquired PCD around the Z–axis with a specified angle. By setting the rotation angle to 1 degree, 360 augmented PCD files can be generated from only one PCD file by using the data augmentation application developed on MATLAB as shown in Fig. 6, resulting in that 72 × 360 = 25,920 PCD files could be prepared.

Figure 6. Data augmentation application for LAS, PCD, and PLY files developed on MATLAB.

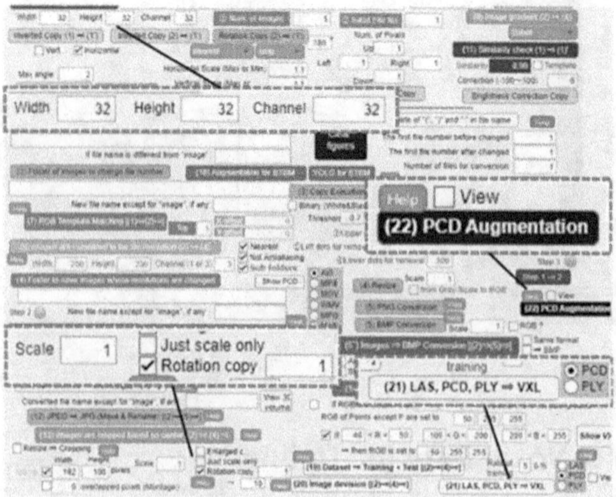

Figure 7. Example of voxel data transformed from a PCD.

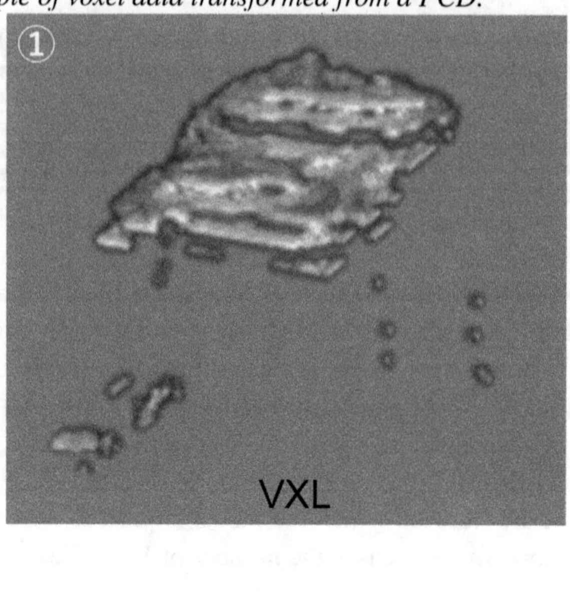

B. Conversion to Voxel Data from PCD

A PCD file containing depth information obtained by Realsense D435 has a large amount of 3D point data. As a result, when the number of these PCD files increases, efficient learning of 3D CNN becomes difficult. Therefore, PCD files are converted to voxel data files, which can be used as dataset for desirable light training of a 3D CNN. In this test trial, a PCD file is converted into 32×32×32 sized voxel data file. We created a training dataset consisting of 1620 voxel data and a test dataset equally consisting of 1620 voxel data for each category. Figure 7 shows an example of voxel data converted from a PCD file of category No. 1.

Design, Training and Evaluation of 3D CNN Using Converted Voxel Data

3D CNN can generally extract features containing temporal and spatial information of data through 3D filters. We design a 3D CNN structured with a total of 12 layers for classification, including two 3D convolutional layers and two fully connected layers as shown in Table 1. The training conditions such as mini-batch size, epoch number, and learning rate are set to 32, 60, and 0.0001, respectively, using our developed 3D CNN design application available on MATLAB. Figure 8 shows an example of the learning process. After the training process, we conducted classification experiments using 12,960 test voxel data (8 categories × 1620 voxel files) to evaluate the generalization performance of the 3D CNN. As a result, the eight categories of models were classified with accuracy of 100% as given by the confusion matrix shown in Fig. 9. In this experiment, we used a PC with an Intel(R) Core(TM) i7-12700K 3.60 GHz CPU and an NVIDIA GeForce RTX 3080 Ti GPU, equipped with 32 GB of main memory.

It has been confirmed from the results that the 3D CNN model built based on voxel data could have a promising generalization ability for objects' identification that conventional image-based CNN could not cope with. However, in this evaluation shown in Fig. 9, the 12,960 test voxel data were made from the same models used for making the training data, so that the focus on accuracy as the primary evaluation metric was valuable but insufficient. Additional metrics such as precision, recall, and F1 score should be incorporated particularly for cases with imbalanced datasets or where certain misclassifications may have greater consequences. In future work, it will be necessary to evaluate generalization performance using data measured from 3D-printed models other than the ones used during training.

Table 1. Structural parameters of the constructed 3D CNN model designed for voxel data classification task.

Type	Filter size	stride	Output size
Input	32×32×32	-	32×32×32×1
Convolution	5×5×5	2	14×14×14×32
Leaky ReLU	-	-	14×14×14×32
Convolution	3×3×3	1	12×12×12×32
Leaky ReLU	-	-	12×12×12×32
MaxPool	-	2	6×6×6×32
Fuly Connected	-	-	128
ReLU	-	-	128
Dropout	0.5	-	128
Fully Connected	-	-	8
Softmax	-	-	8
Output	-	-	8

Figure 8. Training progress to optimize the 3D CNN model.

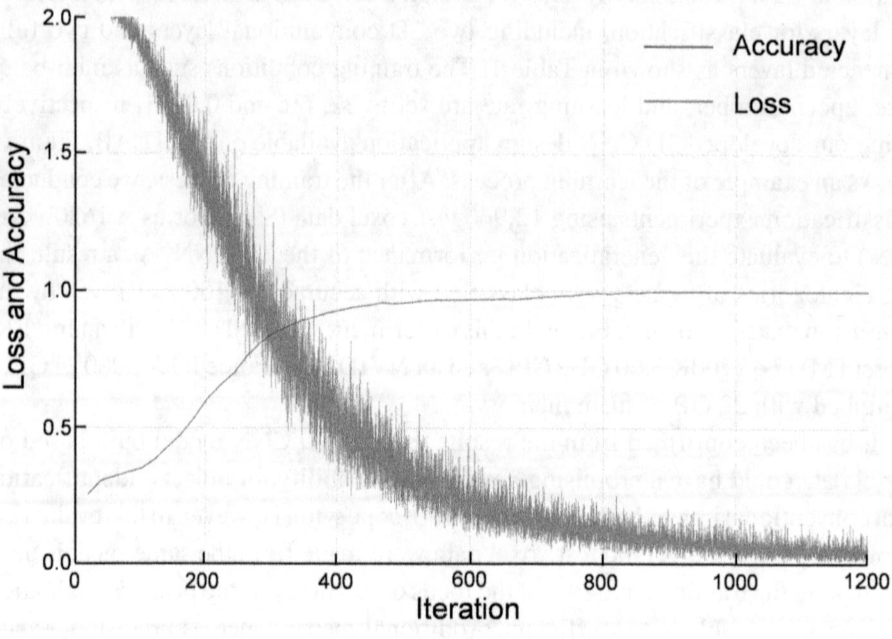

Figure 9. Classification result (confusion matrix) of test voxel data predicted by the built 3D CNN model.

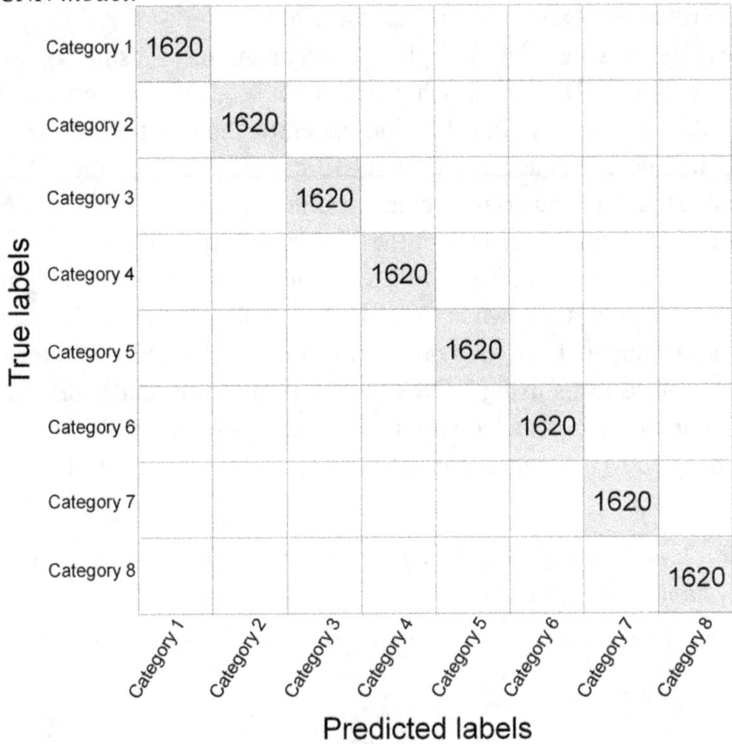

Pick and Place Experiment Using an Industrial Robot

In this final section, a simple pick and place experiment using an industrial robot MG400 incorporated with the trained 3D CNN is demonstrated. Figure 10 shows three forms of a soft gripper driven by a small compressor which can pick workpieces with various types of contours and surface shapes. Three modes, i.e., grip, relax, and open, can be easily switched by setting the values for digital IO ports. The soft gripper is attached to the tip of the robot arm. Figure 11 illustrates the system configuration of the pick and place robot incorporated with the 3D CNN, whose motion sequence can be described with hyper cutter location source (HCLS) data statements. Figure 12 shows the beginning part of the HCLS data for the pick and place task, in which the statement "3DCNN_SHAPE" is set for the identification of workpieces' surface shapes. When the "3DCNN_SHAPE" is processed, a PCD captured by the RealSense D435 is converted into a voxel data, and then the 3D CNN extracts and classifies the feature of the voxel. The concept of HCLS data was proposed by Nagata et al. (2024), in which not only the basic statement for position data but also advanced statements for calling AI models such as CNN and YOLO can be described and executed (Nagata et al., 2024).

Figure 13 shows the pick and place experiment scene using the MG400 incorporated with the 3D CNN, in which the robot sequentially works on following the HCLS data as shown in Fig. 12. Consequently, the robot places workpieces to pre-designated positions according to classification results by the 3D CNN. It is demonstrated that the industrial robot incorporated with the trained 3D CNN is able to conduct the pick and place task of the 3D-printed workpieces.

This single system introduced in Figs. 11 and 13 can be integrated into an IoT-enabled smart factory as shown in Fig. 14. The number of the single system can be increased according to the need of production line. The anomaly monitoring system for pick-and-place tasks using RGB cameras is currently under development, in which if an anomaly is detected based on image processing and CNN techniques, it will be notified to the central control room via each robot control application.

Figure 10. Three forms of the soft gripper which can pick workpieces with various types of contours and surface shapes.

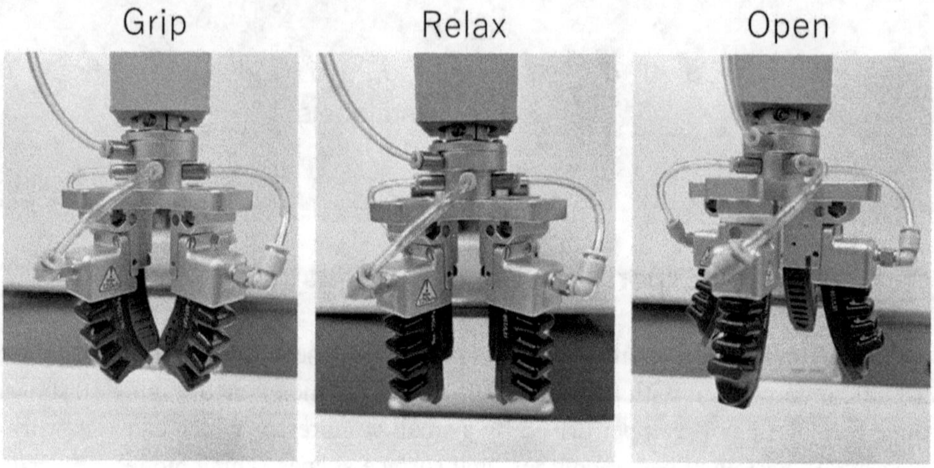

Figure 11. System configuration of the pick and place robot incorporated with the 3D CNN shown in Fig. 13. The soft gripper of the robot is shown in Fig. 10.

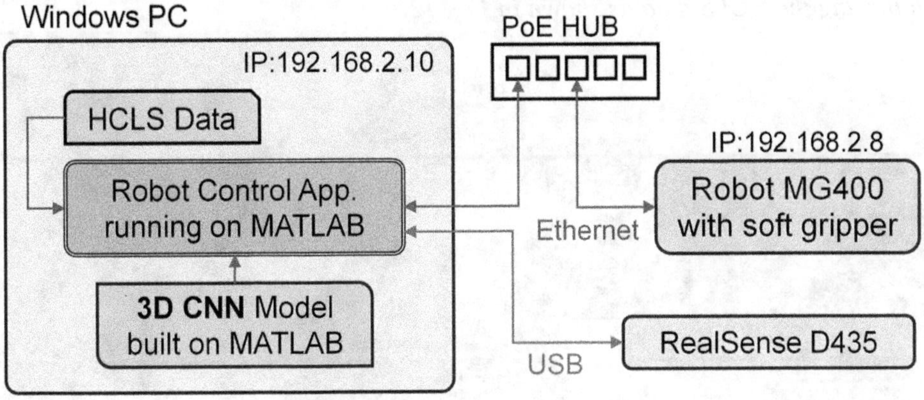

Figure 12. The beginning part of HCLS data for the pick and place task, in which the statement "3DCNN_SHAPE" is set for capturing a PCD and identifying of the target workpiece's surface shape.

```
Absolute
3DCNN_SHAPE
PAUSE 3
GOTO/248. 62, 1. 32, 30. 05, 139. 78, 0, 0, 1
GOTO/248. 62, 334. 35, 30. 05, 139. 78, 0, 0, 1
DO 1 1
PAUSE 2
GOTO/248. 62, 334. 35, -51. 77, 178. 85, 0, 0, 1
GOTO/248. 62, 334. 35, -104. 55, 178. 85, 0, 0, 1
PAUSE 1
DO 1 0
PAUSE 1
DO 2 1
PAUSE 3
GOTO/248. 62, 334. 35, 30. 05, 139. 78, 0, 0, 1
GOTO/248. 62, 1. 32, 30. 05, 139. 78, 0, 0, 1
PLACE
PAUSE 1
DO 2 0
PAUSE 1
DO 1 1
PAUSE 1
MOVZ 30. 0
GOTO/248. 62, 1. 32, 30. 05, 139. 78, 0, 0, 1
DO 1 0
```

PCD capturing and shape identification by 3D CNN

Gripper open

Picking a workpiece

Placing to each target position

Return to initial position

Figure 13. Pick and place experiment scene using an industrial robot MG400 incorporated with the trained 3D CNN, in which the robot sequentially works on following the HCLS data as shown in Fig. 12.

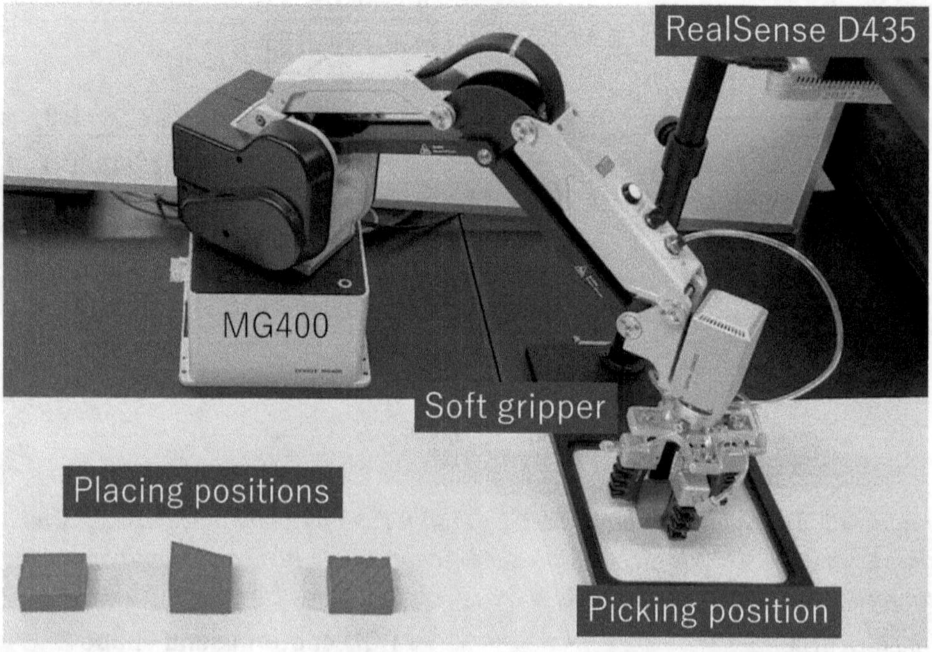

Figure 14. Image of IoT-enabled smart factory using multiple robots, which is under development.

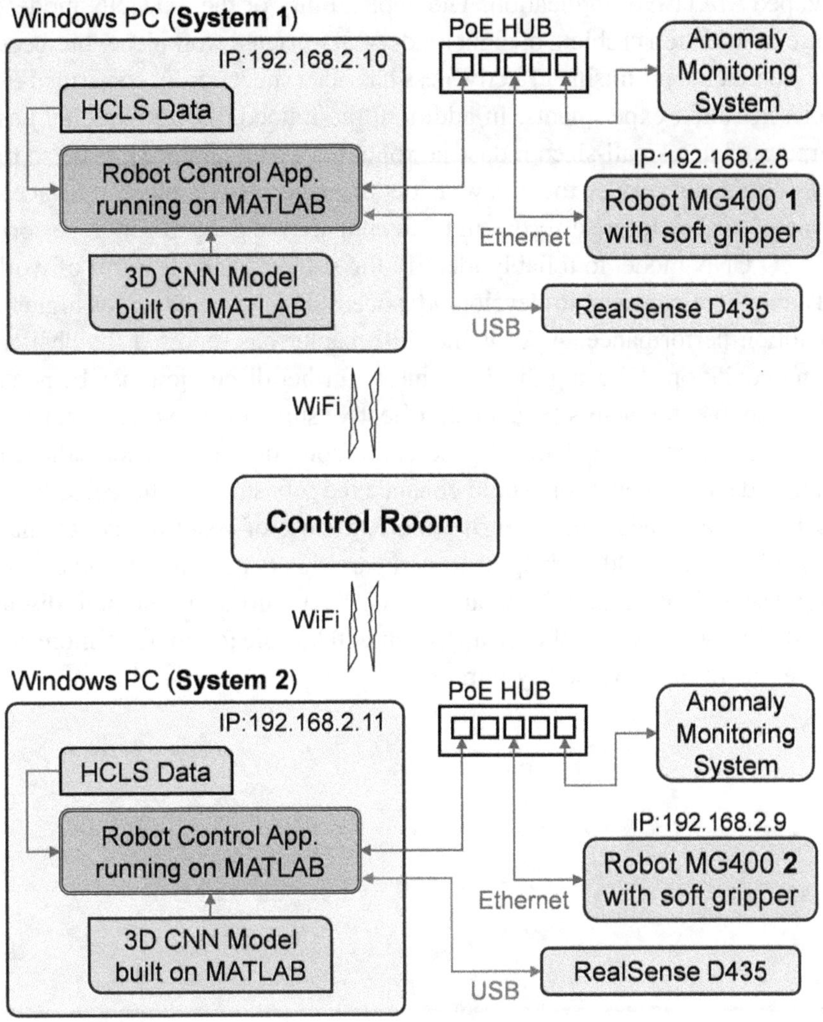

CONCLUSIONS AND FUTURE WORK

3D cameras such as RealSense can measure the distance to an object by irradiating near-infrared rays and obtain point cloud data (PCD), so that it is effective in measuring object shape without being affected by lighting conditions. In this chapter, voxel data have been converted from PCD measured by a 3D camera RealSense D435.

Then, a compact 3D CNN model, which can identify workpieces' surface shape based on voxel data, has been designed, trained, and evaluated using the authors' developed MATLAB application. The applicability of the 3D CNN model to the shape classification problem of eight kinds of 3D-printed workpieces has been evaluated, so that the promising effectiveness has been successfully confirmed through the classification experiments. In addition, the trained 3D CNN model has been incorporated in a small-sized industrial robot for a pick and place task, so that the robot was able to work on the task with identifying each workpiece's surface shape.

One of the key future efforts will be to enhance the generalization performance of the 3D CNN model to reliably identify the surface shape features of workpieces. The authors continue to develop advanced 3D CNN models for higher shape recognition performance. Also, while this chapter has reported the challenge of lighting conditions by using the 3D camera, further discussion may be needed on how environmental factors (e.g., dust, reflective surfaces) can affect data accuracy and model performance. Including potential solutions like sensor calibration or advanced data augmentation would enhance the robustness of the 3D CNN model for surface shape recognition. Furthermore, the use of voxel-based data has been explained, but it would be helpful to compare this approach with other common 3D representations (e.g., point clouds or meshes). Further research to discuss the strengths and weaknesses of each approach will be able to provide a more comprehensive role of the proposed system.

REFERENCES

Bello, S. A., Yu, S., Wang, C., Adam, J. M., & Li, J. (2020). Review: Deep Learning on 3D Point Clouds. *Remote Sensing (Basel)*, *12*(11), 1729. DOI: 10.3390/rs12111729

Choudhary, K., DeCost, B., Chen, C., Jain, A., Tavazza, F., Cohn, R., Park, C. W., Choudhary, A., Agrawal, A., Billinge, S. J. L., Holm, E., Ong, S. P., & Wolverton, C. (2022). Recent Advances and Applications of Deep Learning Methods in Materials Science. *npj Computational Materials*, *8*(1), 59. DOI: 10.1038/s41524-022-00734-6

Fang, S., Zhao, Q., Zhao, K., & Yang, L. (2025). Research on Bolt Loosening Detection Method Based on PointNet. Procs. of 2025 IEEE 8th Information Technology and Mechatronics Engineering Conference (ITOEC), Chongqing, China, 2025, pp. 1331-1335, DOI: 10.1109/ITOEC63606.2025.10967991

Li, Y., Zhang, R., Li, H., & Shao, X. (2020). Dynamic Attention Graph Convolution Neural Network of Point Cloud Segmentation for Defect Detection. Procs. of 2020 IEEE International Conference on Artificial Intelligence and Information Systems (ICAIIS), pp. 18–23. DOI: 10.1109/ICAIIS49377.2020.9194799

Matsuyama, H., Nagata, F., Miki, K., Kato, H., & Watanabe, K. (2021). Evaluation System Using 3D Convolutional Neural Network for Spark Test. Procs. of 22nd SICE SI Conference, pp. 328–331 (in Japanese).

Maturana, D., & Scherer, S. (2015). VoxNet:A 3D Convolutional Neural Network for Real-Time Object Recognition. Procs. of 2015 IEEE/RSJ International Conference on Intelligent Robots and Systems (IROS), Hamburg, Germany, pp. 922–928.

Nagata, F., Abe, R., Sakata, S., Watanabe, K., & Habib, M. K. (2024). Hyper CLS Data-based Robotic Interface and Its Application to Intelligent Peg-in-Hole Task Robot Incorporating a CNN Model for Defect Detection. *Machines (Basel)*, *12*(11), 1–14. DOI: 10.3390/machines12110757

Nagata, F., Hirono, M., Yonehara, H., Kato, H., & Watanabe, K. (2022). Steel Spark Test System Using Video Processing and 2D CNN. Procs. of The Institute of Industrial Applications Engineers (IIAE) Annual Conference 2022, pp. 41–42 (in Japanese).

Nakata, T. (2012). Development of Automated Spark Testing Technique by Image Processing to Measure Carbon Content in Steel Materials. Procs. of IFAC Workshop on Automation in the Mining, Mineral and Metal Industries, Gifu, Japan, pp. 118–119. DOI: 10.3182/20120910-3-JP-4023.00029

Patel, S., Kovarovics, A., More. A., Chun, C., Fletcher, M., Tanveer, M.H. (2025). Towards Autonomous Robotic Grasping: 3D Point Cloud Analysis with Stereo Cameras in Unpredictable Environments. SoutheastCon 2025, Concord, NC, USA, pp. 520–525, .DOI: 10.1109/SoutheastCon56624.2025.10971505

Prunella, M., Scardigno, R. M., Buongiorno, D., Brunetti, A., Longo, N., Carli, R., Dotoli, M., & Bevilacqua, V. (2023). Deep Learning for Automatic Vision-Based Recognition of Industrial Surface Defects: A Survey. *IEEE Access : Practical Innovations, Open Solutions*, *11*, 43370–43423. DOI: 10.1109/ACCESS.2023.3271748

Qi, C. R., Su, H., Mo, K., & Guibas, L. J. (2017). PointNet: Deep Learning on Point Sets for 3D Classification and Segmentation. Procs. of the IEEE Conference on Computer Vision and Pattern Recognition (CVPR), pp. 652–660.

Tran, D., Bourdev, L., Fergus, R., Torresani, L., & Paluri, M. (2015). Learning Spatiotemporal Features with 3D Convolutional Networks. Procs. of the IEEE International Conference on Computer Vision (ICCV2015), pp. 4489–4497.

Usamentiaga, R., Lema, D. G., Pedrayes, O. D., & Garcia, D. F. (2022). Automated Surface Defect Detection in Metals: A Comparative Review of Object Detection and Semantic Segmentation Using Deep Learning. *IEEE Transactions on Industry Applications*, *58*(3), 4203–4213. DOI: 10.1109/TIA.2022.3151560

Xu, M., Luo, Z., Liu, B., Wang, Z., Tang, X., & Song, T. (2025). Automatic Landmark Detection in Severe Craniomaxillofacial Deformities via Three-Dimensional (3D) Point Cloud Deformation Model and Deep Learning Networks. *IEEE Access : Practical Innovations, Open Solutions*, *13*, 32724–32740. DOI: 10.1109/ACCESS.2025.3541601

Yamashita, H., & Kobayashi, H. (2014). Automatic Identification of Carbon Steel on Spark Analysis -Automation of Spark Examination-. Procs. of the 2014 JSME Conference on Robotics and Mechatronics, 2A1-V03, 3 pages (in Japanese).

Ye, J., Liu, X., Madhusudanan, H., Wang, Y., Zhu, J., Wang, Y., Ru, C., Liu, X., & Sun, Y. (2025). Automatic Point Cloud Clustering for Surface Defect Diagnosis. *IEEE Transactions on Automation Science and Engineering*, *22*, 12538–12547. DOI: 10.1109/TASE.2025.3542076

Yoshioka, T., Ohba, Y., & Kobayashi, H. (2019). Development of a Steel Identification System by Spark Image Analysis Method [in Japanese]. *Sanyo Technical Report*, *26*(1), 23–27.

Chapter 2
AI-Based Solutions for Future Trends in IoT and Remote Sensing Integration for Precision Agriculture

V Dankan Gowda
https://orcid.org/0000-0003-0724-0333
Department of Electronics and Communication Engineering, BMS Institute of Technology and Management, Bangalore, India

R. Saranya
Department of Information Technology, V.S.B. Engineering College, Karur, India

S.V. Ramanan
https://orcid.org/0009-0005-3477-6039
Department of Electronics and Communication Engineering, PPG Institute of Technology, Coimbatore, India

N M G Kumar
https://orcid.org/0000-0003-1494-5737
Department of Electrical and Electronics Engineering, Mohan Babu University, Tirupathi, India

Priya SACHIN Jadhav
https://orcid.org/0000-0002-5439-5437
Symbiosis Institute of Technology, Symbiosis International University, Pune, India

DOI: 10.4018/979-8-3693-7994-3.ch002

ABSTRACT

This chapter focuses on the use of AI, IoT, and remote sensing technologies in precision agriculture. This explores how the technologies can be used in collaboration to benefit the improvement of yield in agriculture, efficiency in resource use and sustainable utilization of resources. Among them, current practices consist on crop health monitoring, yield prediction, precise irrigation and correct soil management are presented where they are currently being used and the latest applied technologies are showed in the following chapter. Also, it defines the difficulties associated with these solutions: their cost, data compatibility issues, and the necessity of qualified staff. The summarization of the potential of AI and IoT as well as the remote sensing in the farming industry is also provided as well as a focus on the future trends like edge computing and 5G connectivity. It outlines the matters that should be considered by the policy makers and the need for collaboration among various fields as a way of enhancing the uptake of these technologies.

INTRODUCTION

Overview of Precision Agriculture

This chapter focuses on the use of artificial intelligence, Internet of things, and remote sensing in precision agriculture. This explores how the technologies can be used in collaboration to benefit the improvement of yield in agriculture, efficiency in resource use and sustainable utilization of resources (Victor, N., et al., 2024). In this chapter, the author briefly covers some current areas of employments of AI in crop health management, yield prediction, precision irrigation, and soil health assessment including some selected AI novel technologies case studies (Ye et al., 2013). Also, it defines the difficulties associated with these solutions: their cost, data compatibility issues, and the necessity of qualified staff. The summarization of the potential of AI and IoT as well as the remote sensing in the farming industry is also provided as well as a focus on the future trends like edge computing and 5G connectivity (Sarjerao, J. S., and Sudhagar, G., 2024). It outlines the matters that should be considered by the policy makers and the need for collaboration among various fields as a way of enhancing the uptake of these technologies.

Precision agriculture revolves around the need of farmers to learn about the variability of production within the field. Some of them are; Soil fertility, availability of water, existence of pest, diseases and other natural factors such as weather conditions (Pincheira et al., 2020). Farmers are then able to acquire data from these sources; sensors, satellite imagery, other remote sensing technologies which help them in perceiving these variabilities. In that regard, PA practices assist in spotting the uncertainty of yield and thus enhance the prediction of yield as well as the appropriate irrigation practices (Martínez et al., 2020), including the rational use of fertilizers and pesticides, thus cutting the cost, enhancing yield and sustainability of the environment.

The rationale this paper is based on is the examination of the importance of precision agriculture in the world to come up with strategies for providing an increased rate of food production to feed the ever expanding global population while at the same time keeping the demand on earth's resources to a minimum (Aravindh et al., 2024). This can be attributed to the fact that precision agriculture promotes the use of inputs such as water, fertilizers as well as pesticides at the right places at the right time hence minimizing their use on the fields. It is beneficial not only to the large scale commercial farming but also for the small holders intending to expand their farming business in the developing world (Sethi, H. S., and Prasanth, P. V., 2023). It is the best opportunity for leveraging growth and breakthroughs in agriculture usage of technologies to increase yield and sustainable productions, decrease costs, and reduce environmental impacts to enhance food production global availability.

However, research shows that the actualisation of the precision agriculture technologies that are especially useful in increasing yields in the agricultural sector is not without socio-economic challenges. There are several reasons why the required technology is currently out of reach of many of the farmers, particularly those in the Third World. Finally, with the help of AI and IoT, farmers may need to acquire certain knowledge and skills that cannot be easy to attain for beginners in technologies.

Role of IoT and Remote Sensing in Precision Agriculture

In precision agriculture, IoT and remote sensors technologies have been applied in order to improve crop production and productivity. IoT can be defined as a system of devices that is capable of capturing data and sending it to other computer systems in the agricultural sector for monitoring of environments (Kudari, J. M., and Samal, A., 2021). These can be arranged in the fields or in the crop area to capture different type of data like moisture condition of the soil, temperature, humidity conditions of the environment, light intensity and nutrient grade in soil etc in the fields and in the irrigation systems as well as various equipments used in the farming sector (Veeramanikandan, P., and Kumar, N. A., 2024). IoT thus assists in making specific parameters that relate to the farming process famous to farmers so as to manage crop situations.

Remote sensing can be defined as an accumulation of information from a distance using satellite image, aerial photograph, and unmanned aerial systems such as drone or unmanned aerial vehicles (UAVs). Electrical equipment helps farmers have timely and actual knowledge about the condition of vast expanses of farmland and information about the state of the crops, soil, and climate. The fusion of remote sensing with IoT systems enhances the precision agriculture by giving both micro and macro level perspective on the status of the farm.

Remotely sensed data can also be applied to monitor variability in respect with the soil health, crop status, and climate in precision agriculture (Singh et al., 2024). For instance, such things as poor health of the soil or infestation by pests can cause some parts of the field to yield little or nothing and this may be revealed through satellite imagery. UAVs and drones overcomes the generalisations made by solutions and they offer high precision images of crops that help farmers isolate problematic regions easily. It also enables the farmer to apply fertilizer or pesticide whenever it is required in particular sections of the field without the need to spread it on the entire field which greatly cuts on the pollution.

IoT and remote sensing enabled farmers monitor and gather detailed information on numerous aspects related to crop farming (Shafi, U., et al., 2020). This integration of real time data collection and remote monitoring hence results to better understanding that permits proper intervention among farmers. Besides, IoT and remote sensing can enhance the forecasting of events including drouths, diseases or pests, whereby the farmers can adequately respond to the occurrences. Therefore, these technologies are considered to be very important in increasing efficiency and sustainability within today's agriculture.

Introduction to AI

AI refers to an extensive capability of a computer system to replicate human cognitive functions such as learning and problem solving. Some of the cognitive process identifies include learning, reasoning, problem solving, perception and language comprehension (Singh et al., 2021). Talking about precision agriculture, AI plays the role of the cognitive functionality in the transformation of the huge amount of data collected by IoT sensors and remote sensing. Of the various subfields of AI, ML and DL are especially helpful in diagnosing the large and complex data sets gathered through different sources, as well as in making a wide range predictions, optimizations, and decisions.

IoT and Remote sensing get a boost from AI in precision agriculture since it makes sense of the large amount of data collected and makes it easy to conduct predictive analysis (Guidotti et al., 2019). For instance, AI can help monitor data collected by sensors in the field to identify initial conditions of disease or pest infection. By means of the facial recognition or pattern recognition, a machine can easily see signs which the plants are revealing that a human may not easily detect. In this respect, applications of AI potentially prevent diseases from becoming widespread, thus requiring the use of pesticides in large quantities.

Another area of use of AI technology in precision agriculture is on the deployment of methods for yield estimation. This is because using historical data, external or current weather conditions, and the quality of the soil among other pertinent factors, have been incorporated into the AI algorithms used in determining the likely yields of crops (Gulay, C. F., and Fajardo, A. C., 2024). These ones enable farmers harvest, manage supplies and resources with efficiency in an attempt to maximize on revenues and minimize loss. They can also help in irrigation scheduling since the determination of the time when they are required for conveying water based on the climate conditions and the moisture content of the soil in addition to other parameters aimed at conserving water.

It can also make a positive contribution to remote sensing applications for the purpose of improving image processing and data analysis. It is worth emphasizing that many applications of satellite and drone imageries produce large amounts of data that must be pre processed and analyzed to be useful. Deep learning is a perfect machine learning technique to use in endeavours such as image classification and segmentation; these help in the identification of the areas of interest in agricultural fields. Another aspect of applying the AI in a farming system is that the system can also help monitor the development stages of crops, identify abnormalities, and suggest which particular practices should be employed given the images collected from remote sensing.

Moreover, it is worth stating that the popularity of AI-associated autonomous systems is growing in agriculture rapidly (Pero et al., 2024). These include self-driving tractors and drones within farms to plant crops; weed, and harvest among other farm activities. These machines are capable of operating on their own and cause little or no dependence on manpower, and the activities can be performed in a very efficient manner. This advancment also eliminates the costs of labour and develops efficiency and effectiveness which are costly in large scale farming.

In conclusion it can be noted that AI has a critical role in boosting the functions of IoT and remote sensing in precision agriculture. AI makes it easier for farmers to improve existing decisions, make better decisions, predict scenarios, manage resources efficiently, and even perform some repetitive tasks. Thus, AI's interaction with IoT and remote sensing will definitely transform the practice of agriculture, increase yields and efficiency, and improve sustainability.

LITERATURE SURVEY

IoT in Agriculture

The use of IoT as an ingredient has led to greatness in how the data is garnered featured, process and apply in the precision agriculture. The practical use of IoT in agritance is using the Internet by having sensors placed on the soil and crops, as well as on farm machinery to capture data in real time (Sawant et al., 2014). These sensors include the moisture, temperature, humidity, pH and nutrient content in the soils. This information assists the farmers in identifying irregular services, diseases and aeration, water and feeding services, and control of pests.

Sustainable use of water in agriculture is a prime characteristic of the developing world; hence, smart irrigation systems, which are considered among the most popular IoT applications, applied in agriculture. Example: These system employ use of moisture sensors to measure the level of water in the soil and therefore come up with the irrigation schedules (Felício et al., 2023). According to the different previous studies, the application of IoT-based smart irrigation system in agriculture has been found that its usage can save up to 50% of water usage while irrigated crops remain healthy and productive. For instance, the research by Zhang et al. (2017) and Mehmood et al. (2020) has shown that there is a shift from the conventional irrigation practices and that IoT based irrigation systems can greatly enhance the efficient use of water.

Also other climate tracking gadgets connected to internet of things have been found very crucial in assessing the weather features for instance temperature, humidity and wind speed which affect crop growth (Alam, S., 2023). These enable organisations to forecast weather alterations in the environment, hence, prepare adequately towards the alterations. The technologies like climate sensors connected through IoT that has been used in the study done by Kurniawan et al. (2021) enable farmers to make decisions based on the current climate resulting in fewer loses when the climate is unfavourable such as during frost or a drought season (Rodriguez-Gallo et al., 2024). These particular IoT systems have made the agricultural practices rely on data in an endeavor that not only increases productivity but also minimizes the use of resources. The availability of real-time data makes it easier for farmers to make proper decisions on how much of the inputs such as fertilizers, water and pesticides to use. Bringing out better decision making boosts productivity of farming hence making it sustainable as well as more profitable.

However, there are some challenges that accompany the integration of IoT in the agricultural sector. However, there are the following challenges which can slows down and hinder the general usage: Cost implication: The cost of installing the technologies is high. Some of the studies also urge socio-economic analysis to understand how these challenges can be addressed so as to make the gains from IoT attainable across all the tiers of agriculture.

Remote Sensing in Agriculture

Satellite remote sensing as in satellite imagery, drones, and unmanned aerial vehicles (UAVs) are crucial innovations to farmers in monitoring their fields. Satellite data has proved to be useful in delivering farmers' high-resolution information about the crops, moisture condition of the soil and the land use. Through remote sensing data and information derived from satellites, the farmers have benefited from information that can effectively illustrate the status of crops even in large scale farming among the large-scale farming industries (Pandey et al., 2024). Remote sensing as pointed out by Cingolani et al. (2019) can also be used to over see vegetation indices, pest identification, and assessment of soil health hence positively contributing to crop management.

The use of Drones and UAV's for aerial photography of crop fields is fast becoming popular due to the high quality of the images obtained. These devices supply information on the state of the plants; field, which allows farmers to detect new infections or the affected specific regions of fields in real-time nearly (Sharma et al., 2024). The multispectral and thermal cameras installed in UAVs allow crops to be checked for water stress, pest and disease infestations in crops. Maimaitiyiming et al., in their research carried out in 2019, have noted that drones can also identify the specific parts of the field that produce low yields and, thus, such areas can be attended to with fertilizers or pesticides.

In addition, the remote sensing also play important role in yield prediction as well as pest control. Climate information integrated with remotely sensed data and crop models has improved the ability to forecast crop production and this was proved in the findings of studies conducted by Gitelson et al. (2018) and Chen et al. (2020). These predictions help the farmers in organizing their farming and marketing their produce in the market (Dubey, A. K., and Chahar, R., 2023). Disease or stress can be detected by the help of remote sensing technologies which identifies such areas possibly to cause pest outbreak. What should be noted about it is this kind of early warning system enables the farmers to reduce crop loss by taking preventative measures.

Technology within the context of artificial intelligence helps predict diseases and pests attack on crops. But the practical results of such systems depend on the ordained field situations, the sensors' calibration level, as well as the actual data acquired. Also, it is necessary to train the AI models for geographical differences in the type of soil and climatic conditions for greater accuracy of the results.

The models for predicting yields are very useful for AI, nevertheless, there are problems. There is an issue with climate fluctuations and the given models with the possibility of distortion depending on the region. This is a problem due to the scarcity of real-life data used in making these predictions especially in small and sometimes less technologically developed farming ventures.

The possibility of using AI for monitoring soil health is quite optimistic but faces many challenges concerning the quality and multiple types of sensors, frequent updates in the systems. Additionally, long-term changes as epitomized by matters such as eroding of the soil and diminishing fertility remain a challenge in modeling.

AI Integration

AI with IoT and remote sensing capability have brought more advancement in the area of precision farming. Therefore, big amounts of data that are gathered from tiny sensors and remote sensing jets can be tackled by AI algorithms, more especially the MLs and the DLs. With the help of AI, one is able to accurately predict, make decision within short span of time and even the best practices that can be employed within the farming.

Computer algorithms allow for the analysis of the accumulating, large datasets and are capable of identifying patterns when it comes to the environmental conditions, the growth stage of crops and the health of the soil (Baranwal, & Pateriya, 2016). Several authors have explained how AI models can predict the best planting time, harvest time or watering time from these data patterns as pointed out by Liu et al. (2020) and Liakos et al. (2018). These models help farmers alleviate the guesswork that affects the usage of resources to benefits acquired by making sound decisions.

In this case, Deep learning, which is a branch of AI, has embraced the powerful image recognition role most especially in crop health monitoring. Convolutional neural networks or CNNs are one of the commonly used deep learning tools that have been used to analyze images taken by drones or satellites' sensors. These are capable of identifying diseases, pests and other factors that stress crops and this can be done even before the same can be easily observed by even our own eyes. Sankaran et al. (2017) have opined that the use of AI in recognizing diseases and pests in crops from remote senors images of the crops is successful, which ensures early mitigation measures for enhancing crop productivity.

It also has its impact on the decision-making systems that are involved in execution of various errands in farm operations. The amount of human intervention in the process of planting or weeding or harvesting has begun to reduce drastically and instead machines which self-controlled by artificial intelligence are being used for planting, weeding and harvesting. For instance, the AI tractors and drones can scan fields and provide Fertilizing, Pesticides spraying and other services according to the real-time data collected from IoT sensors and remote sensing instruments.

AI for Crop Health Monitoring

AI has indeed played a great role in monitoring the health of crops in terms of disease and pest identification through images and Sensors. A few of such related works include the works by Mohanty et al. (2016) and Kamilaris et al. (2018) which have employed deep learning algorithms for crop disease identification from images captured by drones or UAVs. AI systems are capable of observing variations of plant health that may normally go unnoticed in the plant's physical condition. They help farmers determine the actual state of crops before the situation worsens and before extent use of pesticides becomes inevitable.

Additionally, models based on AI have been able to address sensor data collected from the field in assessing environmental factors that impact on crops. For instance, temperature, humidity, and soil moisture amount as factors, once fed into AI algorithms, are useful in determining stress factors such as dry and flooded soils that are unfavourable for plant germination. This ability enables farmers to plan on the best manners to avoid the loss of crops hence enhancing productivity.

AI for Yield Prediction

Yield prediction models use on big data acquired from IoT sensors, remote sensing technology as well as a weather forecasting system. These models are used to approximate all key factors like the state of the ground, the weather, and the growth milestones of the crops in order to estimate production output of certain regions. This is important to the farmers since it enables them to properly schedule the harvesting of crops and allocation of resources as well as the appropriate marketing strategies that go in place. Yang et al. (2019) and Kamble et al., (2020) have supported the idea of AI to raise the degree of yield prediction with the use of different data on the environment and crop.

AI in Precision Irrigation

Artificial intelligence is becoming one of the innovation tools that is making significant impacts in the irrigation systems in agriculture and use of water. Weather data, moisture content of the soil, and water needed by crops to grow are some of the inputs in an AI model that helps in the determination of the right time for irrigation as well as the required quantity. This assists in avoiding cases where a plant is overwatered or underwatered and may lead to wastage of water resources. Sadeghi et al. (2021) have also shown that the use of AI-based irrigation system helps in conserving water and yet at the same time nourishing the crops. Currently, data from IoT sensors and weather conditions mean that AI algorithms can determine the irrigation schedule that is efficient in terms of water and energy consumption.

AI-Based Soil Health Management

AI assists in soil management through monitoring data from several sensors on the soil, soil's Detecting services, and other remote sensing technologies. The existing models can detect the lack of nutrients in the soil and suggest the approach to apply fertilizers and monitor the condition of the soil at different stages. These models actually help the farmers in making decisions and come up with a better way of fertilizing so as to enhance the fertility of their soils. For instance, a study conducted by Ray et al, in 2020, and Zhang et al. in 2021 has explained how the AI algorithms may be used to estimate the nutrient demands and suggest the right amounts to apply in the fields, to avoid over-fertilization and deterioration of the soil.

Challenges and Limitations

However, some issues still arise in implementing IoT, remote sensing, and AI in agriculture. However, since the beginning of implementation of these technologies entails high costs within developing areas, the utilization of such technologies is still somewhat restricted. However, there is the need to ensure that farmer data is protected from data breaches that may be witnessed in the collection and processing of farm-related data. This integration of IoT, remote sensing, and AI systems also have challenges with human capacity especially on skilled personnel in implementing the systems that may prove hard to come across in some rural or developing regions. Moreover, the data should be collected continuously so that the AI models stay up to date, which is not always possible in the long run.

Some of the challenges in relation to AI and the IoT in precision farming are more centered on data ownership of farmers. There is also a problem of digital differential, due to which the stable connectivity of internet and technologies which are used in these technologies are not available across the globe.

Future Directions

The current and future trends of IoT, remote sensing, and AI for agriculture have some progressive aspects as follows. The concept of mobile networks which undertakes data processing locally and not in the cloud will also help perform real time uses of data. The adoption and integration of the 5G technology will also enhance the transfer of real time data that will increase the precision of intervention by specialized doctors. Moreover, innovations in utilization of artificial intelligence technology is in self-driven tractors and harvesters to enhance the farming operations by minimizing on the use of human labour. These advancements will prepare for the further shifts of the farming that will be more sustainable and efficient at some point.

PROPOSED METHODS

The devised strategies in this study pertain to the use of AI, IoT and remote sensing to improve automatic agriculture. These methods tend to solve the problem of optimal allocation of resources, early identification of diseases affecting crops, the probability of yielding and management of water in farming.

IoT Sensor Networks for Data Collection

IoT sensors will be installed in the agricultural fields that will monitor and provided real time information about moisture level, temperature, humidity and nutrient of the fields. These sensors will be mounted at various point on the field to give the variability of agriculture environment needed for control. All the collected data will be extracted and transferred to another site for the purpose of analysis. This sensor data will be used as the basis for artificial intelligence and remote sensing solutions that will aid farmers to enhance agriculture activities.

Remote Sensing for Field Monitoring

Satellites, UAVs, and drones will be used for surveillance of large farmlands where information gathering from close distances is not feasible. These drones will be fitted with multispectral and thermal cameras that will be used to take high-quality pictures that will assist in evaluating crop status, mineral nutrient content and water deficit. There are at least two reasons for choosing satellite imagery over drone data in addition to the custom the former provides a wider view of the fields, and the latter enables the monitoring of temporal changes of crop conditions. This triplet of remote sensing data will help determine disease, pest infection and nutrient deficiencies of crops at their preliminary stages.

Figure 1. Proposed Methods for Integrating IoT, Remote Sensing, and AI in Precision Agriculture

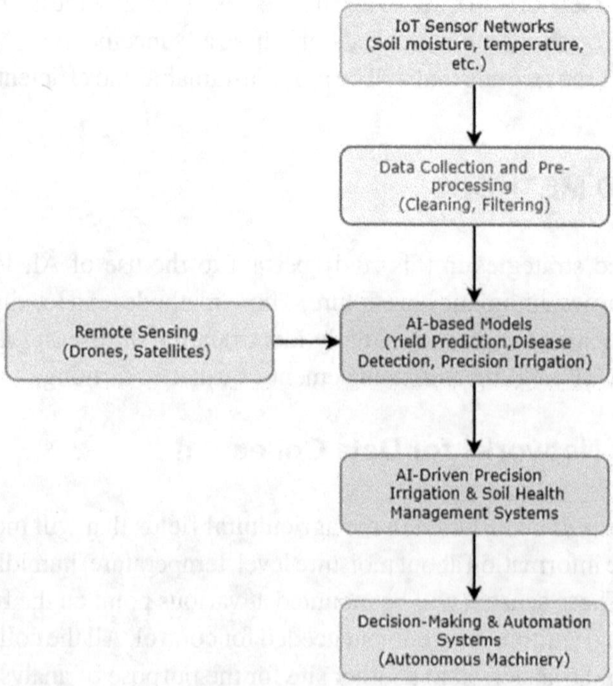

The proposed method for implementing IoT, remote sensing as well as AI into the advancement of Precision Agriculture is depicted as follows Figure 1 below. This starts with IoT senor networks which gather realtime readings of the prevailing environmental conditions such as moisture, temperature and humidity of the soil. They are then cleaned and filtered to remove any other data that is not useful in the analysis of independent variable. Other possible methods include the use of drone technologies and satellite imagery for additional field monitoring, monitoring plant health, the quality of the soil, and the existence of threats. Information gathered from IoT and remote sensing are processed through models for yield estimations, disease or pest determination, or growing requite in yield automation. AI systems are also important in the management of soils as they provide suggestions on fertilization regime and quality of the soil. The last of them concerns decision-making and automation systems with the help of which real-time decisions can be made and there is also the use of self-driving tractors for farming activities. This approach also increases efficiency, utilization of resources and ensures practices that are sustainable within the farming activities.

Several factors have been found affecting the degree of using AI models for predictive analytics in precision agriculture, despite the models' great potential. A current drawback is the absence of public data to verify most of this models. I think, certain pilot works and field trials are to be carried out to assure that those models work in different conditions and situations, such as climate difference or difference in crops.

Use of smart watering techniques, particularly automatic control using AI, have been proved to have potential in efficient use of water. Still, the implementation of such systems within areas that lag behind in terms of rather digital development lacks the quality. However, fixed costs of these technologies and the recurrent cost needed to maintain the systems limits the adoption of technologies by small-scale farmers.

The combination of all the artificial intelligence based solutions for field tracking, watering and crop health is necessary to develop a automatic, environment friendly farming system. But it do calls for adequate and sound data management platform and enhanced conformity of interoperability. Without these technologies, it will be possible to get the best out of each technology individually without really getting the intended efficiency of the farming process.

AI Models for Predictive Analytics

The key characteristic of the proposed method is that it is based on the use of AI models for creating predictive models. Supervised and unsupervised learning approaches will be used to mine the data generated from the IoT sensors and the remote sensing devices. They will include models that will help in estimating the yields of different crops, disease or pest outbreak predictions, and the right time and frequencies for irrigation. For yield prediction, it is going to use the crop information that it will receive to give accurate outcome that pertains to the yield that is expected to be produced. For the health of crop, images captured from the drone and satellites will be play in the deep learning model especially the Convolutional Neural Network (CNN). These models shall observe and from the unaltered visuals of crops and real-time information, detect diseases or pest infestations in its early stages.

AI-Driven Precision Irrigation System

An automatic and accurate irrigation system will be implemented to apply water in the field based on artificial intelligence. It will draw instantaneous information from the moisture in the soil, weather predation, and water necessary for crops. The AI models will also ensure that they supervise the amount of water on the soil and adjust the irrigation time to ensure that the crops do not get either too much or too little water, yet, save water.

AI-Based Soil Health Management

Under the soil health management, AI algorithms will use data from the soil sensors to determine on its nutrient content, ph level, and organic matter. In this case, fertilization recommendation for the AI models will be developed to give best measures to fertilize and, in addition, monitoring the changes in the status of the soil over a period. They will enhance farmers' ability to make reasonable decision on use of the soils, ways of soil management, ways of avoiding the whats more use of fertilizers amongst others.

Integration and Automation

The incorporation of these methods will also enable easy coordination between IoT sensor networks, remote sensing systems, and the used AI models. The technology would facilitate timely decisions concerning the use of the inputs like water, fertilizers, and pesticides. The advanced technology will also be applied to increase automation in crop production through its self-driving tractors and drones. The proposed methods therefore intend to cause a round approach to precision agriculture to collect real time data for the analysis of process and automation in farmers' yield and usage of farming resources with an intention of increasing on the overall productivity of the farm.

RESULTS AND DISCUSSIONS

Impact of AI, IoT, and Remote Sensing on Agricultural Efficiency

The advancing technologies such as Artificial Intelligence, Internet of Things, and remote sensing have given some rich guidance on how the agricultural efficiency can be improved, the resources can be utilized more effectively, and even the yield prediction is expected to increase. For instance, using historical crop science information and using data coming from IoT sensors, AI models have also enhanced accuracy when it comes to crop yields' forecasts. These help in decision making as regards to the time of planting, period of incorporating water and time to harvest crops. Cortana uses IoT sensors to give farmers insight on the real world data on the moisture, temperature and humidity of the soil or the conditions that impact crop grow. The integration of remote sensing, for instance, satellite either through aerial imagery or use of unmanned aerial vehicles (UAVs) or drones) enables larger areas in agriculture to be monitored closely and have details information on the status of the crops, the soils, water stress among others.

Figure 2 shows the variation of the soil moisture content with time using IoT sensors mounted on the soil. Hence, the line plot used to identify the soil moisture, demonstrates variation in the amount of water present in the surfaces of soil through seasons. In particular, daily rainfall increases in the rainy season while during dry season the amount of moisture in the soil is considerably low. IoT sensors into the soil monitoring enable the farmers to regulate the usage of water in irrigation without affecting the quality of the crops. This speaks volumes about the importance of using IoT in order to deal with water resources in the best way possible.

In Figure 3, a heatmap for crop health check using remote sensing is produced. Here is an example of how the satellite or drone imagery can help to check vast areas of agricultural fields. This means that an application of AI-based image processing system on RS can help in recognizing various stress indicators including nutrient deficiencies, diseases or pests. Apply the advances in remote sensing technologies with the help of AI in farm management so that the interventions are targeted towards precise areas of the field thus making it economical as well practical.

The AI technology for crop yield predictions continuative the crop yield estimations in figure 4 for different types of crops. The bar chart is used to show the performance of the AI models by comparing the predicted yields which has been produced by the AI models to the actual yields obtained at the time of harvest. Shown in the chart, it is evident that the application of AI in helping farmers in forecasting production can be useful in predicting yield, utilizing resources, and marketing crops. The small deviations established between the actual and the predicted yields suggest that AI can great help with advancement of forecast accuracy leading to improvement of operation decisions.

In Figure 5 below, the line plot displays water usage prior to introducing the use of AI in managing irrigation and after it was included as a tool. The graph below also shows that, every season, the usage of water reduced after the integration of the AI reducing the usage of water marked by the 'after AI optimization line being below that of the 'before AI integration' line. It also suggests that by incorporating real-time data from the IoT sensors, such as moisture level of the soil and weather predictions, the dirty work of providing a more efficient AI-based irrigation model can be carried out. This specific irrigation system plays a critical role in water sparing which is scarce in some areas of the globe.

Figure 6 is a scatter plot for pest infestation monitoring, based on images captured by the drone as well as the aid of AI to process the image. Pest infestation can be identified when AI algorithms analyse drone images and identify indications of pests that would be otherwise difficult to discern. From these images, the AI is able to see other areas that are infested and this helps the farmers to come up with other ways and means of dealing with the issue. The timely identification of pests and their control reduces crop loss, makes it unnecessary for large-scale pesticides application, which can be very beneficial for the sustainable agriculture.

Figure 7 is a bar chart that shows the amount of fertilizer used before and after the use of artificial intelligence. The above figure Clear graph indicating that, there is reduction on use of fertilizers after implementation of artificial intelligence ' smarter soil management'. They also predict fertilizer requirements with the aid of results got from the soil sensors hence avoiding over application that causes negative effects on the soil environment. This change leads to improved use fertilizers and hence improved soil quality and less impact on the environment from the use of fertilizer.

The central focus of this paper, and the overall capability that the proposed AI solution enables, can be summarized with a flowchart presented in Figure 8. The process starts with the observations from the IoT sensors as well as the remote sensing approaches which include the soil water content, the weather conditions of the area and crops' health status. This information is then channeled through artificial intelligent systems which act as decision making tools that produce suggestions which may include irrigation, application of fertilizers or use of pesticides. The effects of these decisions are then implemented by automatic equipment including self-driving tractors and drones. The flowchart highlights the continuous process of data processes and decisions, and automation whose adoption by the AI driven solution improves farm operations because human interferences are limited, and the processes are well managed.

Figure 2. Soil Moisture Variation Over Time

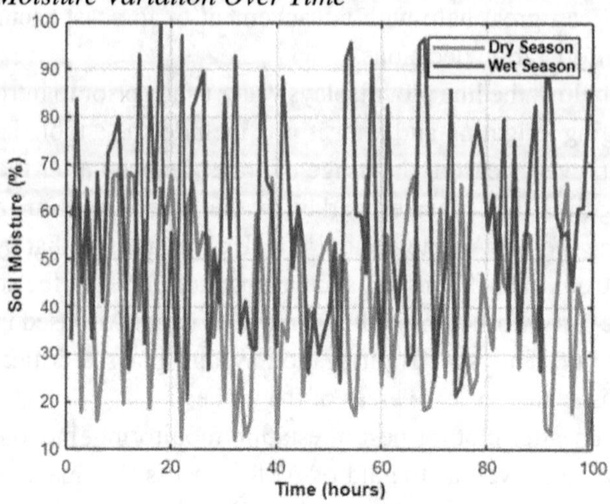

Figure 3. Crop Health Monitoring Using Remote Sensing

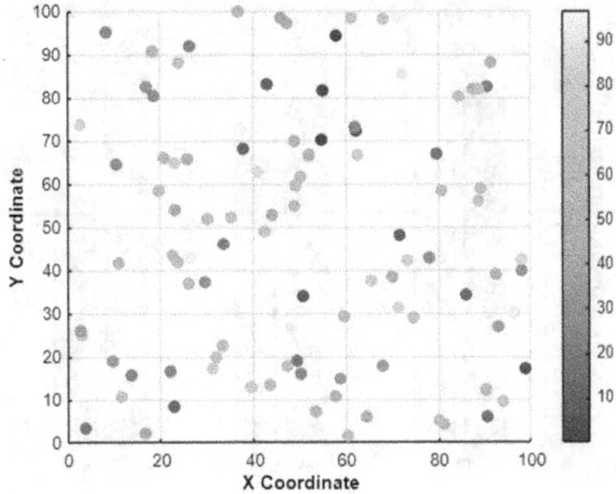

Figure 4. Comparison of Crop Yield Predictions vs. Actual Yields

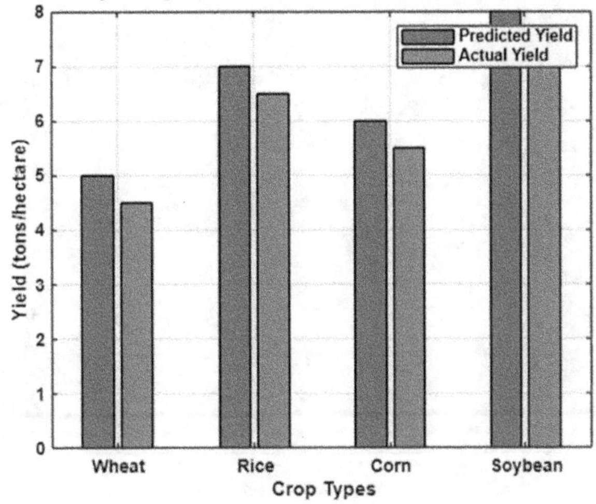

Figure 5. AI-Driven Irrigation Optimization

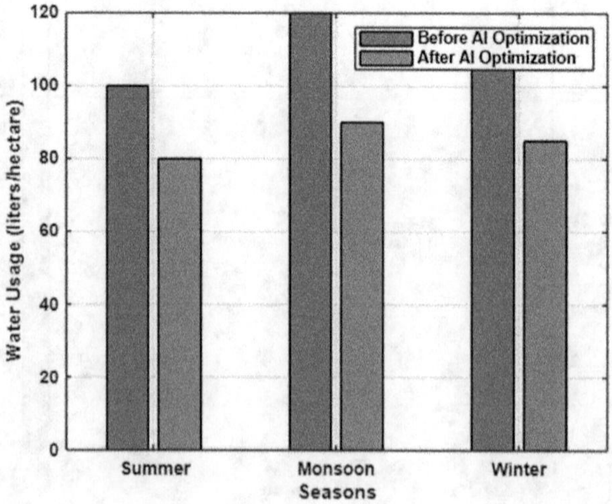

Figure 6. Drone-Based Pest Detection

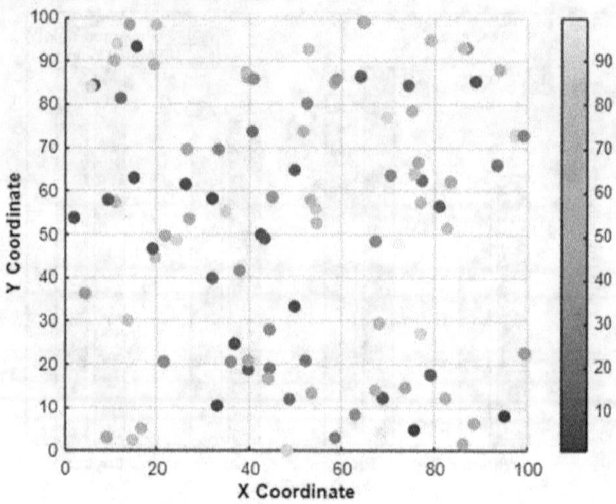

Figure 7. AI-Powered Fertilization Optimization

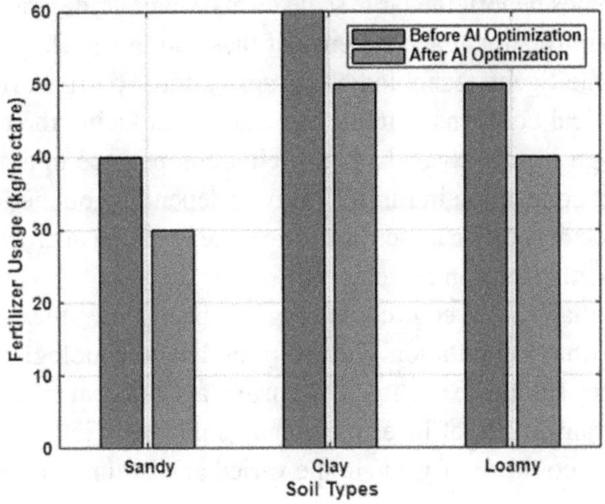

Figure 8. Autonomous Farm Operations Using AI

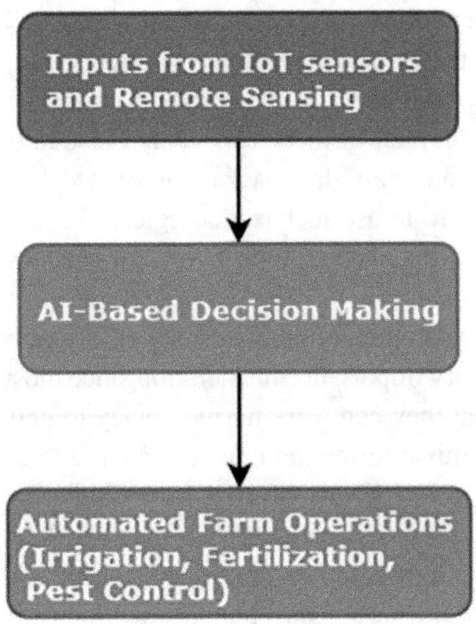

Thus, as can be seen in pilot studies, the effect of AI and IoT improves agricultural efficiency through adaptive labor cost, improved yield of crops, and better crop management. Such technologies are however shown to have a major problem of scalability based on these studies. It is for this reason that while elaborately developed technologies benefit the large scale farms, small holder farmers are unable to bear the capital investment and expense of these innovations.

The use of intelligent technologies in agriculture, particularly, AI and IoT face certain practical concerns. Among the challenges facing the expansion of e-commerce are high capital costs, lack of digital competence in a population, and restricted internet connection in rural. However, depending on the IoT devices and sensor calibration are specific issues that have to be resolved in order to meaningful continuous data gathering can be achieved.

It is predicted that 5G and edge computing will bring down the expense of relaying data and real-time computation. However, for these technologies to be effective and beneficial, they should be anchored on current agricultural infrastructures. This would involve a fair degree of investment and some appreciation of how some of these technologies could be of value in the varied agricultural environme

Challenges in Implementation

However, there are various factors that demerit the advancement of AI, IoT, and remote sensing as follows; location of health facilities is one of the major challenges that arise when implementing the system, mainly because it is expensive to start such projects. The Internet of Things sensor networks, remote sensing devices, and AI systems that would apply in the automation of the system may be costly, especially for the smallholder farmers, especially those in the developing regions. However, there are also certain drawbacks, one of which is a cost implication in educating farmers on how to use such technologies. The second problem is related with data integration where data from different sources have to be used together. As mentioned earlier, data from IoT sensors and remote sensing devices contrast in the type of data they generate and thus the current format of the two is not easily compatible. This is a very important consideration since most of these systems need to be integrated so that they can work harmoniously to deliver intended services. Moreover, privacy of this information must be ensured to avoid any misuse of the information of the farmers. The use of these technologies also implies that special skills are needed in order to use and maintain them. If there is no such specialist, which may be the case in rural or less developed countries, the application of such technologies remains quite slow. Thus, policies that focus on training delivery and capacity development should be adopted to ensure farmers receive better returns from these innovations.

Scalability of Solutions

AI, IoT, and remote sensing have been used in the large-scale farming system, but the issue arises on how to implement them in smallholder farmers in the developing world. This is true because the cost of technology is high, while other requirements such as infrastructure and human capital for technology implementation are very scarce. Depending on the requirements, several interventions are already being proposed as possible solutions include cheap IoT sensors, low-cost drones as well as cloud models for AI that are preferred due to the reduction in need for expensive hardware. Moreover, existing efforts to enhance internet connection as well as access to mobile technology in the rural areas are creating opportunities on how smallholder farmers can adopt precision agriculture. This is because as these technologies become more cheap and within the reach of most farmers, they will acquire them, thus improving on productivity and sustainability of farming in the third world countries.

Impact of 5G and Edge Computing

5G along with the edge computing will contribute towards improving the future of precision agriculture through improving the communication between the IoT sensors and remote sensing devices with the help of AI systems. This will enhance a real time transmission and processing of data that will enable farmers to get feedback on their practices in the farms thereby making decisions in real time. Edge computing where data are processed locally instead of placing them in the cloud will build on this feature to reduce the decision-making time. In view of this, edge computing will lead to more efficient use of bandwidth and relieve centralized cloud computing systems. This will prove advantageous in areas that may have poor internet connectivity in the rural areas that practice farming. Thus, 5G, in combination with EC, will let the industry provide crop growers with accurate, nearly real-time data analysis and automated decision-making. They will help in implementation of advance farming techniques to conserve the results grow in a sustainable manner.

CONCLUSION

This chapter has discussed about the use of Artificial Intelligence (AI), Internet of Things (IoT) and remote sensing technologies in precision farming. These technologies are bringing radical changes in the agricultural practices that are making them more productive, eco-friendly, and based on analysis of a large number of variables. AI solutions based on IoT sensors and remote sensing devices today have enhanced yields, resource utilization and work output estimations. Soil moisture and climate sensors offer current information of environmental conditions while the drones and satellite images give the big picture on the state of the land. From this case of integrated technologies it has led to proper operation of irrigation, right use of fertilizers, early identification of pests and diseases, and better treatment of crops that in turn lead to higher yields as well as conservation of the environment. The integration between AI, IoT and remote sensing technologies has made farming more integrated, automatic and intelligent system so farmers can easily develop better strategies for the improvement of efficiency and efficacy.

AI is still in the process of growing its application in theories of precision agriculture as we look to the future. Some of the developments that may occur in the future are robot controlled tractors and vehicles that can plant, weed, or harvest crops without assistance of human beings. The use of real-time data in farming systems will increase as well as artificial intelligence algorithms becoming a part of setting up the operations to satisfy the efficient use of the available resources. In addition, the use of modern learning techniques in 'big data analytics will help in improving crop yield prediction, pests control, and the right time for irrigation. However, the adoption of 5G along with edge computing will enhance the data rates and time for processing the changes in the agricultural conditions. They will also improve the efficiency of operations, increase sustainability, and expand precision farming systems' accessibility for farmers.

In conclusion, AI, IoT, and remote sensing mean a tremendous increase in agricultural productivity and sustainability but carry a set of socio-economic and logistics' adjustment predictors. Its challenges like cost of technology, ability to use the new technology, and ownership of data have to be solved so that these technologies that are an innovation would be of benefit to all farmers, big or small. However, further, the agricultural sector requires ethic and environment impacts to decide that these technologies are helping the society to progress in the right direction.

REFERENCES

Alam, S. (2023). "Security Concerns in Smart Agriculture and Blockchain-based Solution," *2022 OPJU International Technology Conference on Emerging Technologies for Sustainable Development (OTCON)*, Raigarh, Chhattisgarh, India, 2023, pp. 1-6. DOI: 10.1109/OTCON56053.2023.10113953

Aravindh, S., Nithisha, C., & Tanguturi, R. C. (2024). "Enhanced Agricultural Methods and Sustainable Farming Through IoT and AI Technology," *2024 Second International Conference on Intelligent Cyber Physical Systems and Internet of Things (ICoICI)*, Coimbatore, India, 2024, pp. 1206-1212.

Baranwal, T., & Pateriya, P. K. (2016, January). Development of IoT based smart security and monitoring devices for agriculture. In 2016 6th international conference-cloud system and big data engineering (confluence) (pp. 597-602). IEEE.

Dubey, A. K., & Chahar, R. (2023). "Revolutionizing agriculture: harnessing IoT for productivity, sustainability, and innovation," 2023 6th International Conference on Recent Trends in Advance Computing (ICRTAC), Chennai, India, 2023, pp. 390-396. DOI: 10.1109/ICRTAC59277.2023.10480861

Felício, T., Postolache, O. A., Rodrigues, M. J., & Sebastião, P. (2023). "Vineyard Thermal Stress Assessment Through the Combination of In-Situ and Remote Sensing Technology," *2023 IEEE Sensors Applications Symposium (SAS)*, Ottawa, ON, Canada, 2023, pp. 1-6. DOI: 10.1109/SAS58821.2023.10254061

Guidotti, D., Marchi, S., Antognelli, S., and Cruciani, A., (2019). "Water Management: Agricolus tools integration," 2019 Global IoT Summit (GIoTS), Aarhus, Denmark, 2019, pp. 1-5.

Gulay, C. F., & Fajardo, A. C. (2024). "Rethinking Soil Diagnostics: Why IoT-Based Soil Sensors are Essential for the Future of Precision Agriculture," *2024 IEEE Asia-Pacific Conference on Geoscience, Electronics and Remote Sensing Technology (AGERS)*, Manado, Indonesia, 2024, pp. 318-323. DOI: 10.1109/AGERS65212.2024.10932884

Kudari, J. M., & Samal, A. (2021). Smart Agriculture and Smart Farming using IoT Technology. *Journal of Physics: Conference Series, 2089*(1), 012038. DOI: 10.1088/1742-6596/2089/1/012038

Martínez, J. V. Y., Skarmeta, A. F., Zamora-Izquierdo, M. A., & Ramallo-González, A. P. (2020). "IoT-based data management for Smart Agriculture," *2020 Second International Conference on Embedded & Distributed Systems (EDiS)*, Oran, Algeria, 2020, pp. 41-46. DOI: 10.1109/EDiS49545.2020.9296443

Pandey, U., Vashisth, A., & Mishra, A. (2024). "AI and IoT Enabled Soil Health Monitoring System," 2024 Second International Conference Computational and Characterization Techniques in Engineering & Sciences (IC3TES), Lucknow, India, 2024, pp. 1-6. DOI: 10.1109/IC3TES62412.2024.10877563

Pero, C., Bakshi, S., Nappi, M., & Tortora, G. (2024). IoT-Driven Machine Learning for Precision Viticulture Optimization. *IEEE Journal of Selected Topics in Applied Earth Observations and Remote Sensing*, *17*, 2437–2447. DOI: 10.1109/JSTARS.2023.3345473

Pincheira, M., Donini, E., Giaffreda, R., & Vecchio, M. (2020). *"A Blockchain-Based Approach To Enable Remote Sensing Trusted Data," 2020 IEEE Latin American GRSS & ISPRS Remote Sensing Conference*. LAGIRS.

Rodriguez-Gallo, Y., Cañas, H., & Cardona, M. (2024). "Agriculture 4.0 in Central America: Current Status, Challenges, and Future Trends," 2024 IEEE 42nd Central America and Panama Convention (CONCAPAN XLII), San Jose, Costa Rica, 2024, pp. 1-6. DOI: 10.1109/CONCAPAN63470.2024.10933593

Sarjerao, J. S., & Sudhagar, G. (2024). "Integration of Remote Sensing and IoT for Real-Time Monitoring of Irrigation in Smart Farming," 2024 1st International Conference on Innovative Sustainable Technologies for Energy, Mechatronics, and Smart Systems (ISTEMS), Dehradun, India, 2024, pp. 1-6. DOI: 10.1109/ISTEMS60181.2024.10560275

Sawant, S. A., Adinarayana, J., & Durbha, S. S. (2014). "KrishiSense: A semantically aware web enabled wireless sensor network system for precision agriculture applications," *2014 IEEE Geoscience and Remote Sensing Symposium*, Quebec City, QC, Canada, 2014, pp. 4090-4093. DOI: 10.1109/IGARSS.2014.6947385

Sethi, H. S., & Prasanth, P. V. (2023). "Optimizing Resource Allocation in Agriculture through IoT and AI," 2023 7th International Conference on Electronics, Communication and Aerospace Technology (ICECA), Coimbatore, India, 2023, pp. 1573-1578.

Shafi, U., Mumtaz, R., Iqbal, N., Zaidi, S. M. H., Zaidi, S. A. R., Hussain, I., & Mahmood, Z. (2020). A Multi-Modal Approach for Crop Health Mapping Using Low Altitude Remote Sensing, Internet of Things (IoT) and Machine Learning. *IEEE Access : Practical Innovations, Open Solutions*, *8*, 112708–112724. DOI: 10.1109/ACCESS.2020.3002948

Sharma, S., Rani, S., Sharma, A., & Dogra, A. (2024). "Advancements in Image Fusion Techniques for IoT Systems: Current Trends and Applications," *2024 Second International Conference on Intelligent Cyber Physical Systems and Internet of Things (ICoICI)*, Coimbatore, India, 2024, pp. 368-373. DOI: 10.1109/ICoICI62503.2024.10696732

Singh, P. N., Saxena, U., & Chaudhary, M., and Pravartan, (2024). "Smart Agriculture: IoT-Enabled Precision Environmental Monitoring and Management," *2024 International Conference on Electrical Electronics and Computing Technologies (ICEECT)*, Greater Noida, India, 2024, pp. 1-6.

Singh, R. K., Berkvens, R., & Weyn, M. (2021). AgriFusion: An Architecture for IoT and Emerging Technologies Based on a Precision Agriculture Survey. *IEEE Access : Practical Innovations, Open Solutions*, *9*, 136253–136283. DOI: 10.1109/ACCESS.2021.3116814

Veeramanikandan, P., & Kumar, N. A. (2024). "Design and Implementation of an IoT-Based Control System for Precision Food Manufacturing," in Rathore, V.S., Piuri, V., Babo, R., S, K. (eds) Universal Threats in Expert Applications and Solutions. UNI-TEAS 2024. *Lecture Notes in Networks and Systems*, vol 1007, Springer, Singapore.

Victor, N., Maddikunta, P. K. R., Mary, D. R. K., Murugan, R., Chengoden, R., Gadekallu, T. R., Rakesh, N., Zhu, Y., & Paek, J. (2024). Remote Sensing for Agriculture in the Era of Industry 5.0—A Survey. *IEEE Journal of Selected Topics in Applied Earth Observations and Remote Sensing*, *17*, 5920–5945. DOI: 10.1109/JSTARS.2024.3370508

Ye, J., Chen, B., Liu, Q., & Fang, Y. (2013). "A precision agriculture management system based on Internet of Things and WebGIS," 2013 21st International Conference on Geoinformatics, Kaifeng, China, 2013, pp. 1-5. DOI: 10.1109/Geoinformatics.2013.6626173

Chapter 3
AI-Powered Defect Detection Using Deep Learning Enhancing Industrial Automation with U-Net and YOLOv3

Fumiya Yamaguchi Yagi
Sanyo-Onoda City University, Japan

Fusaomi Nagata
https://orcid.org/0000-0002-7128-2151
Sanyo-Onoda City University, Japan

Maki K. Habib
https://orcid.org/0000-0001-8088-8043
The American University in Cairo, Egypt

Keigo Watanabe
Okayama University, Japan

ABSTRACT

This chapter explores the application of deep learning models, specifically U-Net and YOLOv3, in detecting crack-type defects in industrial materials. The research addresses key challenges, such as the limited availability of defective data samples, the difficulty of distinguishing fine defect details, and the need for precise quantitative evaluations in automated inspection systems. The chapter introduces the significance of defect detection in the context of smart manufacturing and discusses the limitations

DOI: 10.4018/979-8-3693-7994-3.ch003

of traditional manual inspection methods. U-Net, known for its superior pixel-level semantic segmentation capabilities, is utilized for detailed defect localization, while YOLOv3, a real-time object detection model, is optimized for fast detection but has trade-offs in precision, particularly for irregular defect shapes. Both models are trained on a custom dataset provided by a material manufacturer incorporating a novel augmentation method that enhances generalization and improves detection accuracy by expanding defective sample diversity.

INTRODUCTION

A. Background on Recent Defect Detection System

The integration of Artificial Intelligence (AI) into industrial automation has significantly been transforming quality control processes, particularly in defect detection. Traditional manual inspection methods are often labor-intensive and prone to undesirable human error, leading to inconsistent product quality. In contrast, AI-driven systems offer enhanced accuracy, efficiency, and the ability to learn and adapt to new challenges automatically.

There are various types of quality control in manufacturing, including visual inspection, dimensional inspection, strength inspection, and performance inspection, but this chapter focuses on visual inspection. Recent advancements in deep learning, especially in CNN (Convolutional Neural Network), SVM (Support Vector Machine), YOLO (You Only Look Once) have further propelled the capabilities of automated visual inspection systems. Prunella et al. reviewed and introduced more than 220 relevant articles from the related literature published until February 2023, covering the recent consolidation and advances in the field of full automatic and deep learning-based surface defects inspection systems, applied to many industrial applications (Prunella et al., 2023). Recent studies have also explored novel deep learning architectures for defect inspection, including hybrid CNN-transformer models and self-supervised learning for anomaly detection (Pei et al., 2023).

The authors also have developed several defect detection systems on MATLAB. Additionally installed options are Computer Vision Toolbox, Deep Learning Toolbox, Global Optimization Toolbox, Image Processing Toolbox, Parallel Computing Toolbox, Statistics and Machine Learning Toolbox, and so on. For instance, Nagata et al. (2021), proposed a defect detection system using a convolutional neural network-based support vector machine (CNN-SVM) technique to enhance feature extraction and classification accuracy. Their approach demonstrated improved generalization ability and robustness in detecting micro-defects in industrial components (Nagata et al., 2021). Similarly, Zhang et al. (2021) introduced a novel automated visual inspection framework integrating generative adversarial networks (GANs) for defect augmentation, thereby overcoming dataset imbalance issues in manufacturing applications (Zhang et al., 2021).

In addition to the essential need such that defective products should be identified based on the presence or absence of a defect, there is another important need from manufacturers to want to classify defects based on quantitative criteria such as area and length. Recent methodologies have explored integrating deep regression models within defect detection frameworks to automate precise defect measurement in industrial quality control (Fang et al., 2023). Carrilho et al. (2024) provided an extensive review of deep learning applications in automated defect detection, with a particular focus on fabric inspection. Their survey highlights the growing use of computer vision-based regression models for precise defect localization and measurement, aligning well with the discussion on quantitative evaluation in industrial quality control. Additionally, the authors emphasize challenges such as dataset standardization and scalability, which are critical factors in real-world defect detection systems.

B. Crack-Type Defect Detection

Many challenges for crack-type defect detection and inspection have been conducted. For example, Yuan et al. used Matlab software for image processing to improve the contrast because the cracks in the test block were too small, and the contrast was not high. It is reported that the length of defects was detected by introducing a scale for quantitative analysis of the iron block (Yuan et al., 2020).

Shaterian and Mrozowski proposed a novel crack detection sensor using a microstrip loaded with a Dumbbell-Shaped Defected Ground Structure (DS-DGS). The proposed sensor exhibited a good sensitivity above 260 MHz shift for a crack with a 200 μm width at the relatively low operating frequency of around 2 GHz (Shaterian and Mrozowski, 2022). Zheng et al. proposed a crack detection method based on attribute weighted naive Bayes improved OTSU algorithm, in which based on the analysis of the surface defects in MEMS devices image, the edge information of the crack defect was extracted by some conventional image processing techniques (Zheng et al., 2023). Their experimental results showed that the proposed crack detection method could obtain the crack defect information of MEMS devices efficiently and accurately. Also, Gong et al. proposed a magnetic circuit model to analyze the magnetic field concentration mechanism. Their experimental results showed that the proposed method could enhance the magnetic leakage field strength resulting from the cracks and improve the signal-to-noise ratio of the MFL signals (Gong et al., 2024).

However, it seems that there is almost no system that can not only determine the presence or absence of defects, but also indicate the extent of the defects using quantitative criteria such as length or area based on pixel-wise quantity. In this paper, to cope with the need for pixel-wise quantitative evaluation, a U-Net (Ronneberger et al., 2015) which is one of typical FCN (Fully Convolutional Network) models, and a YOLOv3 (Redmon and Farhadi, 2017, The MathWorks, Inc., 2025-1, The MathWorks, Inc., 2025-2, Yanan et al., 2018) which is one of YOLO (You Only Look Once) models, are designed, trained and applied to the detection of crack-type defects in an industrial material as shown in Fig. 1, respectively. As indicated by red arrows, the defects appearing in the manufacturing process of this industrial material are thin long clacks. Defect detection performances, i.e., generalization ability, of U-Net and YOLOv3 are evaluated and compared using test images of an industrial material. Also, easiness of making gTruth dataset, designability of model, training time, forward calculation time, and ability of quantitative evaluation are compared. In addition, image processing techniques for data augmentation are also introduced for the U-Net model and the YOLOv3 model.

Figure 1. Image samples used for training U-Net models, in which crack-type defects to be detected are seen in the images of NG products. Note that upper half parts are not products and covered with blue sheets but lower half parts are target products. Also, handwritten blue lines are not defects, but thin clacks Indicated by a red arrow are the target defects.

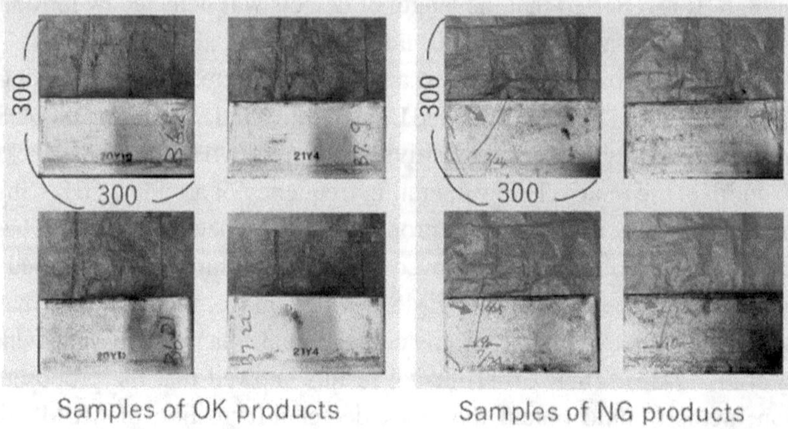

DESIGN OF FCN MODEL AND APPLICATION TO DEFECT DETECTION

C. Semantic Segmentation

In this study, we have focused on Semantic Segmentation (SS) which has potential ability of quantitative evaluation for defect detection. SS is a deep learning algorithm that associates labels and categories with all pixels in an image. It is used to recognize an object consisting of pixels that form a characteristic category. If an image can be divided into multiple areas at the pixel level, it is possible for SS to detect irregularly shaped objects compared to the conventional object detection methods which have to fit the target object within the bounding box (Chen et al., 2018).

D. Ground Truth (gTruth) Dataset for Training FCN

The gTruth dataset for training a U-Net model consists of one gTruth file, multiple image files (.jpg) with a defect at least and their pixel-labelled files (.png). Also, the gTruth file has full path names of all the image files and pixel-labelled files. Figure 2 shows the structure of the gTruth dataset.

In this study, SS is applied to the detection of clack-type defects that occur in the vertical direction in an industrial material as shown in Fig. 1. Although the manufacturing process of this material is currently produced on an automation line, images for inspection are manually taken by a worker. In addition, not only the upper part of the product contains a blue sheet, but also the product part includes handwritten letters and lines that resemble scratches. Therefore, it is not easy for general CNN models to accurately detect the target clack-type defects.

A total of 103 images consisting of 31 OKs and 72 NGs were provided by a cooperative manufacturer for evaluation. 38 NG images out of 72 were pixel-wisely labeled on defective areas needed for learning, and the remaining 34 images were used for testing. We used an annotation application called Image Labeler available on MATLAB for labeling the 38 NG images with three types of features, i.e., background, product, and clack. A gTruth file for training was created through the labeling process. Note that the dataset A com- posed of the gTruth file and the included image files is used for training an SS model. Also, the dataset B is augmented from the dataset A by horizontally flipping the images and reconstructing the gTruth file, consequently total of 38×2=76 image files (.jpg), their pixel-labelled files (.png), and the gTruth file, were made into the dataset B.

Figure 2. Structure of gTruth datasets A and B for U-Net.

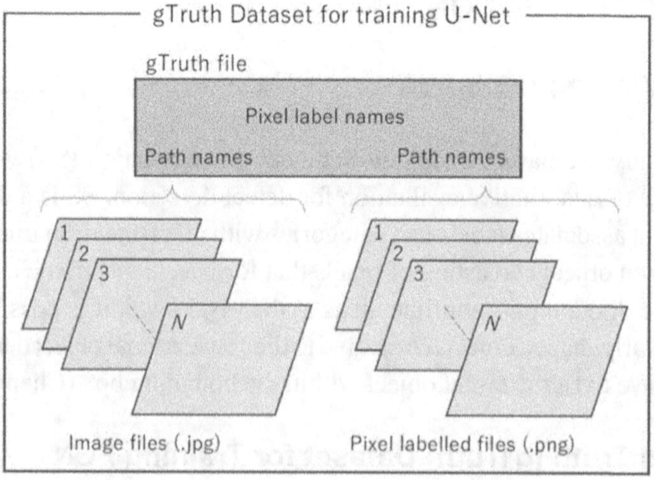

E. Augmentation of gTruth Dataset for U-Net

Since the defect occurrence rate of the product focusing on in this paper is low, some kind of augmentation method of gTruth dataset for efficiently training an FCN model should be considered to enhance the generalization ability. Several researches, in which detection accuracy was improved with the increase of the number of images used in CNN training, have been reported. However, depending on industrial products, the occurrence frequency of defective products tends to be often low. In such a case, it is not easy for makers to collect sufficient number of images including defects. To cope with the problem, the authors propose a new automatic augmentation method of training data, i.e., gTruth dataset for U-Net as shown in Fig. 2. The effectiveness of the method is evaluated through the classification accuracy and the reliability of detection of defective areas after training.

The proposed augmentation method is explained using Figs. 3 and 4. First, only the product area consisting of 300×150 pixels is cropped from the original image as shown in Fig. 3 that has been pixel-wisely labeled in advance. Then, the cropped image as shown in Fig. 3(b) is pasted on an image randomly selected from 75 background images (300×300 pixels) prepared in advance, while moving it by an arbitrary number of pixels from the top to the bottom. Through this process, multiple training images can be automatically augmented from one original image. Figure 4 shows examples of the augmentation of which pasted positions are 1, 50, and 100 pixels.

Actually, a total of 380 images were generated from dataset B by using the proposed augmentation method, in which five images are augmented from one cropped image while putting it on the bottom position in the randomly selected 5 background images. By adding these 380 images to the original dataset B, dataset C composed of 76+380=456 images were obtained. Furthermore, within dataset C, only images with the defects that could be more clearly identified through human visual inspection were reselected, and a thorough relabeling process was conducted to certainly cover the defects. Dataset D was obtained through this process.

Figure 3. Cropping of a target material area.

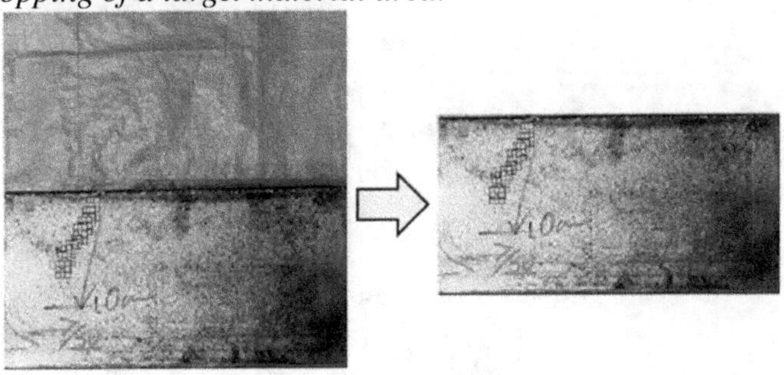

(a) Product image (b) Cropped image

Figure 4. Examples of the augmented images by moving the cropped image with 1, 50, and 100 pixels.

1 pixel 50 pixels 100 pixels

Figure 5. The authors' originally constructed U-Net type model to detect crack-type defects.

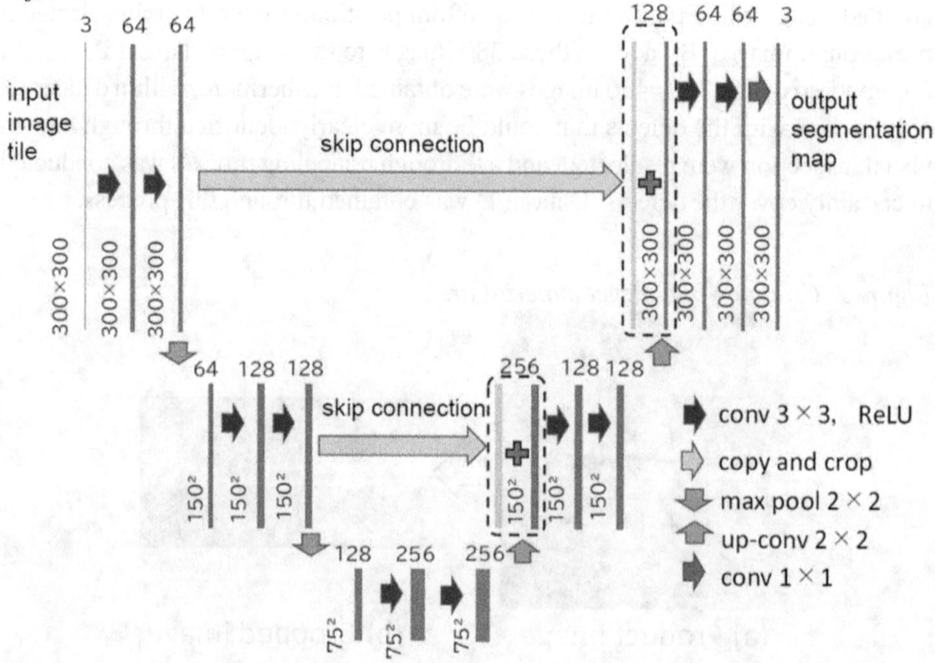

F. Training of U-Net

We built a U-Net on MATLAB as shown in Fig. 5 which is one of typical SS models (The MathWorks, Inc., 2025-3). U-Net consists of an encoder network and a corresponding decoder network. U-Net also has skip structures called skip connections that connect the encoder feature map to the decoder feature map. Skip connection is a direct connection from a layer at the encoder part to the corresponding layer at the decoder part as shown in Fig. 5. The depth of the encoder and decoder in the U-Net is set to 2. The depth is determined by how many times an input image is down-sampled by the encoder and up-sampled by the decoder. In the U-Net, feature maps encoded in the former half layers are skipped to the latter half layers, and copied as they are, and synthesized. Low level features in input-side layers and middle level features in hidden layers can be propagated to the output-side layers by long skip connections, so that they can be effectively utilized for final prediction (The MathWorks, Inc., 2025-4). By shortening the processing by the middle layers and directly propagating features to the terminal layer, the U-Net has an advantage that the features from the former layers can be actively used for prediction.

Models trained using each dataset and the Adam optimizer (Adaptive Moment Estimation) (Kingma and Ba, 2015) were saved as U-Net A, U-Net B, U-Net C, and U-Net D, respectively. The performances of these four models are compared with 31 OK product images and 34 NG product images for testing. We applied an 8-connectivity method to images generated from the output layer of U-Nets to segment pixels identified as defects. Classification accuracies are assessed by measuring the areas (number of pixels) of the objects formed by these segmented pixels. In this experiment, a threshold value of 140 is determined by trial and error, and if the area of the connected defect region is 140 or more, it is classified as a defective product. Note that the output from each U-Net is a map with the same resolution as the input image. Each pixel has one of the values of 1, 2, or 3 based on the predicted result, corresponding to background, product, and defect, respectively. As a result, by changing the values of pixels estimated as defects from 3 to 255 and also setting the values of other pixels to 0, a binary image clearly showing defect regions can be obtained. By calling the function named 'regionprops()' provided by MATLAB, which enables the measurement of properties of image areas, it becomes possible to extract regions with defect features using the 8-connectivity method and calculate several attributes such as area and COG.

G. Classification Experiment Using 4 Types of U-Nets

After comparing the classification results of four U-Net models, it was confirmed that U-Nets C and D were able to output all the three categories. It was also confirmed that U-Net D achieved a classification accuracy of over 96.9% for both OK and NG products. Tables 1 and 2 represent the classification results by confusion matrices when 34 NG test images and 31 OK test ones were given. Figure 6 shows six image examples output from U-Net D, in which they are classified as defect products. Since yellow, blue, and light blue represent background, product, and clack-type defects, respectively, we can confirm that all three categories are well segmented. It is expected from these results that quantitative assessment based on thresholds for the areas and lengths of defects will be possible, which is an advantage compared with YOLO models.

Table 1. Classification results by U-Net C model (row: true labels, column: predicted label), in which accuracy, precision, recall, and F value are 93.9, 94.1, 94.1, and 94.1, respectively.

	Anomaly	Normal
Anomaly	32	2
Normal	2	29

Table 2. Classification results by U-Net D model (row: true labels, column: predicted label), in which accuracy, precision, recall, and F value are 96.9, 97.1, 97.1, and 97.1, respectively.

	Anomaly	Normal
Anomaly	33	1
Normal	1	30

Figure 6. Image samples of industrial materials generated by U-Net D model, in which crack-type defects are visually predicted with light blue color.

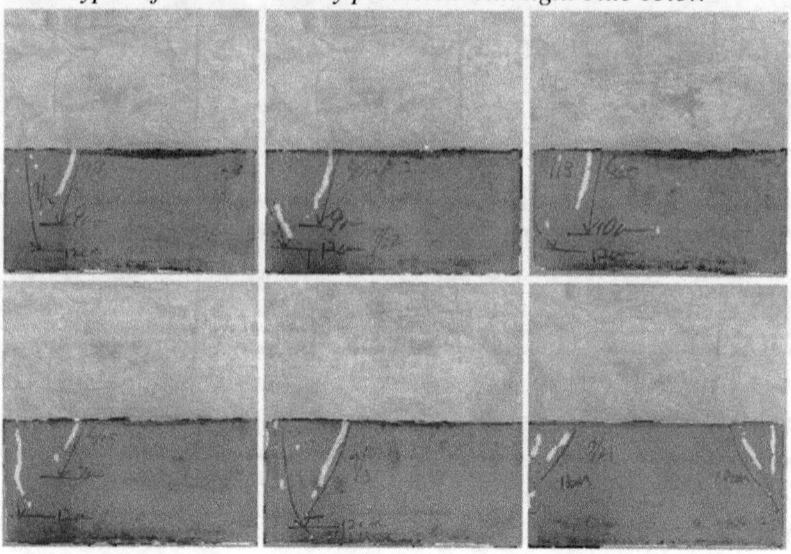

DESIGN OF YOLO MODEL AND APPLICATION TO DEFECT DETECTION

A. Systematic Classification Using a YOLOv3 Model

YOLOv3 is an AI model that uses CNN to detect the positions and classes of objects included in the input image (Wang et al., 2022). YOLOv3 object detection model runs based on a CNN to generate network predictions from multiple feature maps obtained through convolution. Specifically, it generates bounding boxes to surround objects and visualize their position, size, and the area. Additionally, it uses anchor boxes to detect the objects' classes within the image, while predicting the following three attributes for each anchor box.

(1) Objectness score of each anchor box (Intersection over Union (IoU))
(2) Anchor box offset used to adjust the position of each box
(3) Class probability assigned to each anchor box

By using anchor boxes, it becomes possible to evaluate predictions for all objects in real time.

YOLOv3 is a multi-scale object detection network which employs a feature extraction network and multiple detection heads to conduct predictions at multiple scales (The MathWorks, Inc., 2025-1). In this chapter, a pretrained ResNet50 is used for feature extraction in a YOLOv3 model. It is reported that the accuracy of object detection compared to YOLOv2 is improved and the detection ability of smaller objects is also enhanced. However, unfortunately, it seems that the speed of object detection has slightly decreased.

In this section, we apply the YOLOv3 to the detection of crack-like defects in the industrial materials as introduced in the previous section, and the effectiveness is assessed. Image Labeler is also used to make gTruth dataset for training YOLOv3, by which annotation is conducted by marking rectangular boxes around the crack-like defect areas within an image. Figure 7 presents annotated image examples that form one part of the gTruth dataset as shown in Fig. 8. Furthermore, the data augmentation method for YOLO model (Arima et al., 2023) is utilized. A cropped material area is pasted on an image randomly selected from the 75 background images as shown in Fig. 4. As a result, three training images can be newly generated from one original image as shown in Fig. 4. Through the proposed augmentation process, a total of 228 images (76×3) are generated, so that 304 images can be totally prepared for training. Note that the information in the gTruth file also has to be reconstructed according to the augmented training images.

After training, generalization ability is evaluated using the test dataset consisting of 31 OK and 34 NG images, of which result is shown in Table 3. As can be seen, only one OK image is over-detected as including defect. On the other hand, for the NG images, all defects were successfully detected in all the test images. This result shows the YOLOv3 model can systematically classify the test images as accurately as U-Net D model.

Figure 7. Annotated image examples with crack-type defects used to train the YOLOv3 model.

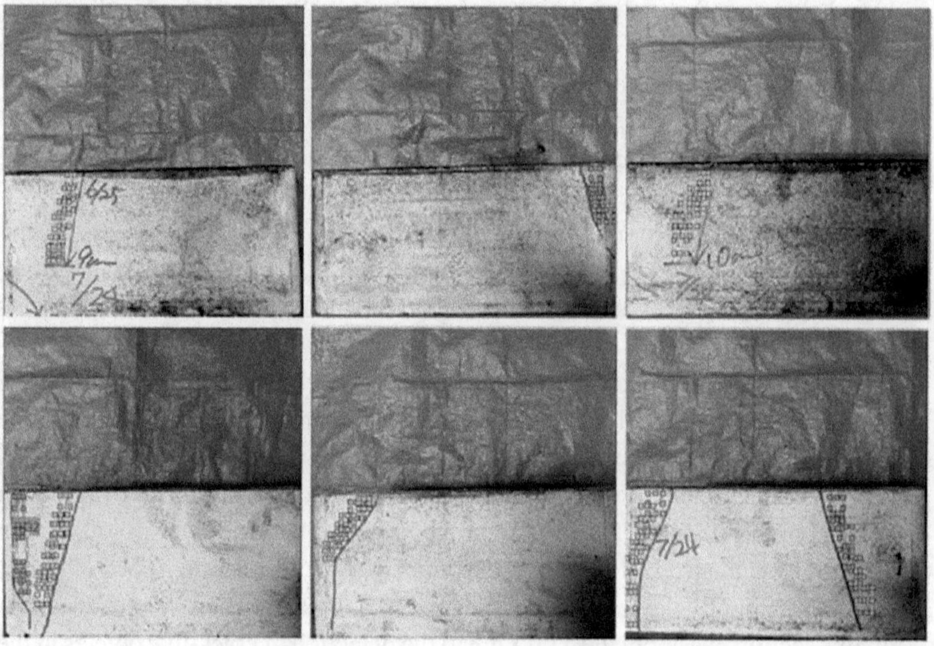

Figure 8. Structure of gTruth dataset for a YOLOv3 model.

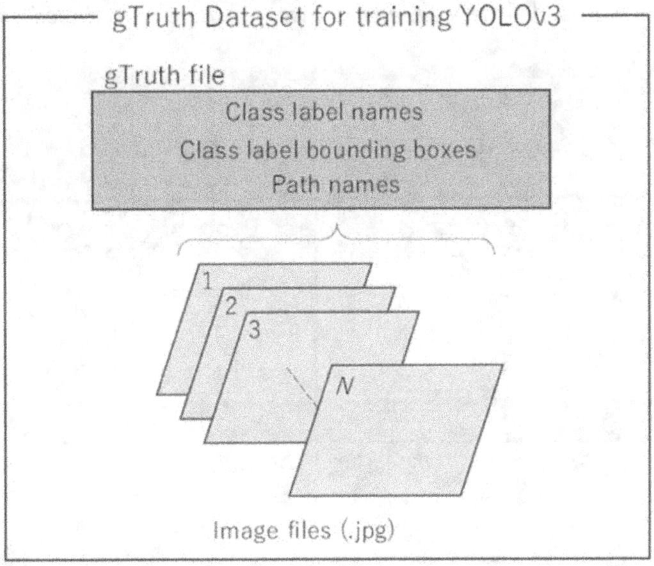

Table 3. Classification result systematically predicted by YOLOv3 model (row: true labels, column: predicted label).

	Anomaly	Normal
Anomaly	34	0
Normal	1	30

B. Deep Verification by Human

Table 3 shows the classification result that the trained YOLOv3 model systematically predicts. As can be seen, almost the same performance with the U-Net D model seems to be achieved. However, just to be sure, 34 test images classified as anomaly in Table 3 were carefully verified based on visual inspection by human eyes, so that it was confirmed that the results of 4 images out of 34 are invalid. Actually, the invalid defect detection by the YOLOv3 means that non-defective areas in the images are incorrectly identified as defects by bounding boxes and scores as shown in Fig. 9. After correcting the Table 3 based on visual inspection by human eyes, the classification result is revised as shown in Table 4.

Figure 9. Test images of the NG category partially misidentified by the YOLOv3 model, in which non-defective areas are over-sensitively predicted as defects with boundaries.

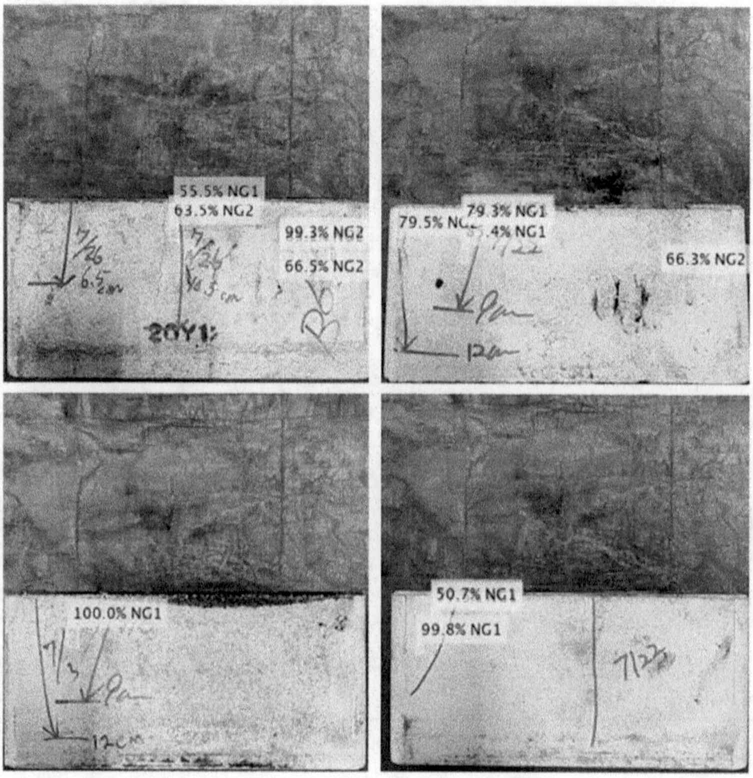

Table 4. Classification results rewritten by revising Table 3 based on visual inspection by human eyes (row: true labels, column: predicted label).

	Anomaly	Normal
Anomaly	30	4
Normal	1	30

C. Comparison of Classification Performance Between U-Net and YOLOv3 Models

We compared the performance concerning four items: working cost required to create gTruth dataset by labeling, learning time, classification performance, and time required for forward calculation. Firstly, regarding the working cost for labeling, the YOLOv3 model outperformed U-Net model in terms of both the easiness of labeling task and the time required for the task. In the case of U-Net, labeling is needed for all the pixels in the image, so that the problem required for more complex and finer works became clear. As for the learning time, both models required about 20 hours to complete. Also, the superior classification performance of the U-Net D model to the YOLOv3 model is observed from Tables 2 and 4.

Additionally, the forward calculation time was compared, i.e., the time from giving a test image to the input layer to obtaining a classification result. The PC's CPU is an Intel(R) Core(TM) i9-10850K with a clock frequency of 3.60 GHz, and the GPU memory is 55.9 GB. The average values of forward calculation times by the U-Net and the YOLOv3 were 0.134 [s] and 0.035 [s], respectively.

CONCLUSIONS AND FUTURE WORK

In this chapter, to quantitatively identify thin, curved, and long crack-type defects included in an industrial material based on physical quantity such as area and length, a U-Net which is one of typical FCN models and a YOLOv3 which is one of YOLO series models, have been designed, trained, and compared. The two trained models have been applied to the detection of crack-type defects in the industrial material. Each defect detection performance, i.e., (1) generalization ability has been evaluated and compared using test images. In addition, (2) easiness of making gTruth dataset, (3) designability of model, (4) training time, (5) forward calculation time, and (6) ability of quantitative evaluation are compared, so that U-Net and YOLOv3 outperform each other in terms of (1), (6) and (2), (5), respectively.

This time, a highly constrained dataset composed of only 103 original images (31 OK and 72 NG) had to be used for training and test due to the limited dataset size provided by the manufacturer. Although data augmentation expanded the dataset, the underlying diversity of real-world defects remained limited. Further experiments using more images and additional criteria by ROC curves should be conducted to evaluate the robustness to variance in defect types, lighting conditions, or material textures.

Also, as for the visualization ability, in spite of less training data, the U-Net D model was able to pixel-wisely identify the defects, and the YOLOv3 model was overall able to predict the defective areas with boundary boxes although it had four oversensitive defect detections. In this experiment, it seems that completed training was not able to be achieved due to the less training data, however, it is expected that the false positive and negative rates will be reduced by increasing training data.

Moreover, although this chapter reports forward calculation times (0.134 [s] for U-Net and 0.035 [s] for YOLOv3), it seems to lack a thorough discussion on latency requirements for real-time industrial deployment. The authors were not able to explore how performance scales with higher image resolutions, multiple camera feeds, or under embedded processing constraints. In practical factory automation, inference under varying environmental conditions, temperature, or hardware configurations is crucial. In future work, these operational constraints should be discussed to evaluate the system's deployability.

ACKNOWLEDGEMENTS

This work was supported by JSPS KAKENHI, Grant Number JP25K07532.

REFERENCES

Arima, K., Nagata, F., Shimizu, T., Otsuka, A., Kato, H., Watanabe, K., & Habib, M. K. (2023). Improvements of Detection Accuracy and Its Confidence of Defective Areas by YOLOv2 Using a Dataset Augmentation Method. *Artificial Life and Robotics*, *28*(3), 625–631. DOI: 10.1007/s10015-023-00885-9

Carrilho, R., Yaghoubi, E., Lindo, J., Hambarde, K., Proença, H. (2024). Toward Automated Fabric Defect Detection: A Survey of Recent Computer Vision Approaches. Electronics, Vol. 13, No. 18, 3728, 20 pages. DOI: 10.3390/electronics13183728

Chen, L. C., Zhu, Y., Papandreou, G., Schroff, F., & Adam, H. (2018). Encoder-Decoder with Atrous Separable Convolution for Semantic Image Segmentation. Procs. of European Conference on Computer Vision 2018 (ECCV2018), pp. 833–851. DOI: 10.1007/978-3-030-01234-2_49

Fang, Q., Ibarra-Castanedo, C., Garrido, I., Duan, Y., Maldague, X. (2023). Automatic Detection and Identification of Defects by Deep Learning Algorithms from Pulsed Thermography Data. Sensors, Vol. 23, No. 9, 4444, 33 pages. doi:DOI: 10.3390/s23094444

Gong, W., Akbar, M. F., Jawad, G. N., Shrifan, N. H. M. M., & Zhang, F. (2024). A Magnetic Field Concentration Method for Magnetic Flux Leakage Detection of Rail-Top Surface Cracks. *IEEE Access : Practical Innovations, Open Solutions*, *12*, 43245–43254. DOI: 10.1109/ACCESS.2024.3377692

Kingma, D. P., & Ba, J. L. (2015). Adam: A method for stochastic optimization. Procs. of the 3rd International Conference on Learning Representations (ICLR2015), pp. 1– 15.

Nagata, F., Habib, M. K., & Watanabe, K. (2021). Design and implementation of convolutional neural network-based SVM technique for manufacturing defect detection. *International Journal of Mechatronics and Automation*, *8*(2), 53–61. DOI: 10.1504/IJMA.2021.115240

Pei, M., Liu, N., Zhao, B., & Sun, H. (2023). Self-Supervised Learning for Industrial Image Anomaly Detection by Simulating Anomalous Samples. International Journal of Computational Intelligence Systems, Vol. 16, No. 152, 15 pages. DOI: 10.1007/s44196-023-00328-0

Prunella, M., Scardigno, R. M., Buongiorno, D., Brunetti, A., Longo, N., Carli, R., Dotoli, M., & Bevilacqua, V. (2023). Deep Learning for Automatic Vision-Based Recognition of Industrial Surface Defects: A Survey. *IEEE Access : Practical Innovations, Open Solutions*, *11*, 43370–43423. DOI: 10.1109/ACCESS.2023.3271748

Redmon, J., & Farhadi, A. (2017). YOLO9000: Better, Faster, Stronger. Procs. *2017 IEEE Conference on Computer Vision and Pattern Recognition*, pp. 6517–6525.

Ronneberger, O., Fischer, P., & Brox, T. (2015). U-Net: Convolutional Networks for Biomedical Image Segmentation, Medical Image Computing and Computer-Assisted Intervention (MICCAI), LNCS, Vol. 9351, pp. 234–241.

Shaterian, Z., & Mrozowski, M. (2022). Crack Detection in Metallic Surfaces Based on Dumbbell-Shaped Defected Ground Structures in Microstrip Technology. Procs. of 2022 24th International Microwave and Radar Conference (MIKON), Gdansk, Poland, pp. 1–4, DOI: 10.23919/MIKON54314.2022.9924958

The MathWorks, Inc. (2025-1). https://www.mathworks.com/help/vision/ug/getting - started-with-yolo-v3.html Accessed on 4 February 2025

The MathWorks, Inc. (2025-2). https://www.mathworks.com/help/vision/ug/object -detection-using-yolo-v3-deep-learning.html Accessed on 4 February 2025

The MathWorks, Inc. (2025-3). https://jp.mathworks.com/help/vision/ref/unetlayers .html Accessed on 4 February 2025

The MathWorks, Inc. (2025-4). https://cvml-expertguide.net/terms/dl/inter-layer - connection/skip-connection/ Accessed on 4 February 2025

Wang, Z., Zhu, H., Jia, X., Bao, Y., Wang, C. (2022). Surface Defect Detection with Modified Real-Time Detector YOLOv3. Journal of Sensors, Vol. 2022, No. 1, Article ID 8668149, 10 pages, doi:DOI: 10.1155/2022/8668149

Yanan, S., Hui, Z., Li, L., & Hang, Z. (2018). Rail Surface Defect Detection Method Based on YOLOv3 Deep Learning Networks. Procs. of 2018 Chinese Automation Congress (CAC), pp. 1563–1568. DOI: 10.1109/CAC.2018.8623082

Yuan, L., Zhu, X., & Hong, K. (2020). Detection of Material Surface Cracks by Infrared Non-destructive Testing. Procs. of 2020 11th International Conference on Prognostics and System Health Management (PHM-2020 Jinan), Jinan, China, pp. 587–591, DOI: 10.1109/PHM-Jinan48558.2020.00114

Zhang, G., Cui, K., Hung, T. Y., & Lu, S. (2021). Defect-GAN: High-Fidelity Defect Synthesis for Automated Defect Inspection. Procs. of 2021 IEEE Winter Conference on Applications of Computer Vision (WACV), Waikoloa, HI, USA, pp. 2523–2533. DOI: 10.1109/WACV48630.2021.00257

Zheng, Y., Li, S., Xiang, Y., & Zhu, Z. (2023). Crack Defect Detection Processing Algorithm and Method of MEMS Devices Based on Image Processing Technology. *IEEE Access : Practical Innovations, Open Solutions*, *11*, 126323–126334. DOI: 10.1109/ACCESS.2023.3331152

Chapter 4
Automation of the Process of Charge Sampling at Metallurgical Enterprises Using Robots

Sergei Jatsun
South-West State University, Russia

Jamil Safarov
https://orcid.org/0009-0004-0392-3860
University of Technology of Azerbaijan, Ganja, Azerbaijan

Egor Tistsov
https://orcid.org/0009-0004-1779-5396
South-West State University, Kursk, Russia

Dmitrii Efremov
https://orcid.org/0009-0004-5408-6986
South-West State University, Russia

Gavriil Fursov
South-West State University, Russia

ABSTRACT

This chapter per the authors reveal the problem of the complexity of selecting the fired charge from the furnace and propose automation of this. They talk about the development of a robot for automatic charge selection, as well as modeling its bucket using a fifth-order polynomial in analog and discrete spaces. The calculation results are presented, as well as graphs of the position, speed, acceleration,

DOI: 10.4018/979-8-3693-7994-3.ch004

torque, and changes in motor current when the bucket is moved. The chapter also contains a description of the sampler job and its main parts, such as a control cabinet, a manipulator arm, and a laser protective barrier. This chapter describes the intended modes of operation of the charge sampling robot, such as timer operation and single sampling.

INTRODUCTION

Metallurgical enterprises offer great prospects for industrial robots. This is easily explained if we take into account the conditions of metallurgical production: high temperatures, the weight of each unit, etc.

In metallurgical production, there are a large number of operations currently performed using various mechanisms and machines with the participation of an operator, which can be successfully assigned to robots.

An analysis of the sites and workshops of metallurgical production to be robotized allows us to conclude that mass-produced industrial robots are not suitable for use in this production, for example, due to the presence in many cases of significant dynamic loads associated with the need not to simply move some part in space, but to perform additional mechanical work. Created mainly for use in mechanical engineering, serial robots are usually not adapted to work in aggressive environments and at high temperatures typical of metallurgical production (Yin, 2021; Arents & Greitans, 2022; Dzedzickis et al., 2021).

The charge is one of the key components in metallurgical production, which plays an important role in the metal production process. Understanding what a charge is and how it is used makes it possible to optimize production processes and improve the quality of the final product. It is a mixture of various materials used in metallurgical production to produce metals and alloys. The charge includes metal ores, fluxes and fuels, which together ensure efficient ore melting and impurity removal. The main problems associated with the preparation of the charge include the complexity of accurate calculations, the need to use specialized equipment and the dependence of the quality of the charge on the quality of the raw materials. To do this, automation of the charge sampling process must be widely implemented.

In many enterprises, the analytical control services are based on the principle of periodic manual sampling of technological products followed by their chemical or hardware analysis.

The sampling frequency is 1-3 hours. If you add 1-2 hours of receiving the test results, then the analysis data may be up to 3-5 hours late relative to the sampling time. If we take into account that the variability of the quality of the feedstock has a characteristic time of 1-3 hours, then the resulting delay in obtaining information practically eliminates the possibility of forming adequate control actions and leads to significant economic losses. Additional losses also arise from the influence of the human factor, associated with the inability to ensure proper representation of samples, and sometimes simply with the dishonesty of performers.

Charge sampling by a mechanized method involves the use of automated samplers. Spot sampling is performed using buckets or trays, while the charge flow is not interrupted. Mass sampling is carried out by shut-off devices, due to the intersection of the charge flow on the conveyor belt.

Sampling the charge manually is a much more time-consuming process. For sampling, the operator uses various scoops and probes, as well as sampling frames.

Currently, the proportion of manual labor in charge sampling is quite high, and process control is carried out manually (a lot of time is spent on obtaining analysis results). Therefore, today the task of automating the control of the technological process of metallurgical plants, in particular, technological processes in charge preparation shops, is urgent.

For example, blast furnaces are equipped with automatic dosing and charge feeding systems. The charge level is measured using special probes. A trolley with a probe tube and supporting carriages is moved along the rails using an electric winch operated by an operator. Similar probe installations are used for temperature measurement and metal sampling from buckets and converters.

The experience of mechanization of transport operations in the warehouses of blanks and finished products of rolling mills and other workshops shows that in many areas of a metallurgical plant, the gantry layout of robots may be advantageous, requiring the least production space and providing free access to the work area (Gonzalez et al., 2022; Kaltsoukalas, Makris, & Chryssolouris, 2014; Cui et al., 2021).

1. MATHEMATICAL MODELING USING ARTIFICIAL INTELLIGENCE

System Parameters:

Number of pulses per revolution P = 66,6.

Rated torque on the motor shaft M=1,91 N*m.

Voltage constant C_{ω}=34 V/1000 rpm

Moment coefficient C_M=0,55 N*m/A

Gear ratio:

Planetary gearbox ;

Desired maximum bucket rotation angle $\beta \in [0;180]$

$T_{signal} \in [2*10^{-3}\ ; 90*10^{-3}]\ c$ – The period of the PWM signal belongs to this range.

1.1. Simulation in the analog space

The law of motion describing the movement of the bucket can be written as a polynomial of the 5th order:

$$\beta^* = \sum_{i=0}^{5} \beta_i t^i$$

, (1.2.1)

where β^* — the value of time travel.

Let's write down the 5th order polynomial and the column vector of the initial conditions:

$$\beta_0 = 0 \; ; \dot{\beta} = 0 \; ; \ddot{\beta} = 0 \; ;$$

$$\mathrm{T} = 5.6 \; ; \beta = 180 \; ; \dot{\beta} = 0 \; ; \ddot{\beta} = 0 \; ;$$

$$\Delta t = 0{,}001$$

A 5th-order polynomial in MATLAB:

$$\mathrm{B} = [\beta = \beta_0 + \beta_1 t_0 + \beta_2 t_0^{\,2} + \beta_3 t_0^{\,3} + \beta_4 t_0^{\,4} + \beta_5 t_0^{\,5} \; ;$$

$$\dot{\beta} = \beta_1 + 2\beta_2 t_0 + 3\beta_3 t_0^{\,2} + 4\beta_4 t_0^{\,3} + 5\beta_5 t_0^{\,4} \; ;$$

$$\ddot{\beta} = 2\beta_2 + 6\beta t_0 + 12\beta_4 t_0^{\,2} + 20\beta_5 t_0^{\,3} \; ;$$

(1.2.2)

$$\beta_к = \beta_0 + \beta_1 T + \beta_2 T^2 + \beta_3 T^3 + \beta_4 T^4 + \beta_5 T^5 \; ;$$

$$\dot{\beta}_к = \beta_1 + 2\beta_2 T + 3\beta_3 T^2 + 4\beta_4 T^3 + 5\beta_5 T^4 \; ;$$

$$\ddot{\beta}_к = 2\beta_2 + 6\beta_3 T + 12\beta_4 T^2 + 20\beta_5 T^3] \; ;$$

Vector-column of initial conditions in the environment MATLAB:

$$\mathrm{C} = [\beta_0; v_0; a_0; \; \beta_k; v_k; a_k] \; ;$$

Finding vectors of coefficients of a 5th order polynomial:

(1.2.3)

Deviation is a value that characterizes how much the current position value differs from the desired position.

Control action is the force with which an object corrects a deviation.

To implement the controller, you need to know the deviation of the actual position from the set one:

$$\Delta\beta = \beta^{*} - \beta$$

, (1.2.4)

where: $\Delta\beta$— deviation value; β^{*}— desired position value; β— the real meaning of the position.

Knowing how the moment changes, you can find the law for changing the acceleration of the arrow using the following formula:

$$\ddot{\beta} = \frac{M_{bucket}(t)}{J}$$

. (1.2.5)

By integrating equation 1.2.5 twice, we obtain the laws of change of the arrow velocity and position:

$$\dot{\beta} = \dot{\beta} + \Delta t\ddot{\beta}$$

; (1.2.6)

$$\beta = \beta + \Delta t\dot{\beta}$$

. (1.2.7)

The gear ratio of the planetary gearbox mounted on the engine is 15. Knowing this, it is possible to calculate the boom moment of the control system:

$$M_{bucket}(t) = M_{engine}(t) \cdot u_1 = 0{,}55\dot{i}(\Delta\alpha) \cdot 15$$

. (1.2.8)

Let's simulate how the boom control system will behave under this control:

$$L\frac{di}{dt} + i(t)R + c_{\omega}\dot{\varphi}(t) = U(\Delta\alpha)$$

, (1.2.9)

where: L = 10.7 mGn; R = 1.93 Om; c_{ω} = 3,4; $U(\Delta\beta)=\lambda(t)*U_0$, U_0 = 220V

Since the inductance of the motor phases is too small, the differential component of the equation can be neglected. Then equation 1.2.9 will take the following form:

$$1{,}93i(t) + 3{,}4\dot{\varphi}(t) = U(\Delta\alpha)$$

. (1.2.10)

From this formula, we express the current strength:

$$i(t) = \frac{U(\Delta\beta) - 3{,}4\dot{\varphi}(t)}{1{,}93}$$

. (1.2.11)

Knowing the current applied to the motor, we can find the torque of the motor shaft using the formula:

$$M_{engine}(t) = C_M i(\Delta\beta)$$

; (1.2.12)

C_M = 0,55;

Having modeled the system, we obtain a graph of changes in the desired position (1), the real position (2), the velocity (3) and acceleration (4) of the bucket (Fig. 1). We conclude that the graph of changes in the desired position (1) and the real position (2) coincide. We will also get graphs of torque changes on the bucket rotary module (Fig. 2), current changes on the motor (Fig. 3), torque changes on the motor (Fig. 4) and changes in the angle of rotation of the motor shaft (Fig. 5).

Figure 1. A graph of changes in the desired position (1), the actual position (2), the speed (3), and the acceleration (4) of the bucket.

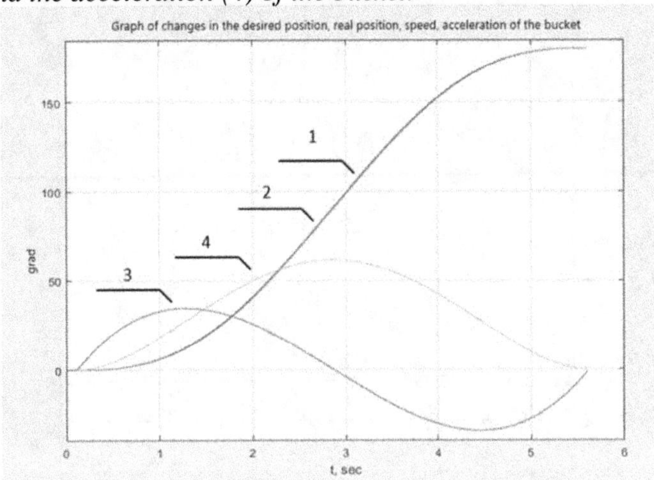

Figure 2. Graph of the torque change on the rotary bucket module

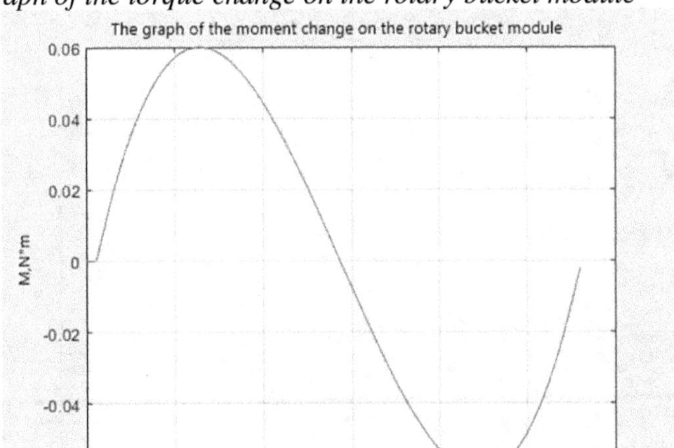

Figure 3. Graph of current changes on the motor

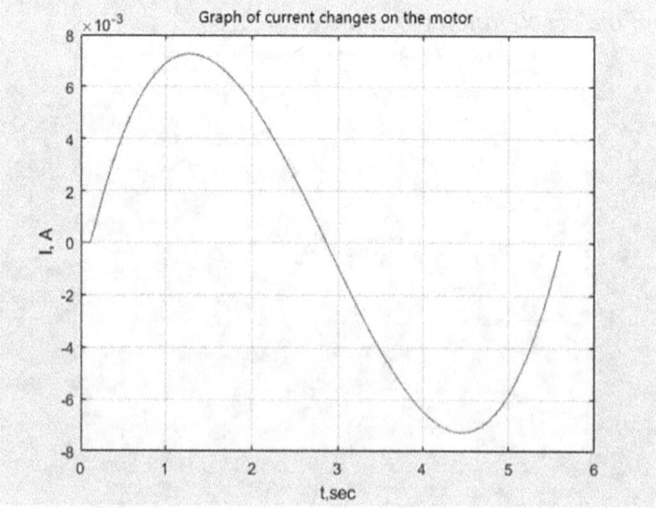

Figure 4. Graph of the torque change on the engine

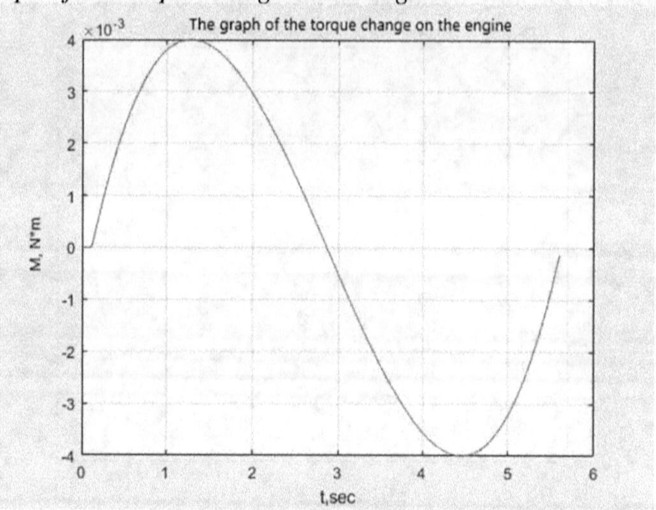

Figure 5. Graph of the angle of rotation of the motor shaft

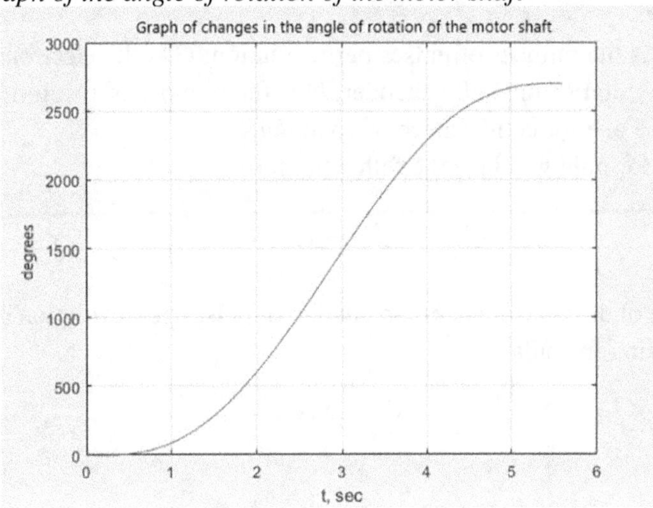

1.2. Modeling in discrete space

To simulate the rotation of the bucket in a discrete space, the following values of the angular velocity and its corresponding periods were selected:

Table 1. Table of values of angular velocity and its corresponding periods

*	T, ms
90	10
75	12
60	15
45	20
30	30
15	60

To control the rotation speed of the motor using a PLC, a certain frequency, frequency and fullness of the PWM signal are set. The maximum possible frequency supplied from the PLC is 500 Hz. In this case, the borehole value can be only 50%. Because $T_{min} = 2*10^{-3}$c, the time of switching the transistor input from the “log. 1” state to the “log.0” state - 1 ms. C = 2500 – encoder resolution. Based on this, the maximum speed using the PLC is determined by the parameters P12 and P13. To determine their values, we use the following formula:

$$PG = 4CN, \quad (1.2.1)$$

where: P is the number of pulses per revolution; G is the electronic gear ratio; C is the resolution of the 2500 encoder; N is the number of revolutions, which is assumed to be one for convenience of counting.

Then the formula will take the following form:

$$PG= 10^4. \quad (1.2.2)$$

The value of the coefficient of the number of pulses per revolution is determined by the following formula:

$$P = \frac{k}{\omega^*_{max}}, \quad (1.2.3)$$

where: ω^*_{max} - the desired speed of rotation of the motor shaft in revolutions per minute; k is the number of pulses per minute.

To determine the number of pulses, use the following formula:

$$k = \frac{60}{T_{min}}, \quad (1.2.4)$$

where: Tmin is the minimum period of supply of the PWM signal, which ensures the maximum speed of rotation of the shaft, which is 2 * 10-3 s.

Using formulas (1.2.2), (1.2.3) and (1.2.4), we express the electronic gear ratio using the formula:

$$G = \frac{10^4 * \omega^*_{max}}{30 * 10^3} = \frac{10\omega^*_{max}}{30}. \quad (1.2.5)$$

Based on the data in the driver configuration table, equation 1.2.5 takes the following form:

$$\frac{P_{12}}{P_{13}} = \frac{10\omega^*_{max}}{30}, \quad (1.2.6)$$

where: P_{12}– the numerator of the electronic transmission; P_{13}– the denominator of electronic transmission;

After installing P12 and P13, further speed control is performed by changing the parameters of the PWM signal through the code uploaded to the PLC.

The motor shaft is positioned based on the value obtained in equation (85). Knowing the number of pulses per revolution, you can calculate how many pulses need to be applied to rotate the motor shaft at a certain angle.:

$$L = \frac{\varphi^*}{360}P \qquad (1.2.7)$$

,

where: L is the number of input pulses; φ^*- the desired angle of rotation of the motor shaft.

Knowing the number of input pulses required to position the motor, we calculate the time during which it is necessary to send a PWM signal to the driver from the PLC.

$$\tau = LT_{signal}, \qquad (1.2.8)$$

where: τ required time; T_{signal} – the period of the PWM signal.

To achieve smooth operation of the system, it is necessary to select the optimal acceleration of the control system. Let's consider the principle of determining the acceleration and time of the transition process using the bucket module as an example.

The smoothness of the bucket movement is ensured by communicating positive acceleration to the system during the first half of the movement and negative acceleration during the second. Knowing this, you can determine the bucket rotation time using the following formulas.

Let's use the formula to determine the rotation angle for the first half of the time:

$$\beta_1 = \frac{\ddot{\beta}\left(\frac{\tau}{2}\right)^2}{2} \qquad (1.2.9)$$

.

The final velocity of the first half of the time is the initial velocity of the second half of the time. We will find the speed using the formula:

$$\dot{\beta} = \ddot{\beta} * \frac{\tau}{2} \qquad (1.2.10)$$

.

Having found out the initial velocity of the second time interval, we will determine the angle of rotation for this period:

$$\beta_2 = \dot{\beta}\frac{\tau}{2} - \frac{\ddot{\beta}\left(\frac{\tau}{2}\right)^2}{2} = \ddot{\beta}\frac{\tau^2}{4} - \frac{\ddot{\beta}\left(\frac{\tau}{2}\right)^2}{2}$$

. (1.2.11)

The required rotation angle is π (radian), from this we obtain the following formulas:

$$\beta + \beta_2 = \pi$$

; (1.2.12)

$$\pi = \ddot{\beta}\frac{\tau^2}{4}$$

; (1.2.13)

From equation (1.2.13), we express the time it takes to make a 90° turn. The equality takes the following form:

$$\tau = \sqrt{\frac{4\pi}{\ddot{\beta}}}$$

. (1.2.14)

Let's simulate the system, display graphs of changes in engine speed (Fig. 6), changes in engine acceleration (Fig. 7), the desired change in engine rotation angle (Fig. 8) and the movement of the horizontal module (Fig. 9). (Antipov, Maslakov, & Yurenko, 2015; Yu & Zhang, 2021; Miková et al., 2014; Csuros & Csuros, 2010b).

Figure 6. Graph of engine speed changes

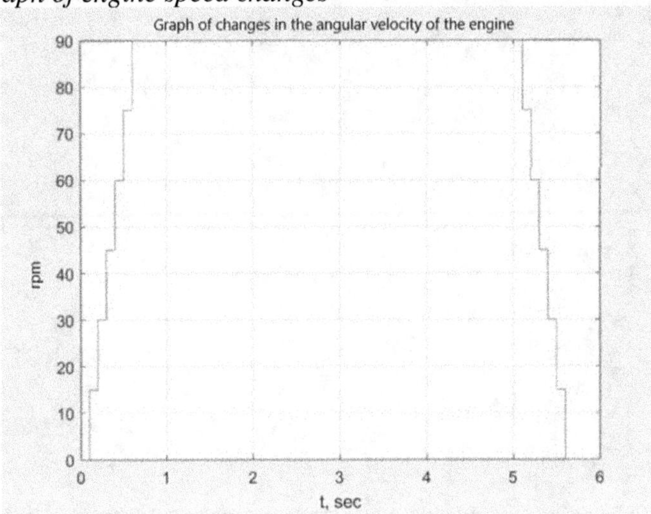

Figure 7. Graph of engine acceleration changes

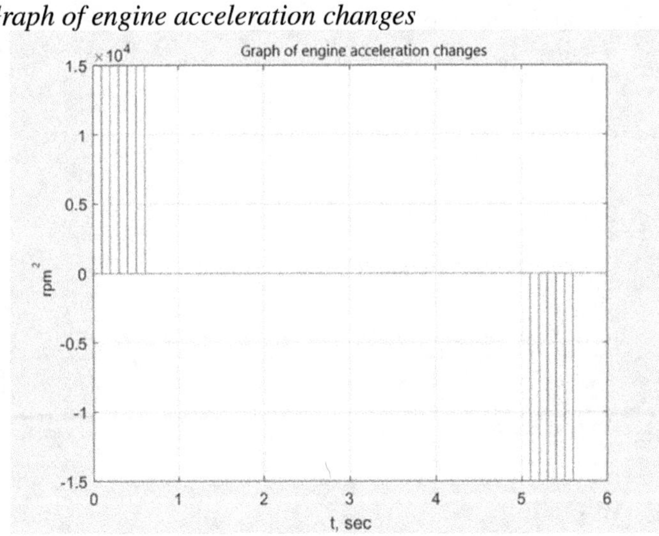

Figure 8. Graph of the change in the angle of rotation of the engine

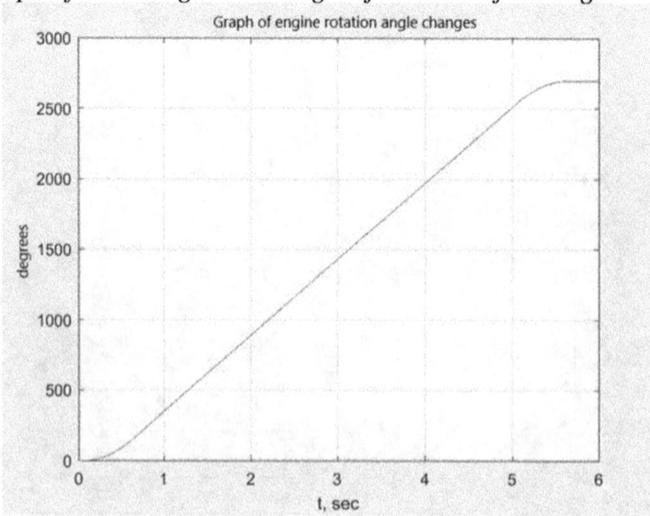

Figure 9. Graph of bucket rotation

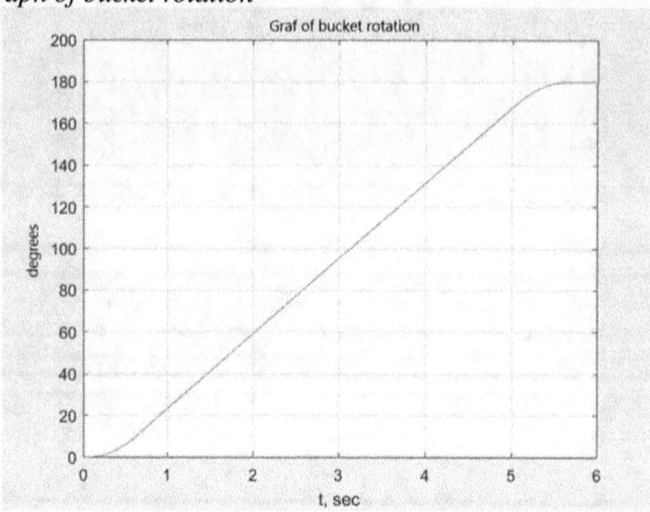

2. DESCRIPTION AND CHARACTERISTICS OF THE ROBOTIC COMPLEX

The robotic complex is designed for automatic sampling of the fired charge directly from the kiln, which greatly simplifies this process. The burnt charge is removed from the furnace using a special heat-resistant ladle and poured into a heat-resistant container, in which it must cool for some time.

The sampler supports 3 modes of operation: timer mode, in which sampling will be performed at preset intervals, one-time sampling mode, and external signal mode, in which the signal will come from an external automated process control system.

Figure 10 shows a diagram of the main components of Probot-2000-300.

Figure 10. Diagram of the main components

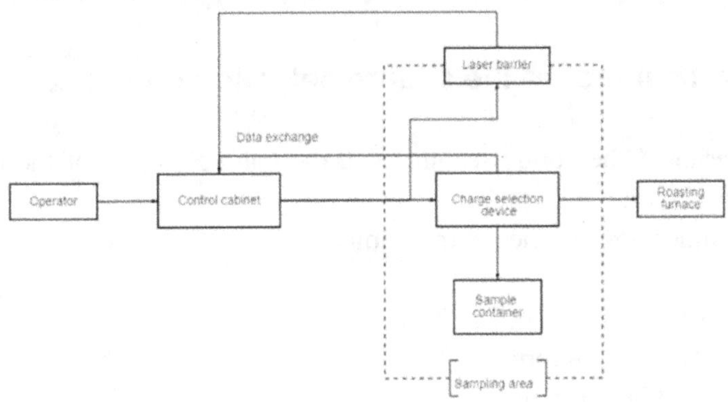

The appearance of Probot-2000-300 and the main elements of the product are shown in Figure 11.

Figure 11. appearance of Probot-2000-300: 1 – control cabinet; 2 – charge sampling device; 3 – sample container; 4 – laser protective barrier

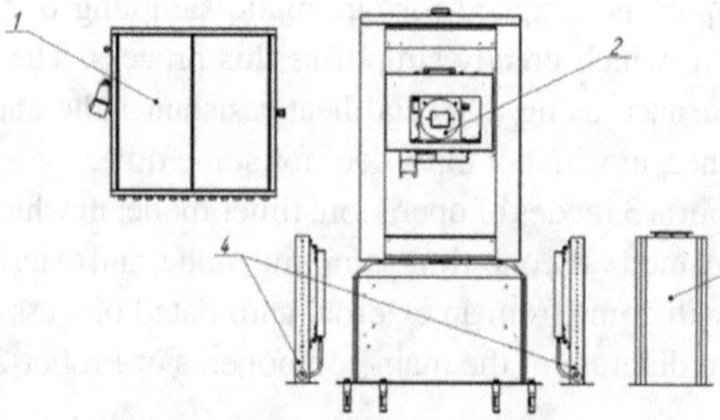

2.1 Control cabinet of the charge selection complex

The exterior of the control cabinet and the main elements of the product are shown in Figure 12.

Overall dimensions of the control cabinet:

- Width: (930 ± 10) mm;
- Height: (815 ± 10) mm;
- Length: (435± 10) mm.

Figure 12. appearance of the control cabinet of the charge selection complex: 1 - housing; 2 - doors; 3 – fixed plug; 4 – stuffing box for power wires; 5 – emergency shutdown button; 6 - oil seal closed; 7 - stuffing box; 8 – plate with connectors

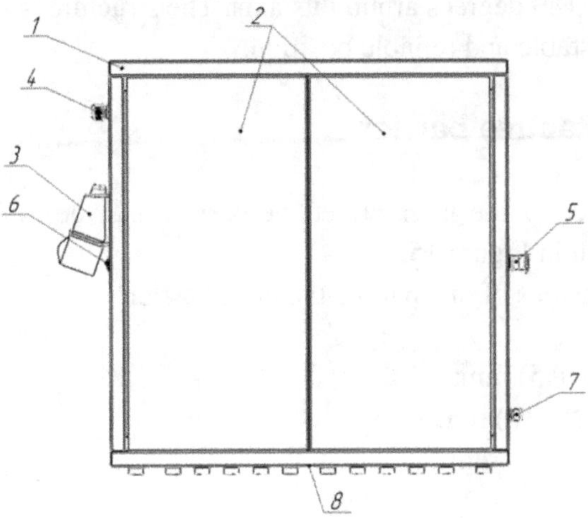

There are 4 buttons on the inner panel of the control cabinet, 3 of which are responsible for selecting the mode and one for canceling the task, 8 green lights, 2 red, a screen and a switch with a rotary knob. All elements are provided with explanatory labels. The panel can also be opened to access the electronics.

2.2 Charge selection device

The appearance of the charge sampling device and the main elements of the product are shown in Figure 13. The general appearance of the sampling robot is shown in Figure 14.

Figure 13. appearance of the charge sampling device: 1 – rotary module; 2 - vertical module; 3 - linear module; 4 – bucket

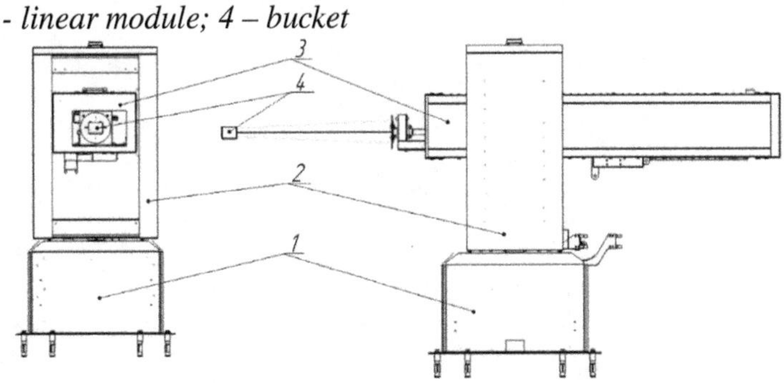

The charge sampling device has 4 degrees of freedom, has a maximum vertical adjustment in the range of 530 mm (working adjustment 370 mm) and can rotate 90 degrees. The linear module can extend the bucket to a distance of 1,500 mm. The bucket can rotate 180 degrees around its axis. The structure is mounted on anchor bolts to give it a stable and reliable position.

2.3 Laser protective barrier

The appearance of the laser protective barrier and the main elements of the product are shown in Figure 15.

Overall dimensions of one part of the laser barrier:

- Width: (190 ± 5) mm;
- Height: (702 ± 10) mm;
- Length: (190 ± 5) mm.

The laser barrier consists of 4 parts located at a distance of 1500 mm from each other.

Figure 14. The appearance of one part of the laser barrier

3. GENERAL TECHNICAL INFORMATION ABOUT PROBOT-2000-300

The total weight of the product is not more than 800 kg.
The recommended weight of the selected charge is not more than 2 kg.
The accuracy of repeated positioning is ± 1 mm.
The degree of protection of electronic components is IP67.
The serial number is 42400.

4. PREPARATION FOR USE

To start working with the device, insert the plug of the device's power cord into a 220V 50Hz outlet. A self-diagnosis mode message will be displayed on the control cabinet screen.

Upon successful power supply to the manipulation system, the "POWER" indicator lights up on the control cabinet, the marker light lights up on the shank, while the mechanical components move to the starting position, after reaching the starting positions, the "READY" indicator lights up.

Samples from-2000-300 supports 3 operating modes: timer operation mode, one-time sampling mode and external signal operation mode.

5. TIMER OPERATION MODE

When the TIMER button is pressed, the "ON" indicator lights up. TIMER SAMPLING", this mode of operation allows sampling after the timer expires, the time of which can be set on the digital display. When sampling is started, the "READY" indicator goes out, the indication of the selected operating mode continues to light, the "START OF SAMPLING" indicator also lights up, the "END OF SAMPLING" indicator lights up at the end of sampling, the manipulation system returns to its original position, if there is a sample in the container, the "SAMPLE AVAILABILITY" indicator lights up. While the sample is in the container, it is forbidden to start a new sampling cycle, which identifies the absence of a GORENJE indicator. To change the operating mode, you need to press the cancel task button and then select a new control mode.

6. ONE-TIME SAMPLING MODE

When the "MANUAL MODE" button is pressed, the "START SAMPLING" indicator lights up, this mode of operation allows you to start a one-time sampling. When sampling is started, the "READY" indication goes out, the indication of the selected operating mode continues to light, the "START OF SAMPLING" indication also lights up, when sampling is completed, the "END OF SAMPLING" indication lights up, the manipulation system returns to its original position, the control mode is reset and the "START OF SAMPLING" indication goes out, if there is a sample in the The "SAMPLE AVAILABILITY" indicator lights up in the container. While the sample is in the container, it is forbidden to start a new sampling cycle, which identifies the absence of a GORENJE indicator.

CONCLUSION

As a result of the research, the concept of a new robotic complex for automatic charge selection was developed and bucket motion was simulated in analog and discrete spaces using artificial intelligence.

REFERENCES

Antipov, K., Maslakov, M., & Yurenko, K. (2015). Improvement of the automated control systems for the development of the metallurgy. *Procedia Engineering*, *129*, 1010–1014. DOI: 10.1016/j.proeng.2015.12.164

Arents, J., & Greitans, M. (2022). Smart Industrial Robot Control Trends, Challenges and Opportunities within Manufacturing. *Applied Sciences (Basel, Switzerland)*, *12*(2), 937. DOI: 10.3390/app12020937

Csuros, M., & Csuros, C. (2010). Environmental sampling and analysis for metals. In *CRC Press eBooks*. DOI: 10.1201/9781420032345

Cui, G., Li, B., Tian, W., Liao, W., & Zhao, W. (2021). Dynamic modeling and vibration prediction of an industrial robot in manufacturing. *Applied Mathematical Modelling*, *105*, 114–136. DOI: 10.1016/j.apm.2021.12.031

Dzedzickis, A., Subačiūtė-Žemaitienė, J., Šutinys, E., Samukaitė-Bubnienė, U., & Bučinskas, V. (2021). Advanced Applications of Industrial Robotics: New trends and possibilities. *Applied Sciences (Basel, Switzerland)*, *12*(1), 135. DOI: 10.3390/app12010135

Gonzalez, M. K., Theissen, N. A., Barrios, A., & Archenti, A. (2022). Online compliance error compensation system for industrial manipulators in contact applications. *Robotics and Computer-integrated Manufacturing*, *76*, 102305. DOI: 10.1016/j.rcim.2021.102305

Kaltsoukalas, K., Makris, S., & Chryssolouris, G. (2014). On generating the motion of industrial robot manipulators. *Robotics and Computer-integrated Manufacturing*, *32*, 65–71. DOI: 10.1016/j.rcim.2014.10.002

Miková, Ľ., Kelemen, M., Virgala, I., & Michna, M. (2014). Simulation model of manipulator for model based design. *Applied Mechanics and Materials*, *611*, 175–182. . DOI: 10.4028/www.scientific.net/AMM.611.175

Yin, R. (2021). Review on the study of metallurgical process engineering. *International Journal of Minerals Metallurgy and Materials*, *28*(8), 1253–1263. DOI: 10.1007/s12613-020-2220-z

Yu, Y., & Zhang, Y. (2021). Collision avoidance and path planning for industrial manipulator using slice-based heuristic fast marching tree. *Robotics and Computer-integrated Manufacturing*, *75*, 102289. DOI: 10.1016/j.rcim.2021.102289

Chapter 5
Electronic Digital Twin in Industry 5.0:
AI-based electronic Industry 5.0 models

Alessandro Massaro
https://orcid.org/0000-0003-1744-783X
Department of Engineering, LUM University, Bari, Italy

ABSTRACT

The proposed chapter provides a methodology to design electronic circuits and Digital Twin (DT) models in Industry 5.0 environments. Different examples are discussed, by combining the electrical and the electronic behavior with the Artificial Intelligence (AI) algorithms upgrading the DT model about the signal prediction. The analyzed DT models are related to important elements of industrial machines such as filters, amplifiers, sensors, switching and control systems. The goal is to illustrate the setting of open source tools able to implement AI-DT complex models simulating real environment industrial conditions including the disturb effects, noises and interferences of the electronic components.

1. INTRODUCTION: DIGITAL TWIN IN INDUSTRY 5.0

Digital Twin (DT) models are important tools able to simulate real industrial scenarios predicting possible failures of machines in manufacturing processes. Advanced DT models take into account the possibility to reply the operational behaviour of the electronic networks characterized by input and output electrical signals. Furthermore, the integration of Artificial Intelligence (AI) algorithms in DT reinforces the decision making about corrective actions. AI based DT models

DOI: 10.4018/979-8-3693-7994-3.ch005

are suitable in many real industrial applications and industrial scenarios including advanced trajectory controller systems (Chaiprabha & Chancharoen, 2023), fault diagnosis methods for denoising autoencoder (Cai et al. 2022), Industrial Internet of Things (IIOT) applications (Xu et al. 2023; Stergiou & Psannis, 2022; Attaran et al. 2024; Koulamas & Kalogeras, 2018), Programmable Logic Controller (PLC) systems (Aniba et al. 2024; Yashika et al. 2024; Thürer et al. 2022; Sun et al. 2022; Werth & Morris, 2023), signal filtering (Jamshidi et al. 2024; Gafurov et al. 2024), power electronics (Bai et al. 2024; Shen et al. 2023; Zhao et al. 2021; Nguyen et al. 2021), cybersecurity (Sasikala et al. 2024; De Azambuja et al. 2024; Krishnaveni et al. 2024; Soudan 2024) manufacturing processes (Soori et al. 2023; Can & Turkmen, 2023; Huang et al. 2023; Alexopoulos et al. 2020; Kharchenko et al. 2020; Scime et al. 2022; Karkaria et al. 2025; Khdoudi et al. 2024) and Industry 5.0 (Bongomin et al. 2025; Zafar et al. 2024; Balog et al. 2023; Verdugo-Cedeño et al. 2023; Coronado et al. 2024; Modad et al. 2024; Frick 2023; Papacharalampopoulos et al. 2025). The state of the art suggests to propose some examples about circuital DT based on the AI algorithms to improve industrial systems. Specifically, the chapter discusses the DT modelling approach in Industry 5.0 environments based on mechatronics and industrial electronics. Specifically, AI algorithms are integrated in DT models to predict possible noises and interferences and supporting the preventive machine control and the circuital parameter corrections.

The used approach is based on the '*black box*' concept useful to classify the circuit disturbs without knowing their origin. Typically, the understanding of the effects of the signal noises or interferences is difficult to interpret because many variable could be involved in the electrical signal deterioration. The DT potentially classifies failures by means the prediction of output electrical signals without identifying the exact point of the disturbance on the electronic circuit under investigation.

Different DT models are analysed in the chapter matching with practical applications in industrial electronics such as:

(i) additive noises characterized by specific carriers acting in amplification systems (Massaro, 2024 a);
(ii) random multiplicative white noises acting on passive electronic filters (Massaro, 2024 b);
(iii) interferences and white noises influencing the whole PLC line including opto-coupler components (Massaro, 2024 c);
(iv) variation of the electrical parameters of sensors under Hardware Trojan (HT) cyber attacks acting on the field level (Massaro et al., 2024 d; Massaro et al. 2025 a): amplification, filtering, PLC, and sensor/actuator systems are modelled by DT suitable to simulate possible failures due to non-intentional and intentional electronic disturbs chancing the circuital output and, consecutively, altering

the monitoring and the actuation of production machines in manufacturing processes.

The analysed DT are multi-dimensional and multi-ports (Massaro, 2024 c) to be addressed on the simulation of complex electronic networks of production lines. The multi-dimensional DT is implemented by considering a large number of input nodes (ports) able to add the electrical disturbs in the electronic network as signals coupled into a specific part of the network, thus simulating a 3D spatial manufacturing environment.

The AI algorithms executed in DT models are the supervised algorithms such as Artificial Neural Networks (ANNs) and Random Forest (RF) suitable for prediction and classification. The AI data processing analyses the disturbed voltage output signals calculated by circuit models simulating the noise and the interference effects acting on specific electronic components. The tools adopted to construct DT models are the LTSpice for circuit simulation, and the Konstanz Information Miner (KNIME) for AI prediction and classification process. LTSpice is an open source tool useful for time domain and frequency domain circuit simulations calculating currents and voltages and modelling the real behaviour of electronic components and their sensitivity responses adopting a parametric setting of the electrical variables. On the other side, KNIME is a low code open source tool used for the AI calculus adopting graphical interfaces to set the AI hyper-parameters. The application of the DT models are explained by means the Business Process Modelling and Notation (BPMN), the Unified Modelling Language (UML), describing the vertical integration of electronic processes (Massaro, 2021; Massaro, 2022; Massaro, 2023 a) and the organizational models following the implementation of DT in real industrial environments. Focusing the attention on BPMN, it is observed that this tool is useful both for mapping operational and management flows, and for the flows of electrical or digital signals describing the network to be simulated. The BPMN is also adopted to represent the fluxes of electrical signals matching with logics of machine control thus improving the Process Mining (PM) modelling in Industry 5.0 scenarios consisting of the application of AI algorithms in operating production processes enabling decisions based on a prediction or a classification.

The goal of the chapter is to show how the application of AI-based DT could be precious to control possible corrective filtering networks or to decide possible interventions on specific parts of the industrial network by means the output voltage classification and prediction. The main advantage of the AI-based DT is to automate the switching of multi-stage circuits selecting specific faulty part of circuit to correct and by sectioning this part from the remaining part of the circuit. This aspect is also important for maintenance and predictive maintenance processes. Furthermore, are proposed different DT circuit parametric models supporting the understanding of the sensitivity of variations of electrical signals disturbed in industrial environments.

The chapter is structures as follows:

- introduction of the proposed DT modelling providing the proof of concept of the DT integrating AI with the circuit model (section 2);
- description of noise and interferences in industrial electronics describing the theory about some critical issue in electronics to be considered in the DT model (section 3);
- circuital DT modelling and simulation and AI prediction of different electronic parts of industrial systems (section 4) illustrating some important and typical applications;
- discussion about AI algorithms, DT integration in industrial processes and human resource skills adopting DT, including ethical, regulatory, global standards, limitations and challenge aspects (section 5).

2. DT MODELLING

DT in industrial systems are based on the modelling of the electronic circuits providing data to be processed by AI algorithms to obtain information about the machine behaviours. The proposed DT model of Fig. 1 is composed by two parts: the first one is the circuital DT simulating the input/output behaviour as a '*black box*' coupled with multiple ports of interferences (intentional and unintentional interferences); the second one is the AI engine analysing the output electrical signals and suggesting possible corrective actions by means the prediction or the classification approach. Furthermore, the circuital DT provides the dataset to be used for the training of the AI supervised algorithms. The corrective actions to perform are typically signal filtering, input power setting, phase shifting corrections. The parametric simulations of the circuit model are important to know the circuit sensitivity to effects of interferences and noises which are difficult to classify in a strong noisy environments. In this direction, the main goal of the circuit modelling and simulation is to analyse the noise and interference effects varying the amplitude of the undesired signal: the parametric analysis allows to model the amplitude modulation of the interferences and noises by varying the resistances of the generators modelling the undesired signals. In the proposed chapter the circuit model and the AI data processing are implemented by the open source tools LTSpice (Kavitha et al. 2025; Asadi 2023) and KNIME (Warr 2012; Acito 2023), respectively. LTSpice is an open source SPICE based electronic circuit simulator estimating all the electrical variables of a simulated circuit. The simulator solves the circuit equations using numerical integration algorithms and calculates the voltages of the nodes and the currents on the branches of a designed circuit layout. KNIME is a single open source platform for advanced end-to-end data analytics including AI data processing. KNIME is based on the implementation of a workflow connecting different nodes to be configured for the different phases of data processing (data pre-processing, data processing, and data reporting and results interpretation). The multitudes of ports of the circuital DT indicates the possibility to model complex '*black boxes*' simulating noises and interferences comping the whole three-dimensional (3D) space. The possibility to apply the AI algorithm allows the modelling suitable for Industry 5.0 applications.

More details about the integration of the DT in production processes are discussed in Section 5 by using the BPMN and the UML standard notations.

Figure 1. AI-based Digital Twin (DT) model in Industry 5.0.

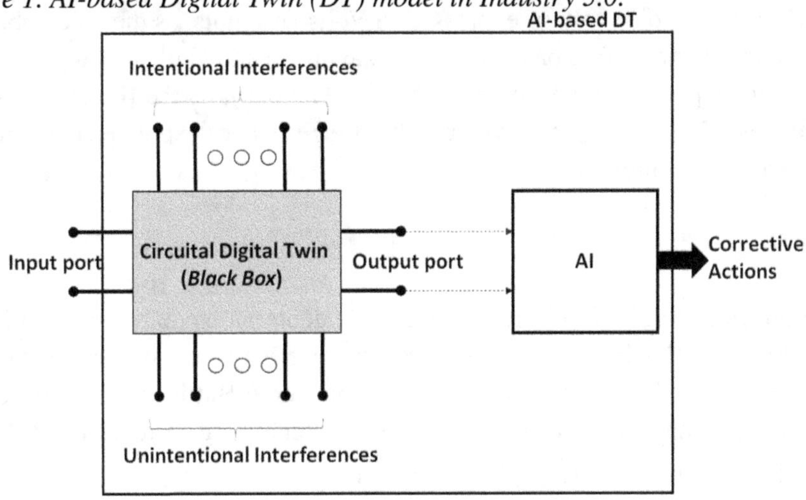

In the next section are provided more details about noise and interferences impinging at the input port of the DT model, and the usefulness of AI in predicting noisy signal (useful for early hardware tuning).

3. NOISES AND INTERFERENCES IN INDUSTRIAL ELECTRONICS

This section describes the main noises and interferences on industrial electronics.

3.1 Electronic Noise

The noise degrades the accuracy of the electrical signals. In electronics are classified the following main noises:

- **Johnson (Nyquist) noise:** noise generated into resistors as random-noise voltage representing a thermal fluctuation (Brownian motion of electrons increasing with temperature and the electrical resistance); this noise is characterized by a flat frequency spectrum (same behaviour for the frequencies); its power density is expressed by (Johnson 1971):

$$v_n^2 = 4K_B TBR \quad (1)$$

, being v_n the noise voltage expressed in Volts *rms* units, as the value obtained driving a totally noise-free band-pass filter having a bandwidth *B*, T is the resistor temperature expressed in Kelvin units (K= °C +273.16), K_B the Boltzmann's constant expressed in Joules per Kelvin, and *R* is the resistance expressed in Ohms; the equivalent current noise is

$$i_n = v_n / R \quad (2)$$

- **Shot noise (noise in diodes):** noise equivalent to a random statistical fluctuations in a flowing current caused by the discrete nature of electrical charge due to a charge quantum behaviour (Beenakker & Büttiker, 1992) by considering the charges acting independently, the fluctuating current's noise density expressed in A/Hz units is given by:

$$i_n^2 = 2qI_{DC} \quad (3)$$

being *q* the electron charge ($1.60 \bullet 10^{-19}$ Coulomb), and I_{DC} is the steady current; this noise characterizes the charges crossing a barrier in junction diodes where the charges move by diffusion;

- **Flicker noise:** additional random noise (Voss & Clarke, 1976) having a power rising as $1/f$ in at low frequencies (prevalent at lower frequencies);its spectral is given by:

$$v_{fn}^2 = \frac{1}{f^a} \quad (4)$$

where f is frequency, and $0 < a < 2$, with exponent a usually equals to 1 ($a = 1$ is the case of the pink noise);

- **Burst noise:** this noise is typically observed at low-frequencies without any spectral peaks in semiconductor devices (Hsu et al. 1970) and it is caused by material device defects; the noise is due to random jumps between a pair of levels in electronic energy levels; it is sometimes called as *popcorn noise*, *bistable noise*, or *random telegraph signal noise*; it happens in scale of milliseconds;
- **White noise:** generalization of noises having a flat response in the frequency domain (Arumugam 2022); it a random noisy process as a random process $K(t)$ having a flat power spectral density $S_K(f)$:

$$S_K(f) = constant\ for\ all\ frequencies \tag{5}$$

3.2 Interferences

Interferences are all the electrical signals disturbing an electronic circuit or a data transmission. Possible interferences are mainly classified as:

- signals characterized by a carrier (limited during the time or continuous) of devices and machines working at specific operating frequency; the signal carrier is defined by the following equation:

$$V(t) = \begin{cases} V_{off} + V_0 \bullet sen\left(\frac{\pi}{180} \bullet \varphi\right), t \le T_0 \\ V_{off} + V_0 \bullet exp\left(- \bullet \left(t - T_d\right) \bullet \varphi\right) \bullet sen\left(2\pi \bullet f \bullet \left(t - T_d\right) + \frac{\pi}{180} \bullet \varphi\right), t > T_0 \end{cases} \tag{6}$$

, being T_0 the time delay, V_{off} the offset voltage, V_0 the voltage amplitude, φ the signal phase in degree, f the signal frequency in Hertz, and ϑ the damping coefficient expressed in s^{-1};

- single pulse or train of pulses overlapping with the useful signal and generating a wrong control output or transmission; the train of pulses is associated to the Internet of Things (IoT) transmission protocol (coding the '0' as the high voltage value and the '1' as the low voltage value);
- intentional and unintentional noises coupled with the electronic circuit;
- coupling of other circuits or devices on the same board as for Hardware Trojan (HT) attacks.

The frequency domain analysis of the interference effects is performed by the Fast Fourier Transform (FFT). The FFT is a faster optimized algorithm of the Discrete Fourier Transformation (DFT). DFT transforms the following sequence of N of time domain samples

$$\{y_N\} = y_0, y_1, \ldots, y_{N-1} \tag{7}$$

into a sequence of another set of complex numbers:

$$\{Y_K\} = Y_0, Y_1, \ldots, Y_{N-1} \tag{8}$$

where

$$Y_k = \frac{1}{n}\sum_{n=0}^{N-1} y(n) \bullet e^{-j\frac{2\pi}{N}kn} \tag{9}$$

3.3 Hardware Trojan Cyber Attacks

HT modelling and analysis is a recent research topic concerning the intentional attacks performed directly on hardware components modifying signals in Industrial Internet of Things IIoT systems (Monjur et al, 2020; Monjur et al, 2022; Monjur et al, 2023; Massaro 2024 c)). In industrial scenarios, HT happen in the Level 1 of an industrial Operational Technology (OT) network disturbing the signal reading or the control of a machine. The attacks are due to the:

- coupling the electronic circuit with other electrical or electronic elements (for example linking other circuital elements on the same electronic board);
- local modification of the environment providing an intentional perturbation of the electric signal (for example by adding humidity, by increasing the temperature or the vibrations);
- adding in proximity other disturbing circuits generating carrier noises or white noises (background noises).

Typically, it is too difficult to find the HT sources. In this direction, the AI-based DT model could support to predict or classify the HT origin finding possible quickly corrective actions. In Fig. 2 are summarized the possible HT attacks due to an intentional perturbation of an integrated chip.

Figure 2. Examples of HT attacks acting on a capacitor-based chip.

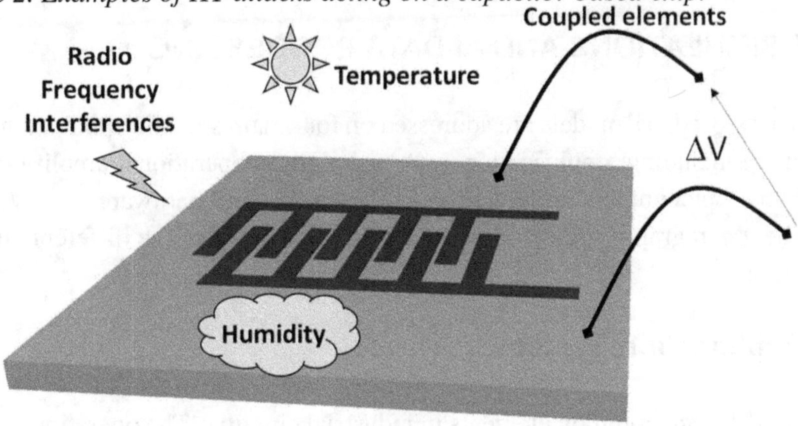

3.4 AI DATA PROCESSING

The DT model of Fig. 1 is improved by the AI engine able to classify and to predict the electrical output signal. This operation allows to understand possible causes of interferences replying a signal which is difficult to classify for example for the superposition of different signal: the AI training model is able to reconstruct the output signal by mans the training of historical signal data. The supervised AI algorithms adopted in literature to process circuital data are (Massaro 2024 (b)):

- Random Forest (RF): decision tree models are constructed analysing different records by splitting the attributes all the attributes; the predicted value for a leaf node is the mean target value (Pu et al. 2021; Parmar et al. 2019; Liu et al. 2012; Genuer et al. 2017);
- Artificial Neural Network (ANN) based on the Multi Layer Perceptron (MLP) model and using the Resilient Backpropagation (RPROP) local adaptation of the weight-updates according to the behaviour of the error function (Riedmiller & Braun, 1993); ANN are applied in different applications in electronics and mechatronics (Zhao et al. 2023; Mokhtar et al. 2021; Alsouda et al. 2018; Alsouda et al. 2019; Rahman et al. 2023; Lo Brano et al. 2014; Shingare et al. 2023; Ogar et al. 2023).

The AI prediction of the output electrical signal is fundamental to execute a corrective machine setting plan thus decreasing drastically the risks of machine failures, component breakages, and product defects.

More details of AI model are discussed in section 5, and in the next section 4 about different AI application cases.

4. DT SIMULATIONS AND AI DATA PROCESSING

The proposed DT models are addressed on four main specific application fields in industry including amplification systems based on operational amplifiers, passive filters, optocouplers integrated in PLC systems, and hardware cybersecurity. In the next paragraphs are detailed the *proof of concepts* of the different circuital applications.

4.1 Amplification System

Amplifiers are common elements in industrial circuitry. The operational amplifier as 'Adder' are suitable to model possible additive noises and interferences in amplification systems.

The modelled circuit is illustrated in Fig. 3 (a): the 'Adder' circuit is in inverting configuration and simulates the additive noise (V_{R4}) and interferences (V_{R2} and V_{R3}) disturbing the input signal (V_{R1}) and coming from the 3D space. The electrical resistances R_2, R_3 and R_4 of Fig. 3 (a) define the amplitude modulation of the interferences and noises behaving as additive noises. By considering all the input resistance equal to the feedback resistance, the output voltage is given by:

$$V_{output}(t) = -\left(V_{R1}(t) + V_{R2}(t) + V_{R3}(t) + V_{R4}(t)\right) \tag{10}$$

being the 'Adder' characterized by the following general output voltage:

$$V_{OUT} = -R_f \sum_{j=1\ldots n} \frac{V_j}{R_j} \tag{11}$$

, where i is the number of the total voltage inputs V_j the input signals, R_i the electrical resistances modulating the input voltage, and R_f the electrical resistance related the negative feedback. The scheme of Fig. 3 (a) is able to model all the circuits characterized by the addition of signals and is representative for the modelling of additive noises and interferences. The simulations are performed assuming $R_f = R_{j=1,2,3} = 10\ \text{k}\Omega$ and changing. Figure 3 (b) illustrates the simulated LTSpice circuit modelling the layout of Fig 3 (a): the input signals are generated by voltage generators able to model the following voltage signals:

- $V_4 = V_{R4}$: white noise (white noise function defined in the LTSpice libraries);
- $V_1 = V_{R1}$: pure sinusoidal signal characterized by a working frequency of f_0 = 1 kHz;
- $V_2 = V_{R2}$: interference defined by eq. 6 with a carrier at frequency f_{C2} = 8 kHz (T_d = 0 s, ϑ = 0 s^{-1}, φ = 0 deg., Number of cycles N = 50, and V_{amp} = 2.4 Volts);
- $V_3 = V_{R3}$: interference characterized by an exponential decay and defined by eq. 6 with a carrier at frequency f_{C3} =8 kHz ($T_d = 8 \bullet 10^{-3}$ s, $\vartheta = 500$ s^{-1}, $\varphi = 45$ deg., and V_{amp} = 2 Volts).

Figure 3. (a) Amplifier and DT model adding a white noise and interferences (interferences 1 and interference 2) characterized by carriers. (b) Equivalent LTSpice simulated circuit.

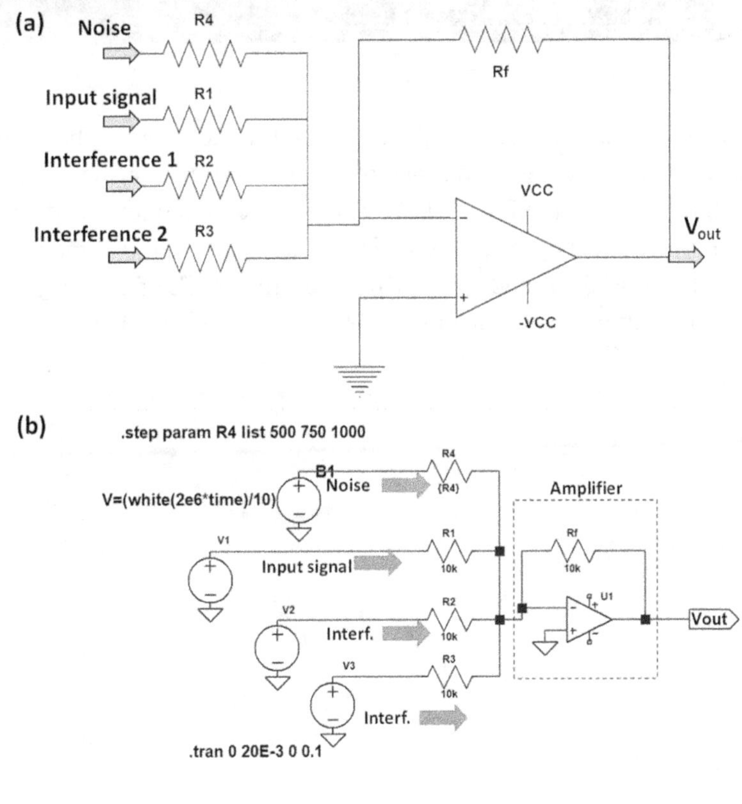

The V_1, V_2, and V_3 signals coupled at the input of the operational amplifier are plotted in Fig. 4 (a), besides the total output voltage V_{out} including the effect of the white noise is illustrated in Fig. 4 (b).

Figure 4. LTSpice simulation: (a) time domain signals V_1, V_2 and V_3 coupled at the input of the operational amplifier; (b) output voltage V_{out} = V(Vout) of the circuit of Fig. 3 adding all the voltage contributions V_1, V_2, V_3 including the white noise V_4 (fixing the resistance R_4 = 1kΩ).

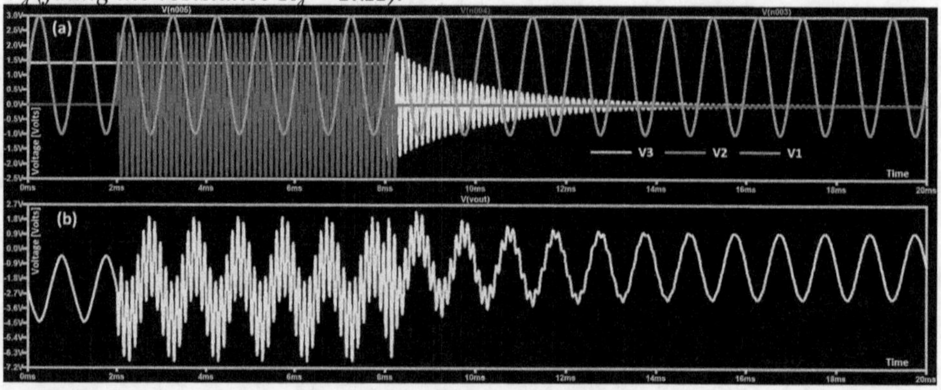

In order to analyse the effect of the amplitude modulation of the white noise due to the R_4 value, it is executed a parametric simulation varying the resistance (R_4 = 500Ω, R_4 = 750Ω and R_4 = 1 kΩ). The time domain output voltage V_{out} and the related noisy frequency response (FFT of the V_{out}) of the parametric analysis are illustrated in Fig. 5 (a) and Fig. 5 (b), respectively. The frequency responses identifies also the interferences having a carrier at $f_{C2} = f_{C3}$ = 8 kHz.

Figure 5. LTSpice parametric simulations varying the resistance R_4 (R_4= 500 Ω, 750 Ω and 1 kΩ): (a) time domain V_{out} = V(vout) signal and (b) related frequency responses (FFT of V_{out}).

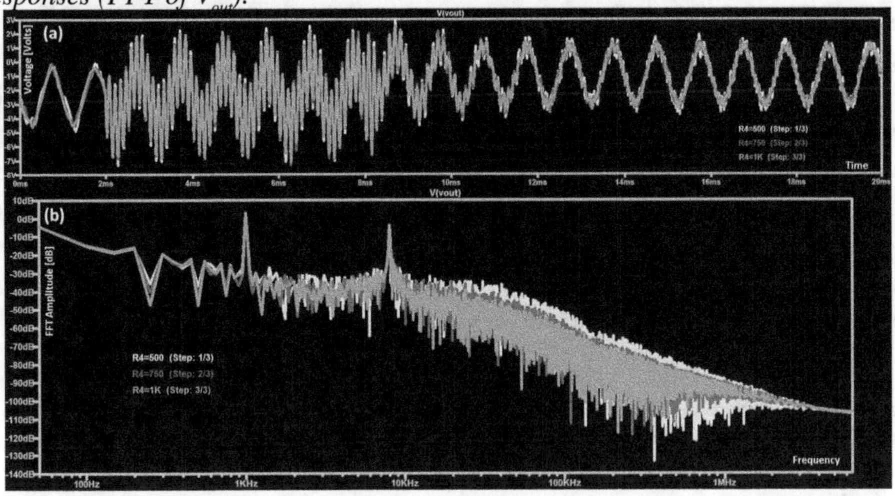

By the Ohm's law, a further decrease of the R_4 resistance corresponds to an increase of the ripple of the electrical current travelling the resistance. This effect is better visible by the parametric analysis of Fig 6 (a) and Fig. 6 (b).

Figure 6. LTSpice parametric simulations varying the resistance R_4 (R_4= 10 Ω, 100 Ω and 200 Ω): (a) time domain V_{out} = V(vout) signal and (b) related frequency responses (FFT of V_{out}).

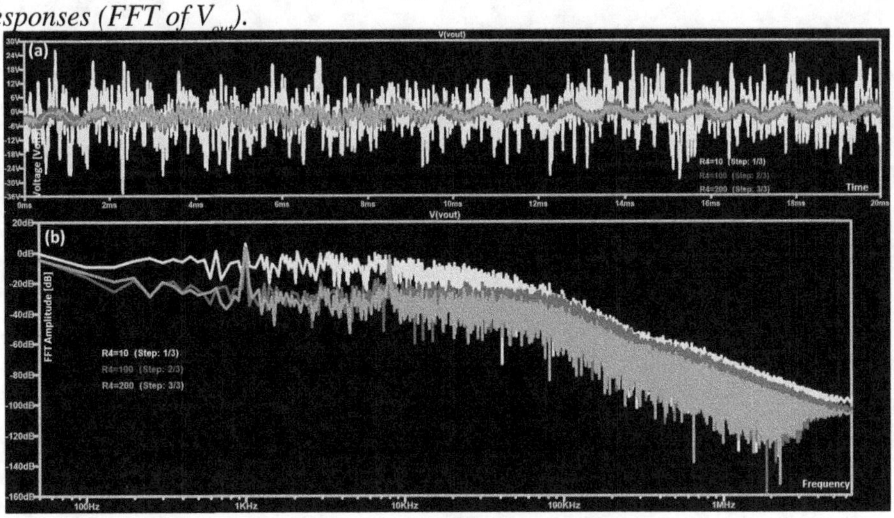

The AI data processing is performed to check the possibility to predict correctly the signal output. The ANN and the RF supervised algorithms provide the results of Fig. 7. The RF approach is closer to the normalized output voltage trend confirming its suitability. The ANN results are convergent in the first time steps but successively approximates the average value of the output voltage. Both the plots of Fig. 7 are obtained by fixing the following hyper-parameters: 70% of training dataset, 30% of testing dataset, 2000 as iteration number (ANN), 4 hidden layer (ANN), 7 neurons per hidden layer (ANN), 10 as tree depth level (RF), minimum node size value equals to 5 (RF), 100 tree models (RF). These results confirms that the choices of the algorithm is a function of the dataset to process, and that it is possible to predict signals characterized by a strong ripple behaviour as for noisy signals composed by interference carriers and white noise.

Figure 7. AI data processing: ANN and RF algorithm predicting the normalized voltage output (Vout of Fig. 3 (b)).

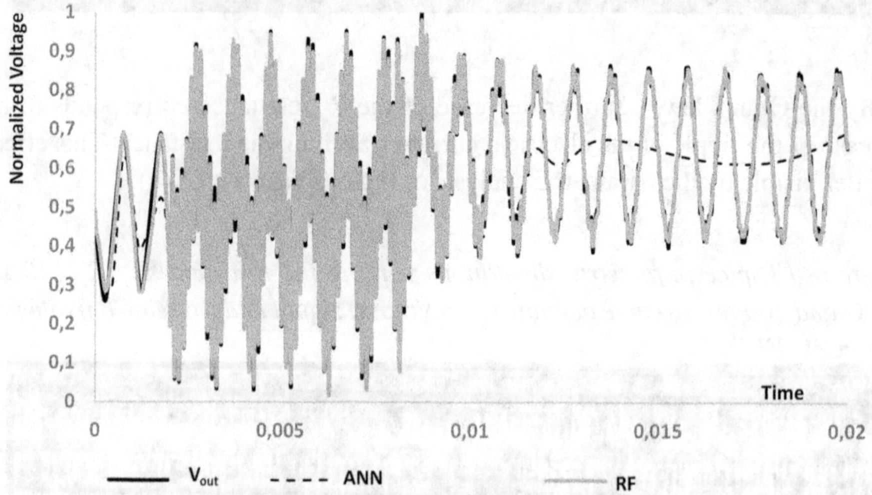

4.2 Passive Filters

Filters are common components of machines and circuits controlling a production. Low pass are typically applied to remove noises (Magsi et al. 2018; Kelemenová et al. 2020) appearing at high-frequencies, to decrease the signal distortion effect, and to limit the bandwidth of signals matching with control system requirements. Other application fields where are implemented Resistance and Capacitance (RC) filters are switching systems of valves and actuators, control, positioning, and electronic energy management systems (Haas et al.. 2014; Mondal et al. 2015; Oshima et al. 2003, Jin et al. 2024; Ramos & Costa-Castell6, 2024; Salari et al. 2018; Lau 2017, Alizadeh et al., 2021, Sun et al. 2019). The low passive RC circuit allows electrical signals with frequencies below a specified cut-off frequency to pass and blocking signals with frequencies higher than the cut-off one. The transfer function $H(s)$ of a single stage RC low passive filter is given by:

$$H(s) = \frac{V_{out}}{V_{input}} = \frac{1}{1 + RC \bullet s} \tag{12}$$

, being $s=j\omega$ the Laplace variable. The model of Fig. 8 (a) illustrates a double stage RC low pass filter coupled with a voltage source (pure sinusoidal signal) and a white noise generator simulating a background noise. The second RC stage behaves as a corrective network cleaning the noisy signal (output of the first RC stage). According with the Kirchhoff's current law (1st Law) the current travelling in the resistance R_4 is the sum of the input source and the white noise currents thus simulating an additive contribution. Figure 8 (a) illustrates the equivalent LTSpice circuit providing the simulation the model of Fig. 8 (a). The performed simulation calculates the currents and the voltages in each section of the circuit: input section (Input + Noise), end of the first filter stage (Output1), and end of the second filter stage (Output2). In order to study the white noise effect, it is performed a parametric analysis varying the R_x parameter representing the white noise amplitude modulation (the Rx are 10Ω, 50Ω and 150Ω).

Figure 8. (a) Example of low pass RC passive filter (two stage) coupled with an input voltage source and a white noise voltage generator. (b) Equivalent LTSpice model of the circuit of Fig. 8 (a) executing a time domain transient analysis.

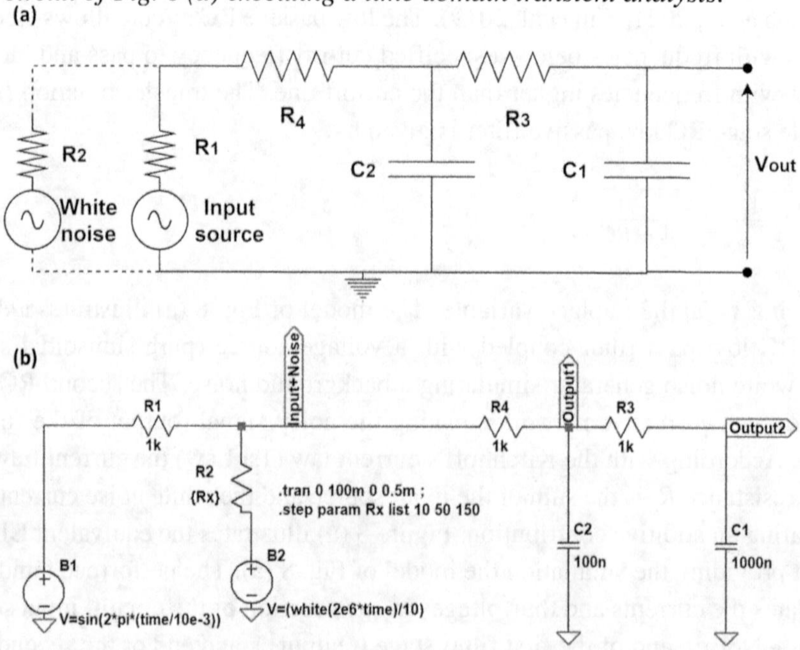

The simulations of Fig. 9 illustrate the effect of the modulated white noise signal on the voltage signal calculated in each section of the circuit of Fig. 8 (b). The effect of the white noise is mainly observed at the input of the simulated circuit (Fig. 9 (a)) for lower values of electrical resistances (according with the Ohm's law $I = V/R$). The first filter RC stage attenuates slightly the signal distortion (Fig. 9 (b)) thus requiring the use of a further corrective network (further stage of the low pass filter). By adding the second filter RC stage, the voltage signal becomes cleaner (Fig. 9 (c)). This circuit simulation is useful to important for the design of robust networks for signal filtering supporting the choice of the electrical parameters (R_3, R_4, C_1 and C_2). In the proposed simulation the filter electrical parameters are fixed as $R_3 = 1 \cdot 10^3\Omega$, $R_4 = 1 \cdot 10^3\Omega$, $C_2 = 100nF$ and $C_1 = 1000nF$.

Figure 9. LTSpice parametric analysis changing the R_x parameter (modulation of the white noise amplitude): (a) voltage at the input port (Input + Noise section); (b) voltage at the output of the first filter stage (Output1 section); (c) voltage at the output of the second filter stage (Output2 section).

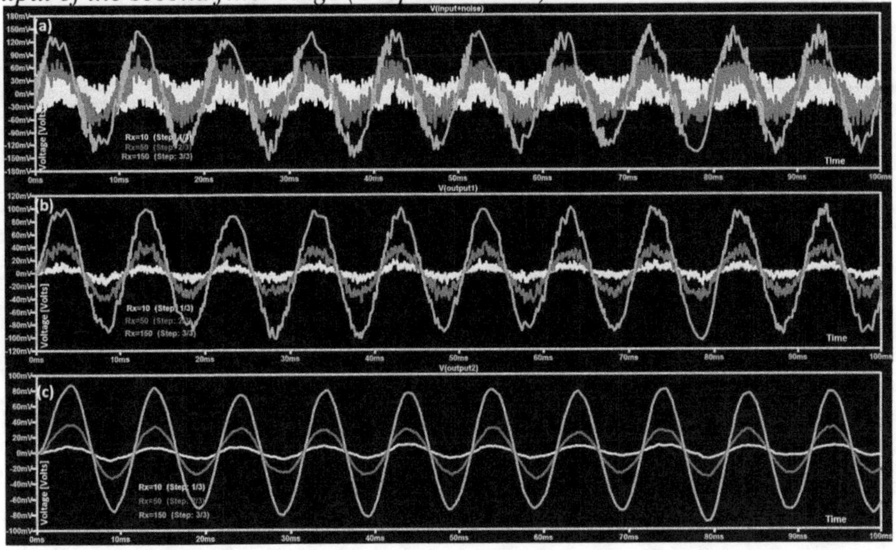

The AI simulation is useful to predict the noisy output signal (output of the first filter stage) when the white noise origin and effect is unknown: the output voltage prediction could provide hypothesized output voltage trends by analysing historical data (historical data characterized by white noise disturbs). The ANN and the RF results of Fig. 10 prove that it is possible to predict the voltage trend (the RF approach is more accurate if compared with the ANN one). The optimised hyper-parameters of results of Fig. 10 are: 70% of training dataset, 30% of testing dataset, 2000 as iteration number (ANN), 15 hidden layer (ANN), 15 neurons per hidden layer (ANN), 10 as tree depth level (RF), minimum node size value equals to 5 (RF), 100 tree models (RF).

Figure 10. **AI data processing: ANN and RF algorithm predicting the normalized voltage output (voltage of the Output1 section ofFig. 8 (b)). For the AI calculus, Rx is fixed at 150 *Ω*.**

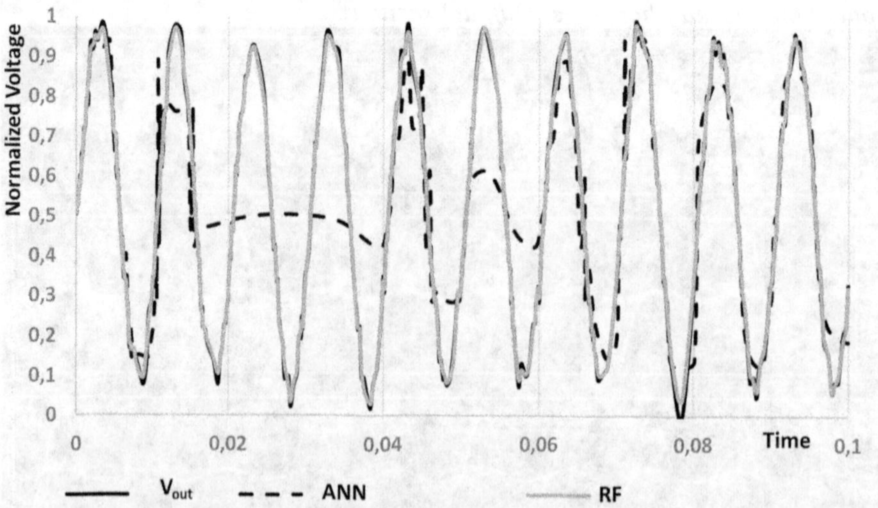

4.3 Optocoupler in PLC Systems

Optocouplers (Kopytov & Ulyanov, 2019; Wu et al.2023; Lai et al. 2022) are typically implemented to electrically isolate a signal source from the PLC circuitry receiving the signal eliminating the risk of damage due to uncontrolled voltage spikes. In optocouplers, light-emitting diodes (LEDs) and phototransistors are coupled to enable an output command such as an electrical current loading a downstream circuit. Another important aspect to study in these systems is the possible interaction of the switching signal of the optocoupler with a background noise. The DT model of Fig. 11 (a) takes into account this interaction by adding a white noise source at the input port coupled with the input signal. The equivalent LTSpice circuit is sketched in Fig. 11 (b) using a 4N25 optocoupler made by a light led and a phototransistor. The circuit simulation is performed by considering a pulse input signal according with the PLC signal transmission protocol (signal of high voltages coding the '1' value and the low voltages coding the '0'value), and a coupled white noise modulated by a resistance (amplitude of the white noise signal modulated by the R_{noise} resistance).

Figure 11. (a) Optocoupler circuit of a PLC system modelling the white noise effect (the optocoupler element is enclosed in the rectangular box and is composed by the led and the phototransistor enabling the output electrical current at the output port). (b) Equivalent simulated LTSpice circuit.

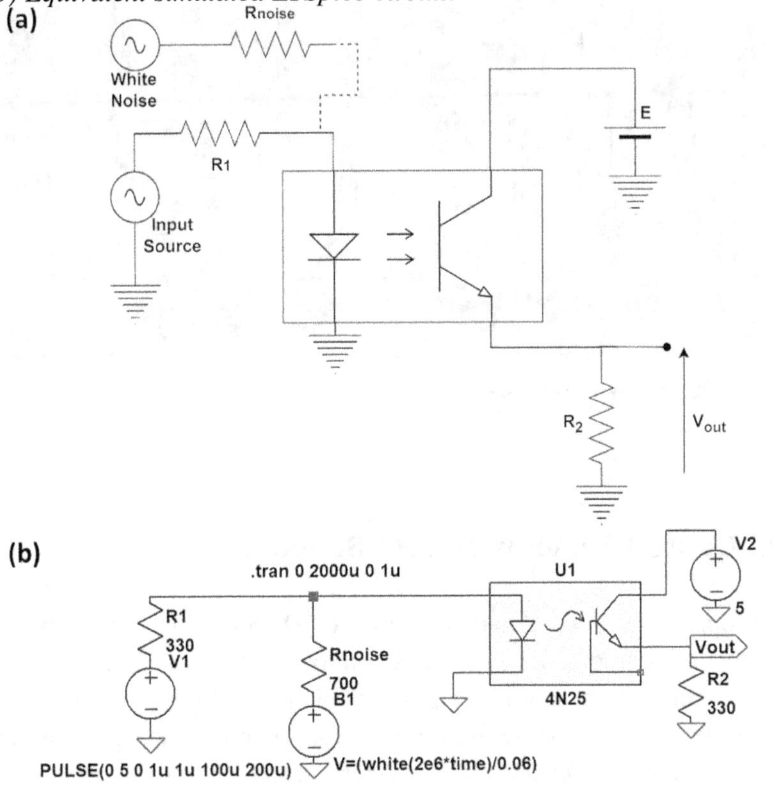

The AI results of the predicted output voltage signal calculated at the output of the circuit of Fig. 11 (b), are plotted in Fig. 12. The ANN and the AI results are follow with a good convergence the output signal proving the accuracy of the approaches. The RF exhibits a closer convergence with the output voltage trend (pulse influenced by the white noise). The optimised hyper-parameters of results of Fig. 12 are: 70% of training dataset, 30% of testing dataset, 2000 as iteration number (ANN), 5 hidden layer (ANN), 10 neurons per hidden layer (ANN), 10 as tree depth level (RF), minimum node size value equals to 5 (RF), 100 tree models (RF). By considering a multi-port model (Massaro 2024 (c)) it is also possible to simulate the effect of other additive or multiplicative noise, such as stationary and impulsive noises.

Figure 12. AI data processing: ANN and RF algorithm predicting the normalized voltage output (voltage of the Vout section of Fig. 11 (b)).

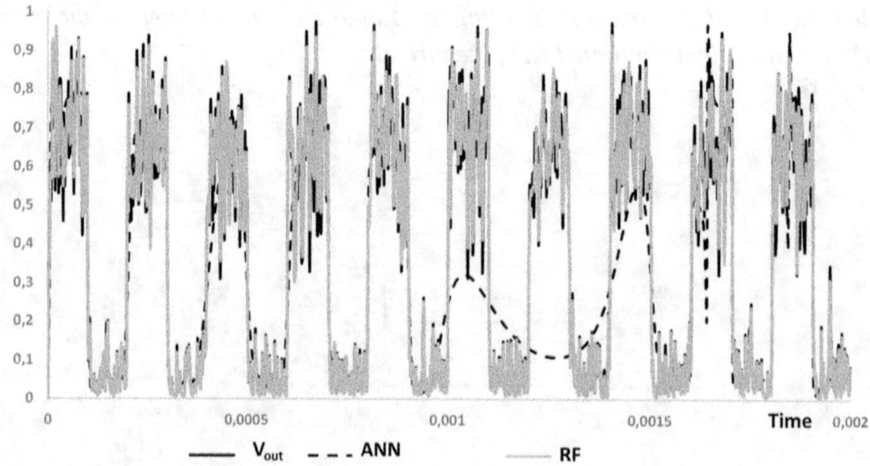

4.4 PID Control Systems and HT Sensors

The Proportional Integral and Derivative (PID) controller is a feedback control system used to set automatically a correction to a control function managing a production machine action. The PID circuit is able to estimate an error function *E(t)* defined as the difference between a desired input set point and a process controlled variable measured by a sensor: for each feedback step, the PID controller estimates the output value function *O(t)* as the sum of the following three terms:

$$O(t) = K_p E(t) + K_i \int_0^t E()d + K_d \frac{d}{dt} E(t) \quad (13)$$

, being K_p, K_i and K_d the constant of the proportional term, integral term and derivative term, respectively.

The PID minimizes gradually the error function optimizing the machine regulation. In Fig. 13 (a) is sketched a basic example of a PID circuit disturbed by a white noise acting on the inverting input of the operational amplifier. The PID circuit is simulated by the layout of Fig. 13 (b). The feedback system is able to measure the variable by means of a sensor (Massaro et al. 2025 b). By considering a temperature as variable to measure, the thermistor model of Fig.13 (c) is suitable to detect the parameter to compare with the set point. In Fig. 13 (c) is sketched the model of a thermistor: the resistance R_1 and the capacitance C_1 values identify a specific value of temperature and relative humidity. Furthermore, the layout of Fig. 13 (d) takes into account a RC circuit coupled to the output of the thermistor and simulating a HT attack: the R_2 resistance the capacitance C_2, and the white noise generator indicate a perturbation of the output of the thermistor (perturbation due to the HT attack). Without the white noise generator, the electrical behavior of the R_2C_2 circuit is the charging and the discharging effect described by the following equations, respectively:

$$V_{out} = V_{input}(1 - e^{-t/R_2C_2}) \tag{14}$$

$$V_{out} = V_{input}(e^{-t/R_2C_2}) \tag{15}$$

Figure 13. (a) Basic PID control system and white noise disturbing the regulation system, and (b) Equivalent PID LTSpice circuit. (c) Thermistor model including HT attack, and (d) equivalent LTSpice circuit.

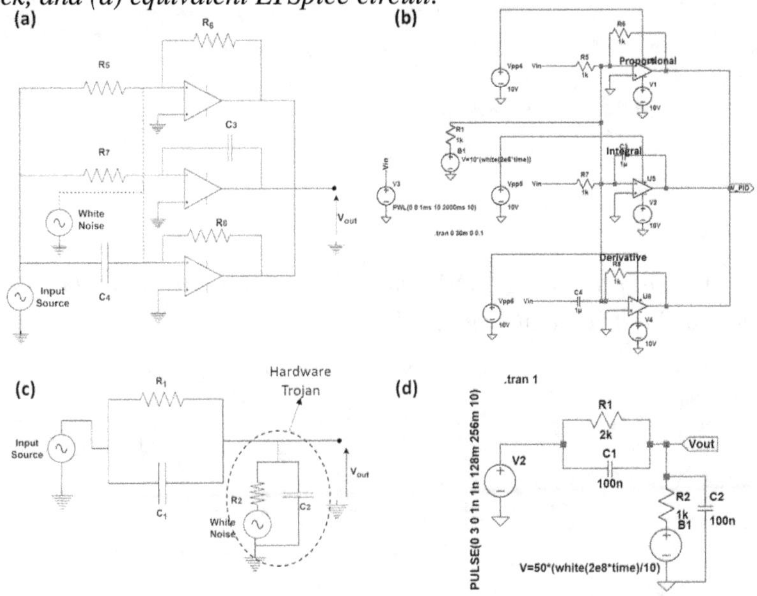

The output voltage V_PID of Fig. 13 (b) and the ANN and RF voltage predictions are plotted in Fig. 14: as observed the ANN is suitable for the average signal trend, besides the RF follows better the ripple behavior of the noisy signal. A more precise convergence of the RF prediction is also observed for the transistor simulation of Fig. 15. The optimised hyper-parameters of results for both Fig. 14 and Fig. 15 are: 70% of training dataset, 30% of testing dataset, 2000 as iteration number (ANN),

5 hidden layer (ANN), 10 neurons per hidden layer (ANN), 10 as tree depth level (RF), minimum node size value equals to 5 (RF), 100 tree models (RF).

Figure 14. AI data processing: ANN and RF algorithm predicting the normalized voltage output (V_PID value of Fig. 13 (b)). Inset: trend of the only ANN prediction.

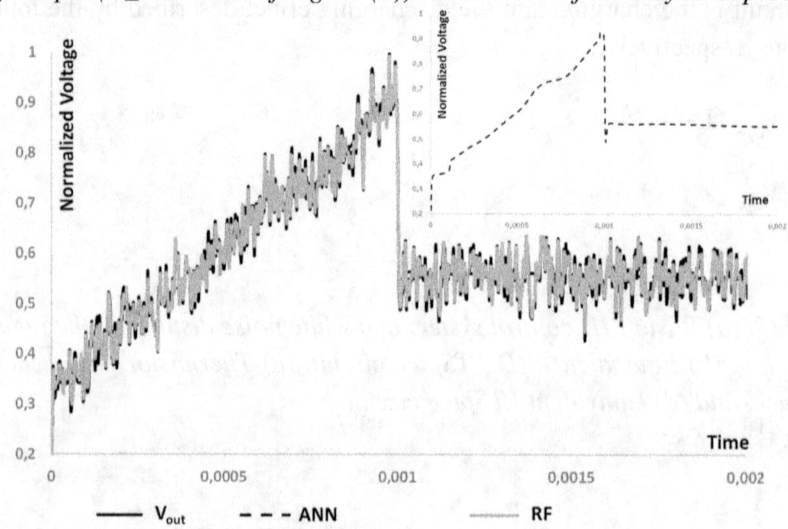

Figure 15. AI data processing: ANN and RF algorithm predicting the normalized voltage output (Vout value of Fig. 13 (d)).

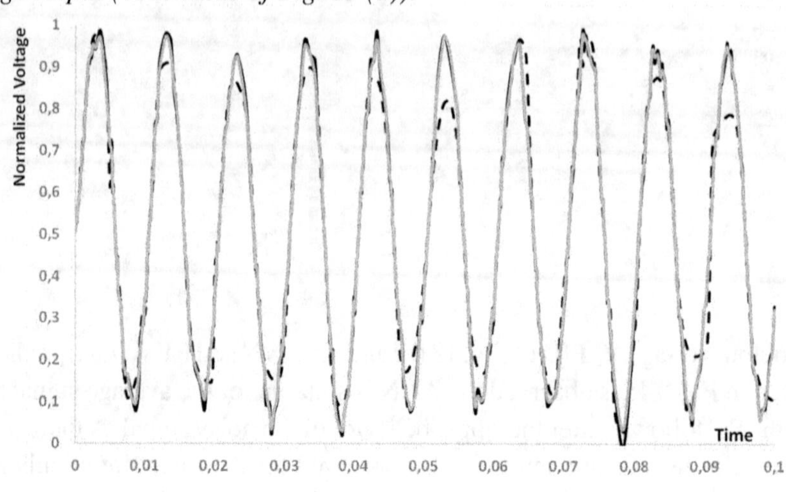

Other implementation aspects about the DT model are discussed in the next section 5 describing organizational processes, AI data workflow, limitations, challenges, and ethics.

5. DISCUSSION

The KNIME workflow of Fig. 16, implemented for the RF and ANN results, is structured in different stages representing the main phases of the AI data processing:

- **Data pre-processing**: phase including data source importing ('Excel Reader' importing the output of the circuit simulations), data normalization ('Normalizer' normalizing the input data using a scale ranging from 0 to 1), data partitioning ('Partitioning' selecting training and testing datasets);
- **Data processing:** training modules ('RProp MLP Learner' training the ANN network, and 'Random Forest Learner' training the RF algorithm), the testing modules ('Multi Layer Perceptron Predictor' testing ANN, and 'Random Forest Predictor' testing RF);
- **Data reporting and results interpretation**: data visualization interfaces as graphical dashboards ('Line Plot'), algorithm performance indicators ('Numeric Scorer'), and data reporting ('Column Appender' merging comparison results of simulated and predicted ones, and 'Excel Writer' exporting the comparison results into an Excel format to be analysed with other graphical dashboards).

Figure 16. KNIME workflow implementing RF and ANN algorithms.

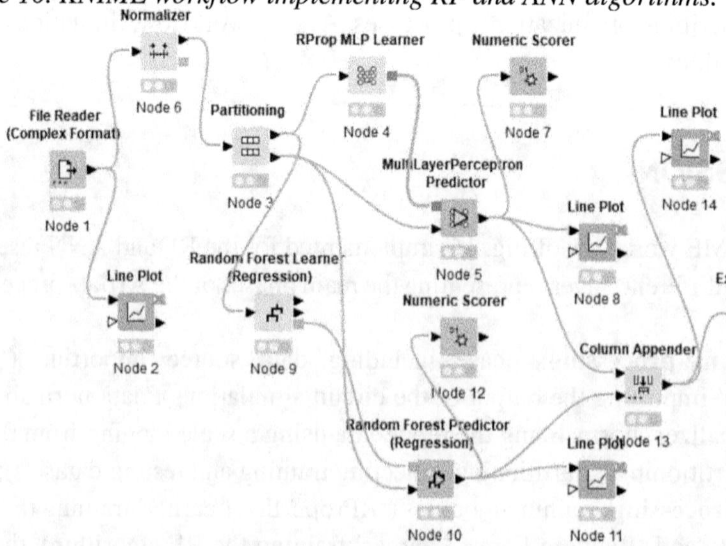

The Mean Absolute Error (MAE) of the ANN data processing ranging from 0.016 to 0.14. The MAE of the RF data processing ranging from ranging from 0.003 to 0.014 (better performance if compared with ANN results). The proposed DT model is integrable in the production Industry 5.0 framework following the scheme of the BPMN workflow of Fig. 17: the diagram includes dynamically the possibility to execute corrective actions in cases of prediction of anomalies and updating the circuital modelling and optimizing the AI data processing. The diagram of Fig. 17 is a process mining representation: the AI-decision making enables the triggering of possible corrective actions based on the prediction of anomalous signal behaviours or of alerting conditions (predicted signal envelopes overcoming thresholds, predicted scattered anomalies due to a superposition of noises and interferences, clustering of electrical variables which are in risk areas of the risk maps, etc.).

Figure 17. BPMN workflows integrating DT data processing into the production process. The AI-decision making is implemented by the 'exclusive event based' gateway.

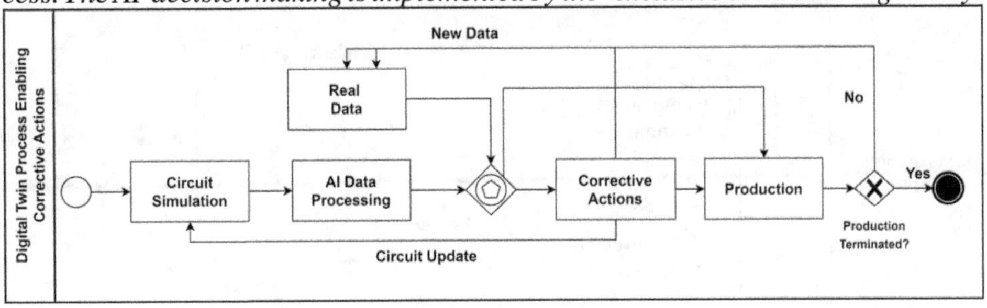

The integration of the DT model into a production process requires a substantial change of the organizational model optimizing some processes such as predictive maintenance and quality monitoring and improvement. The UML Use Case Diagram (UCD) of Fig, 18 explains better this concept: the new skills of data scientist, quality manager and production responsible collaborates to optimize the production where DT are useful to perform corrective actions and to collect digital data improving the AI training models. Specifically, the use cases of the actors are the following:

- **Data Analyst or Data Scientist:** setting of the DT, update and optimization of the AI training model, construction of the dataset to process by merging and comparing real data with simulation ones, check of the corrective actions to execute;
- **Quality Manager:** control of production (monitoring defects and quality of the manufacturing process);
- **Production Responsible:** control of production (real time control of machines), application of the predictive maintenance procedures according with the DT results.

The UML UCD model of Fig. 18 is representative of a human resource organizational model integrating AI DT model in Industry 5.0 scenarios and mapping skills and roles following the introduction of new technologies. This scheme is also useful to implement change management processes fundament to the staff reorganization.

Figure 18. UML UCD diagram of the organizational model using DT in the Industry 5.0 scenario.

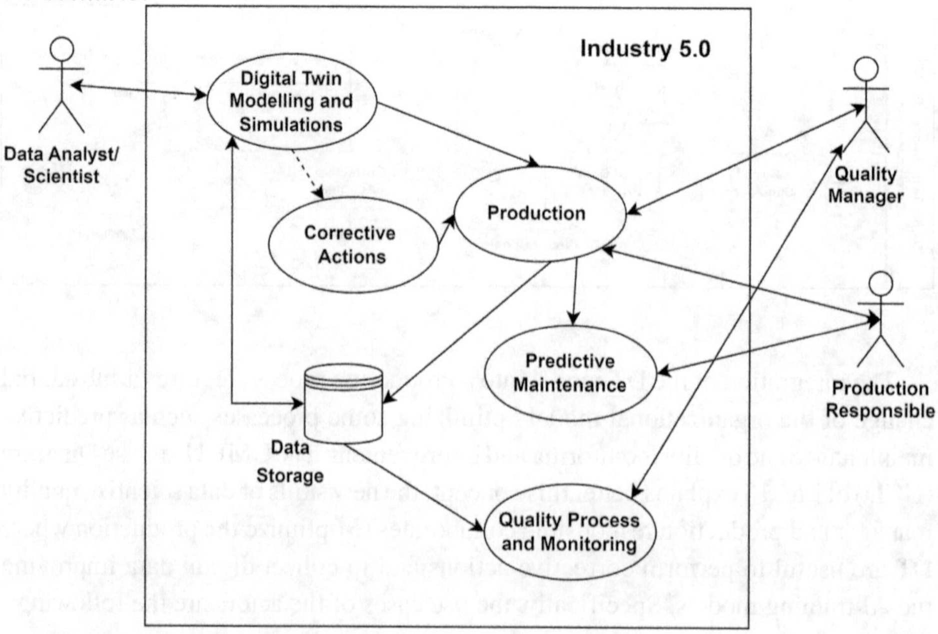

Other aspects to be considered for the implementation of the electronic DT models are related to:

- **Ethics**: ethical aspect are mainly important about the industrial workforce, human resource workload management and the AI decision making (Owolabi et. al 2024; Milosevic & van Schalkwyk, 2024; Bankins 2021; Ferrel et al. 2024). AI should not override respect for human work, rights, and the decision on the setting of the machines should be assigned to the specific worker, ensuring a training plan suitable for figures with different skills. Furthermore, AI could have potential negative impacts on aspects such as job security, job productivity, cybersecurity, safety, and privacy (Liang et al. 2024; Jada & Mayayise, 2024; Damioli et al. 2021; De Azambuja et al. 2023).
- **Regulatory and global standards**: the international standards BPMN (ISO/IEC 19150:2013) and UML (ISO/IEC 19505-1/2:2012) applied in this chapter are useful for the verticalization of the electronics in industrial processes and for the mapping of the new organizational models integrating DT models (Massaro et al. 2025 b). The DT technology must be integrated in industrial protocols including Ethernet networks (Kim & Choi, 2025; Devagiri et al. 2024; Sung et al. 2024) implementing Modbus (Elamanov et al. 2024; Găitan

et al. 2024; Kirdan & Waedt, 2024; Le et al. 2024), Profibus (Rodrigues et al. 2025; Babel 2022) and Open Platform Communications Unified Architecture (OPC-UA) (Ventuneac & Gaitan, 2024; Ji & Xu, 2025; Ladegourdie & Kua, 2022; Martins et al. 2023). Other general international regulatory compliance for industrial machinery and standards are: EU Machinery Regulation (EU) 2023/1230 (European regulatory), IEC 60204-1 (electrical equipment of machines), EN ISO 12100 (general principles for design), NFPA 79 (North America), OSHA guidelines (workplace safety), NFPA 70E (electrical safety in industrial environments), IEEE 802.11 (standards defining wireless networking technology, commonly known as Wi-Fi), IEC 62443 (industrial cybersecurity), NIST Cybersecurity Framework (guidelines and best practices designed to help organizations manage and reduce cybersecurity risks), ISO 27001 (industrial cybersecurity), ISO55000 (framework for the effective management of physical assets).

- **Limitations:** main limitations about the integration of DT in industrial processes are mainly in the construction of the digital knowledge base associated to a specific machine fundamental for the construction of the AI training model, and the computational cost which could be high. Other limitations are in cost of the hardware technology implementing AI-DT, cybersecurity risks due to due to the preparation of the test environments and of hardware testbeds, data privacy when are processed workers data, and workforce adaptation implementing change management procedures.
- **Challenges:** new challenges in Industry 5.0 are in the integration and in the synchronization of many DT modelling entire manufacturing production lines (Massaro 2023 c) thus optimizing maintenance and product quality. Other challenges are in the DT modelling of energy smart metring systems addressed on the energy balancing of production lines or in the energy management of production lines, and on the modelling of advanced electronic components such as memresistors (Tian et al. 2025; Huang et al. 2024; Yu et al. 2024) and nanotechnology-based devices implementing intelligent materials (Massaro 2023 d).

6. CONCLUSION

The goal proposed chapter is to provide theoretical elements to design advanced DT models combining circuit simulation with AI data processing. The circuit modelling and simulation approaches are useful to understand the effect of possible interferences or noises disturbing a measurement or a control signal: this is possible by simulating the coupling at the input port of intentional or unintentional interferences or noises. Furthermore, the circuit simulations provide the sensitivity of the output responses by adopting the parametric analysis approach modulating the amplitude of the undesired signals. The chapter proves the possibility to use the ANN and the RF voltage output predictions as alternative solutions to understand the response behavior of circuit: the different discussed examples of circuits prove that the simulated signals can be adopted for the training of the AI algorithms. The AI predictions proves that for the specific analysed dataset the RF results are closer to the real behaviour of the output signals.

The chapter highlights the process integration aspects of the DT into the production workflow and the organizational model following the integration of the DT in production processes. Finally, the discussion is moved on aspects concerning global standards, ethics, limitations and challenges of DT models providing information about AI-DT perspectives for advanced industrial scenarios. The AI-Based DT models are applicable to complex Industrial of Things (IIOT) control systems and can be used for predictive maintenance of production machinery in Industry 5.0 applications, including the possibility to formulate risk maps based on prediction results. The use of the DT is fundamental also for the design of robust production lines working according with high levels of security and efficiency.

REFERENCES

Acito, F. (2023). *Predictive Analytics with KNIME Analytics for Citizen Data Scientists*. Springer., DOI: 10.1007/978-3-031-45630-5

Alexopoulos, K., Nikolakis, N., & Chryssolouris, G. (2020). Digital Twin-Driven Supervised Machine Learning for the Development of Artificial Intelligence Applications in Manufacturing. *International Journal of Computer Integrated Manufacturing, 33*(5), 429–439. DOI: 10.1080/0951192X.2020.1747642

Alizadeh, M., Moghaddam, M. M., & HosseinNia, S. H. (2021). A Novel Zero Delay Low Pass Filter: Application to Precision Positioning Systems. *ISA Transactions, 111*, 231–248. DOI: 10.1016/j.isatra.2020.11.013 PMID: 33250212

Alsouda, Y., Pllana, S., & Kurti, A. (2018) A Machine Learning Driven IoT Solution for Noise Classification in Smart Cities. *arXiv*, arXiv:1809.00238.

Alsouda, Y., Pllana, S., & Kurti, A. (2029) IoT-Based Urban Noise Identification Using Machine Learning: Performance of SVM, KNN, Bagging, and Random Forest. *InProceedings of the International Conference on Omni-Layer Intelligent Systems*, Crete, Greece, 5–7 May 2019; ACM: New York, NY, USA.

Aniba, Y., Bouhedda, M., Bachene, M., Seddiki, M., Tobbal, A., Hamrani, A., & Benyezza, H. (2024). Enhanced Digital Twin Development for a Conveyor Belt System: Integrating PLC Control, CNN Decision-Making, Gradio-based HMI, and Fuzzy Logic. *Brazilian Journal ofTechnology, 7*(4), 1–18. DOI: 10.38152/bjtv7n4-035

Arumugam, S. (2022). Investigations on Using White Noise as a Test Signal for Performing Frequency Response Measurements on Transformers. *Electric Power Systems Research, 202*, 107586. DOI: 10.1016/j.epsr.2021.107586

Asadi, F. (2023). *Essential Circuit Analysis using LTspice*. Springer Nature., DOI: 10.1007/978-3-031-09853-6

Attaran, S., Attaran, M., & Celik, B. G. (2024). Digital Twins and Industrial Internet of Things: Uncovering operational intelligence in industry 4.0. *Decision Analytics Journal, 10*(100398), 100398. DOI: 10.1016/j.dajour.2024.100398

Babel, W. (2022). Fieldbuses, Communication Protocols, User Interfaces, Hardware ASICs. In *Industry 4.0, China 2025, IoT*. Springer., DOI: 10.1007/978-3-658-37852-3_5

Bai, H., Kuprat, J., Osório, C., Liu, C., Liserre, M., & Gao, F. (2024). Digital Twins for Modern Power Electronics: An investigation into the enabling technologies and applications. *IEEE Electrification Magazine*, *12*(3), 50–67. DOI: 10.1109/MELE.2024.3423111

Bankins, S. (2021). The Ethical Use of Artificial Intelligence in Human Resource Management: A Decision-Making Framework. *Ethics and Information Technology*, *23*(4), 841–854. DOI: 10.1007/s10676-021-09619-6

Beenakker, C. W. J., & Büttiker, M. (1992). Suppression of shot noise in metallic diffusive conductors. *Physical Review. B*, *46*(3), 1889–1892. DOI: 10.1103/PhysRevB.46.1889 PMID: 10003850

Bongomin, O., Mwape, M. C., Mpofu, N. S., Bahunde, B. K., Kidega, R., Mpungu, I. L., Tumusiime, G., Awuor Owino, C., Goussongtogue, Y. M., Yemane, A., Kyokunzire, P., Malanda, C., Tigalana, D., Gumisiriza, O., & Ngulube, G. (2025). Digital Twin Technology Advancing Industry 4.0 and Industry 5.0 Across Sectors. In *Social Science Research Network*. https://doi.org/DOI: 10.2139/ssrn.5072457

Cai, W., Zhang, Q., & Cui, J. (2022). A Novel Fault Diagnosis Method for Denoising Autoencoder Assisted by Digital Twin. *Computational Intelligence and Neuroscience*, *2022*, 1–8. DOI: 10.1155/2022/5077134 PMID: 35909837

Can, O., & Turkmen, A. (2023). Digital Twin and Manufacturing. In Karaarslan, E., Aydin, Ö., Cali, Ü., & Challenger, M. (Eds.), *Digital Twin Driven Intelligent Systems and Emerging Metaverse*. Springer., DOI: 10.1007/978-981-99-0252-1_8

Chaiprabha, K., & Chancharoen, R. (2023). A Deep Trajectory Controller for a Mechanical Linear Stage Using Digital Twin Concept. *Actuators*, *12*(91), 1–14. DOI: 10.3390/act12020091

Coronado, E., Ueshiba, T., & Ramirez-Alpizar, I. G. (2024). A Path to Industry 5.0 Digital Twins for Human–Robot Collaboration by Bridging NEP+ and ROS. *Robotics (Basel, Switzerland)*, *13*(2), 28. DOI: 10.3390/robotics13020028

Damioli, G., Van Roy, V., & Vertesy, D. (2021). The Impact of Artificial Intelligence on Labor Productivity. *Eurasian Bus Rev*, *11*(1), 1–25. DOI: 10.1007/s40821-020-00172-8

De Azambuja, A. J. G., Giese, T., Schützer, K., Anderl, R., Schleich, B., & Almeida, V. R. (2024). Digital Twins in Industry 4.0 – Opportunities and challenges related to Cyber Security. *Procedia CIRP*, *121*, 25–30. DOI: 10.1016/j.procir.2023.09.225

De Azambuja, A. J. G., Plesker, C., Schützer, K., Anderl, R., Schleich, B., & Almeida, V. R. (2023). Artificial Intelligence-Based Cyber Security in the Context of Industry 4.0—A Survey. *Electronics (Basel)*, *12*(8), 1920. DOI: 10.3390/electronics12081920

Devagiri, J. S., Khanmohammadi Hazaveh, P., Rawashdeh, N., Dudipala, S., Desmuhk, P., & Karmarkar, A. (2024). PLC Multi-robot Integration via Ethernet for Human Operated Quality Sampling. Proceeding of 2024 ASEE Annual Conference & Exposition, Portland, Oregon. https://digitalcommons.mtu.edu/michigantech-p2/903

Elamanov, S., Son, H., Flynn, B., Yoo, S. K., Dilshad, N., & Song, J. (2024). Interworking Between Modbus and Internet of Things Platform for Industrial Services. *Digital Communications and Networks*, *10*(2), 461–471. DOI: 10.1016/j.dcan.2022.09.013

Frick, J. (2023). Future of Industrial Asset Management: A Synergy of Digitalization, Digital Twins, Maintenance 5.0 / Quality 5.0, Industry 5.0 and ISO55000. [IJBMM]. *International Journal of Business Marketing and Management*, *8*(4), 93–99.

Gafurov, A. N., Kim, J., Kim, I., & Lee, T.-M. (2024, September 19). Web Tension AI Modeling and Reconstruction for Digital Twin of Roll-to-Roll System. *Journal of Intelligent Manufacturing*. Advance online publication. DOI: 10.1007/s10845-024-02488-y

Găitan, N. C., Zagan, I., & Găitan, V. G. (2024). Proposed Modbus Extension Protocol and Real-Time Communication Timing Requirements for Distributed Embedded Systems. *Technologies*, *12*(10), 187. DOI: 10.3390/technologies12100187

Genuer, R., Poggi, J.-M., Tuleau-Malot, C., & Villa-Vialaneix, N. (2017). Random Forests for Big Data. *Big Data Research*, *9*, 28–46. DOI: 10.1016/j.bdr.2017.07.003

Haas, R., Lukachev, E., & Scheidl, R. (2014). An RC Filter for Hydraulic Switching Control with a Transmission Line Between Valves and Actuator. *Int. J. Fluid Power*, *15*, 139–151.

Hsu, S. T., Whittier, R. J., & Mead, C. A. (1970). Physical model for burst noise in semiconductor devices. *Solid-State Electronics*, *13*(7), 1055–1056. DOI: 10.1016/0038-1101(70)90102-4

Huang, Y., Ando, T., Sebastian, A., Chang, M.-F., Yang, J. J., & Xia, Q. (2024). Memristor-Based Hardware Accelerators for Artificial Intelligence. *Nat. Rev. Electr. Eng.*, *1*(5), 286–299. DOI: 10.1038/s44287-024-00037-6

Huang, Z., Fey, M., Liu, C., Beysel, E., Xu, X., & Brecher, C. (2023). Hybrid Learning-Based Digital Twin for Manufacturing Process: Modeling Framework and Implementation. *Robotics and Computer-integrated Manufacturing*, *82*(102545), 102545. DOI: 10.1016/j.rcim.2023.102545

Jada, I., & Mayayise, T. O. (2024). The Impact of Artificial Intelligence on Organisational Cyber Security: An Outcome of a Systematic Literature Review. *Data and Information Management*, *8*(2), 100063. DOI: 10.1016/j.dim.2023.100063

Jamshidi, M., Lotfi, S., Siahkamari, H., Bleca, T., & Talla, J. (2024). An intelligent digital twinning approach for complex circuits. *Applied Soft Computing*, *154*, 111327. https://doi.org/. DOI: 10.1016/j.asoc.2024.111327

Ji, T., & Xu, X. (2025). Exploring the Integration of Cloud Manufacturing and Cyber-Physical Systems in the Era of Industry 4.0 – An OPC UA Approach. *Robotics and Computer-integrated Manufacturing*, *93*(102927), 102927. DOI: 10.1016/j.rcim.2024.102927

Jin, G., Wu, H., Yin, Y., Zheng, L., & Zhuang, Y. (2024). A High-Accuracy RC Time Constant Auto-Tuning Scheme for Integrated Continuous-Time Filters. *Micromachines*, *15*(1), 166. DOI: 10.3390/mi15010166 PMID: 38276865

Johnson, J. B. (1971, February). Electronic noise: The first two decades. *IEEE Spectrum*, *8*(2), 42–46. DOI: 10.1109/MSPEC.1971.5217957

Karkaria, V., Tsai, Y. K., Chen, Y. P., & Chen, W. (2025). An Optimization-Centric Review on Integrating Artificial Intelligence and Digital Twin Technologies in Manufacturing. *Engineering Optimization*, *57*(1), 161–207. DOI: 10.1080/0305215X.2024.2434201

Kavitha, S., Visalini, D. S., Subasree, P., & Vedhika, A. L. (2025) Assessing Low Power Adder with Carbon Nanotube FET using LTSPICE simulations. *Proceeding of 2025 International Conference on Electronics and Renewable Systems (ICEARS)*, Tuticorin, India, 2025, pp. 406-409, DOI: 10.1109/ICEARS64219.2025.10940106

Kelemenová, T., Benedik, O., & Koláriková, I. (2020). Signal Noise Reduction and Filtering. *Acta Mechatronica*, *5*(2), 29–34. DOI: 10.22306/am.v5i2.65

Kharchenko, V., Illiashenko, O., Morozova, O., & Sokolov, S. (2020). Combination of digital twin and artificial intelligence in manufacturing using industrial IoT. *Proceeding of 2020 IEEE 11th International Conference on Dependable Systems, Services and Technologies (DESSERT)*, 196–201. DOI: 10.1109/DESSERT50317.2020.9125038

Khdoudi, A., Masrour, T., El Hassani, I., & El Mazgualdi, C. (2024). A Deep-Reinforcement-Learning-Based Digital Twin for Manufacturing Process Optimization. *Systems*, *12*(2), 38. DOI: 10.3390/systems12020038

Kim, D., & Choi, J.-Y. (2025). Development of a Common API for Multiple Ethernet Fieldbus Protocols in Embedded Slave Devices. *Electronics (Basel)*, *14*(3), 613. DOI: 10.3390/electronics14030613

Kirdan, E., & Waedt, K. (2024). Modbus Covert Channels Within Industrial Automation and Control Systems. *Proceeding of IECON 2024 - 50th Annual Conference of the IEEE Industrial Electronics Society*, 1–6. DOI: 10.1109/IECON55916.2024.10905882

Kopytov, S. M., & Ulyanov, A. V. (2029) Development and Research of Marker Isolation Schemes from Low-Frequency PLC Signal. In *Proceedings of the 2019 International Multi-Conference on Industrial Engineering and Modern Technologies (FarEastCon)*, Vladivostok, Russia, 1–4 October 2019; IEEE: Piscataway, NJ, USA, 2019, 15, 1–4.

Koulamas, C., & Kalogeras, A. (2018). Cyber-Physical Systems and Digital Twins in the Industrial Internet of Things [cyber-physical systems]. *Computer*, *51*(11), 95–98. DOI: 10.1109/MC.2018.2876181

Krishnaveni, S., Chen, T. M., Sathiyanarayanan, M., & Amutha, B. (2024). CyberDefender: An Integrated Intelligent Defense Framework for Digital-Twin-Based Industrial Cyber-Physical Systems. *Cluster Computing*, *27*(6), 7273–7306. DOI: 10.1007/s10586-024-04320-x

Ladegourdie, M., & Kua, J. (2022). Performance Analysis of OPC UA for Industrial Interoperability towards Industry 4.0. *IoT*, *3*(4), 507–525. DOI: 10.3390/iot3040027

Lai, P., Gonzalez, D., Madhusoodhanan, S., Sabbar, A., Ahmed, S., Dong, B., Wang, J., Mantooth, H. A., Yu, S.-Q., & Chen, Z. (2022). Development of LTCC-Packaged Optocouplers as Optical Galvanic Isolation for High-Temperature Applications. *Scientific Reports*, *12*(1), 1–11. DOI: 10.1038/s41598-022-15631-7 PMID: 35804033

Lau, L. (2017). Wavelet Packets Based Denoising Method for Measurement Domain Repeat-Time Multipath Filtering in GPS Static High-Precision Positioning. *GPS Solutions*, *21*(2), 461–474. DOI: 10.1007/s10291-016-0533-1

Le, L. H., Ngo, T. Q., Toan, N. D., Nguyen, C. C., & Phong, B. H. (2024). Integration of Modbus-Ethernet Communication for Monitoring Electrical Power Consumption, Temperature, and Humidity. [JRC]. *Journal of Robotics and Control*, *5*(6), 1618–1625. DOI: 10.18196/jrc.v5i6.22456

Liang, C. Y., Le, T. H., Ham, Y., Mantha, B. R. K., Cheng, M. H., & Lin, J. J. (2024). Ethics of Artificial Intelligence and Robotics in the Architecture, Engineering, and Construction Industry. *Automation in Construction*, *62*, 105369. Advance online publication. DOI: 10.1016/j.autcon.2024.105369

Liu, Y., Wang, Y., & Zhang, J. (2012). New Machine Learning Algorithm: Random Forest. In Liu, B., Ma, M., & Chang, J. (Eds.), Lecture Notes in Computer Science: Vol. 7473. *Information Computing and Applications. ICICA 2012*. Springer., DOI: 10.1007/978-3-642-34062-8_32

Lo Brano, V., Ciulla, G., & Di Falco, M. (2014). Artificial Neural Networks to Predict the Power Output of a PV Panel. *International Journal of Photoenergy*, *2014*, 193083. DOI: 10.1155/2014/193083

Magsi, H., Sodhro, A. H., Chachar, F. A., & Abro, S. A. K. (2018) Analysis of Signal Noise Reduction by Using Filters. In Proceedings of the 2018 International Conference on Computing, Mathematics and Engineering Technologies (iCoMET), Sukkur, Pakistan, 3–4 March 2018; 1–6. DOI: 10.1109/ICOMET.2018.8346412

Martins, A., Lucas, J., Costelha, H., & Neves, C. (2023). CNC Machines Integration in Smart Factories Using OPC UA. *Journal of Industrial Information Integration*, *34*(100482), 100482. DOI: 10.1016/j.jii.2023.100482

Massaro, A. (2021). *Electronics in Advanced Research Industries: Industry 4.0 to Industry 5.0 Advances*. Wiley., DOI: 10.1002/9781119716907

Massaro, A. (2022). Advanced Control Systems in Industry 5.0 Enabling Process Mining. *Sensors (Basel)*, *22*(22), 8677. DOI: 10.3390/s22228677 PMID: 36433272

Massaro, A. (2023 a). Process Mining in Production Management, Intelligent Control, and Advanced KPI for Dynamic Process Optimization: Industry 5.0 Production Processes. In *Perspectives and Considerations on the Evolution of Smart Systems* (pp. 1–17). IGI Global., DOI: 10.4018/978-1-6684-7684-0.ch001

Massaro, A. (2023 b). Process Mining in Production Management, Intelligent Control, and Advanced KPI for Dynamic Process Optimization: Industry 5.0 Production Processes. Chapter 1, Perspectives and Considerations on the Evolution of Smart Systems, IGI global. DOI: DOI: 10.4018/978-1-6684-7684-0.ch001

Massaro, A. (2023d). Intelligent Materials and Nanomaterials Improving Physical Properties and Control Oriented on Electronic Implementations. *Electronics (Basel)*, *12*(18), 3772. DOI: 10.3390/electronics12183772

Massaro, A. (2024a). ANNs Predicting Noisy Signals in Electronic Circuits: A Model Predicting the Signal Trend in Amplification Systems. *AI*, *5*(2), 533–549. DOI: 10.3390/ai5020027

Massaro, A. (2024b). Artificial Intelligence Enabling Denoising in Passive Electronic Filtering Circuits for Industry 5.0 Machines. *Machines (Basel)*, *12*(8), 551. DOI: 10.3390/machines12080551

Massaro, A. (2024c). Artificial Intelligence Signal Control in Electronic Optocoupler Circuits Addressed on Industry 5.0 Digital Twin. *Electronics (Basel)*, *13*(22), 4543. DOI: 10.3390/electronics13224543

Massaro, A., Epicoco, N., Loseto, G., & Ardito, C. (2024 d), "IIOT Cyberattacks in Industrial Field Control Systems: PID Attack and Hardware Trojan Effects," *IEEE 20th International Conference on Automation Science and Engineering (CASE)*, Bari, Italy, 2024, pp. 1702-1707, DOI: 10.1109/CASE59546.2024.10711839

Massaro, A., Epicoco, N., Loseto, G., & Ardito, C. A. (2025a). AI-Based Detection and Simulation of Hardware Trojan Attacks in Industrial Applications. *IEEE Access : Practical Innovations, Open Solutions*, *13*, 50967–50979. DOI: 10.1109/ACCESS.2025.3552112

Massaro, A., Santarsiero, F., & Schiuma, G. (2025b). Advanced Electronic Controller Circuits Enabling Production Processes and AI-driven KM in Industry 5.0. *Journal of Industrial Information Integration*, *45*, 10841. DOI: 10.1016/j.jii.2025.100841

Milosevic, Z., & van Schalkwyk, P. (2024). Towards Responsible Digital Twins. In Sales, T. P., de Kinderen, S., Proper, H. A., Pufahl, L., Karastoyanova, D., & van Sinderen, M. (Eds.), *Enterprise Design, Operations, and Computing. EDOC 2023 Workshops. EDOC 2023. Lecture Notes in Business Information Processing* (Vol. 498). Springer., DOI: 10.1007/978-3-031-54712-6_8

Modad, O. A. A., Ryska, J., Chehade, A., & Ayoub, G. (2024). Revolutionizing sheet metal stamping through industry 5.0 digital twins: A comprehensive review. *Journal of Intelligent Manufacturing*. Advance online publication. DOI: 10.1007/s10845-024-02453-9

Mokhtar, M., Robu, V., Flynn, D., Higgins, C., Whyte, J., Loughran, C., & Fulton, F. (2021). Prediction of Voltage Distribution Using Deep Learning and Identified Key Smart Meter Locations. *Energy AI*, *6*, 100103. DOI: 10.1016/j.egyai.2021.100103

Mondal, C., Azam, M. K., Ahmad, M., Hasan, S. M. K., & Islam, M. R. (2015) Design and Implementation of a Prototype Electrooculography Based Data Acquisition System. *InProceedings of the 2015 International Conference on Electrical Engineering and Information Communication Technology (ICEEICT)*, Savar, Bangladesh, 21–23 May 2015; IEEE: Piscataway, NJ, USA; 1–6. DOI: 10.1109/ICEEICT.2015.7307395

Monjur, M., Calzadillas, J., & Yu, Q. (2023). Hardware Security Risks and Threat Analyses in Advanced Manufacturing Industry. *ACM Transactions on Design Automation of Electronic Systems*, *28*(5), 1–22. DOI: 10.1145/3603502

Monjur, M. M. R., Heacock, J., Calzadillas, J., Mahmud, M. D. S., Roth, J., Mankodiya, K., Sazonov, E., & Yu, Q. (2022). Hardware Security in Sensor and Its Networks. *Frontiers in Sensors*, *3*, 850056. DOI: 10.3389/fsens.2022.850056

Monjur, M. R., Sunkavilli, S., & Yu, Q. ADobf (2020) Obfuscated Detection Method against Analog Trojans on I2C Master-Slave Interface. *Proceedings of the 2020 IEEE 63rd International Midwest Symposium on Circuits and Systems (MWSCAS)*, Springfield, MA, USA, 9–12 August 2020; pp. 1064–1067. DOI: 10.1109/MWSCAS48704.2020.9184680

Nguyen, S.; Abdelhakim, M. and Kerestes, R. (2021). Survey Paper of Digital Twins and their Integration into Electric Power Systems. *Proceeding of 2021 IEEE Power & Energy Society General Meeting (PESGM)*, 01–05. .DOI: 10.1109/PESGM46819.2021.9638011

Ogar, V. N., Hussain, S., & Gamage, K. A. A. (2023). The Use of Artificial Neural Network for Low Latency of Fault Detection and Localisation in Transmission Line. *Heliyon*, *9*(2), e13376. DOI: 10.1016/j.heliyon.2023.e13376 PMID: 36816249

Oshima, T., Maio, K., Hioe, W., Shibahara, Y., & Doi, T. (2003) Automatic Tuning of RC Filters and Fast Automatic Gain Control for CMOS Low-IF Transceiver. *InProceedings of the IEEE 2003 Custom Integrated Circuits Conference*, San Jose, CA, USA, 24 September 2003; 5–8. DOI: 10.1109/CICC.2003.1249348

Owolabi, O. S., Uche, P. C., Adeniken, N. T., Ihejirika, C., Islam, R. B., & Chhetri, B. J. T. (2024). Ethical implication of artificial intelligence (AI) adoption in financial decision making. *Computer and Information Science*, *17*(1), 49. DOI: 10.5539/cis.v17n1p49

Papacharalampopoulos, A., Sabatakakis, K., Karagianni, O. M., & Stavropoulos, P. (2025). Digital Twins of Manufacturing Processes Under Industry 5.0. In Alexopoulos, K., Makris, S., & Stavropoulos, P. (Eds.), *Advances in Artificial Intelligence in Manufacturing II. ESAIM 2024. Lecture Notes in Mechanical Engineering*. Springer., DOI: 10.1007/978-3-031-86489-6_1

Parmar, A., Katariya, R., & Patel, V. (2019). A Review on Random Forest: An Ensemble Classifier. In: Hemanth, J., Fernando, X., Lafata, P., Baig, Z. (eds) International Conference on Intelligent Data Communication Technologies and Internet of Things (ICICI) 2018. ICICI 2018. *Lecture Notes on Data Engineering and Communications Technologies*, vol. 26. Springer, Cham. https://doi.org/DOI: 10.1007/978-3-030-03146-6_86

Pu, S., Zhang, H., Mao, C., & Yang, G. (2021) A Classification Based on Random Forest for Partial Discharge Sources. *InProceedings of the 2021 33rd Chinese Control and Decision Conference (CCDC)*, Kunming, China, 22–24 May 2021; IEEE: Piscataway, NJ, USA; 2307–2311. DOI: 10.1109/CCDC52312.2021.9602056

Rahman, M. S., Ghosh, T., Aurna, N. F., Kaiser, M. S., Anannya, M., & Hosen, A. S. M. S. (2023). Machine Learning and Internet of Things in Industry 4.0: A Review. *Measurement. Sensors*, *28*, 100822. DOI: 10.1016/j.measen.2023.100822

Ramos, G. A., & Costa-Castelló, R. (2022). Energy Management Strategies for Hybrid Energy Storage Systems Based on Filter Control: Analysis and Comparison. *Electronics (Basel)*, *11*(10), 1631. DOI: 10.3390/electronics11101631

Riedmiller, M., & Braun, H. (1993) A direct adaptive method for faster backpropagation learning: theRPROP algorithm, *Proceedings of the IEEE International Conference on Neural Networks (ICNN),* vol. 16, pp. 586-591, Piscataway, 1993, NJ: IEEE. DOI: 10.1109/ICNN.1993.298623

Rodrigues, C. H. A., Maia, O. B., & Bessa, A. R. R.; W. Júnior S. S. and Carvalho, C. B. (2025) Communication Protocols for Digital Twins. Proceeding of 2025 IEEE International Conference on Consumer Electronics (ICCE), Las Vegas, NV, USA, pp. 1-5, DOI: 10.1109/ICCE63647.2025.10929808

Salari, O., Zaad, K. H., Bakhshai, A., & Jain, P. (2018) Filter Design for Energy Management Control of Hybrid Energy Storage Systems in Electric Vehicles. In *Proceedings of the 2018 9th IEEE International Symposium on Power Electronics for Distributed Generation Systems (PEDG)*, Charlotte, NC, USA, 25–28 June 2018; IEEE: Piscataway, NJ, USA; 1–7. DOI: 10.1109/PEDG.2018.8447608

Sasikala, M., Mahaboob John, Y. M., & Jothi, B. (2024) Integrating Digital Twins with AI for Real-Time Intrusion Detection in Smart Infrastructure Networks. *Proceeding of 2024 International Conference on Intelligent Algorithms for Computational Intelligence Systems (IACIS)*, Hassan, India, 2024, pp. 1-6, DOI: 10.1109/IACIS61494.2024.10721892

Scime, L., Singh, A., & Paquit, V. (2022). A Scalable Digital Platform for the Use of Digital Twins in Additive Manufacturing. *Manufacturing Letters*, *31*, 28–32. DOI: 10.1016/j.mfglet.2021.05.007

Shen, Z., Arraño-Vargas, F., & Konstantinou, G. (2023). Artificial Intelligence and Digital Twins in Power Systems: Trends, Synergies and Opportunities. *Digital Twin*, *2*, 11. DOI: 10.12688/digitaltwin.17632.2

Shingare, S. S., Khampariya, P., & Bakre, S. M. (2023). Efficient Fault Detection and Location in Extra High Voltage Networks: An Artificial Neural Network (ANN)-Based Approach. *Int. J. Intell. Syst. Appl. Eng.*, *11*, 1051–1060.

Soori, M., Arezoo, B., & Dastres, R. (2023). Digital Twin for Smart Manufacturing, A Review. *Sustainable Manufacturing and Service Economics*, *2*(100017), 100017. DOI: 10.1016/j.smse.2023.100017

Soudan, B. (2024) Cybersecurity of Digital Twins in Industrial IoT Environments. *Proceeding of Advances in Science and Engineering Technology International Conferences (ASET)*, Abu Dhabi, United Arab Emirates, 2024, 1-6, DOI: 10.1109/ASET60340.2024.10708640

Stergiou, C. L., & Psannis, K. E. (2022). Digital Twin Intelligent System for Industrial Internet of Things-based Big Data Management and Analysis in Cloud Environments. *Virtual Reality & Intelligent Hardware*, *4*(4), 279–291. DOI: 10.1016/j.vrih.2022.05.003

Sun, M., Cai, Z., & Zhao, N. (2022). Design of Intelligent Manufacturing System Based on Digital Twin for Smart Shop Floors. *International Journal of Computer Integrated Manufacturing*, *36*(4), 542–566. DOI: 10.1080/0951192X.2022.2128212

Sun, Y., Tang, X., Sun, X., Jia, D., Cao, Z., Pan, J., & Xu, B. (2019). Model Predictive Control and Improved Low-Pass Filtering Strategies Based on Wind Power Fluctuation Mitigation. *Journal of Modern Power Systems and Clean Energy*, *7*(3), 512–524. DOI: 10.1007/s40565-018-0474-5

Sung, G.-M., Li, Z.-Y., & Yu, C.-P. (2024). Ethernet–USB Bridge Application-Specific Integrated Circuit Incorporating the User Datagram Protocol and Address Resolution Protocol. *IEEE Access : Practical Innovations, Open Solutions*, *12*, 106874–106883. DOI: 10.1109/ACCESS.2024.3436564

Thürer, M., Li, S. S., & Qu, T. (2022). Digital Twin Architecture for Production Logistics: The Critical Role of Programmable Logic Controllers (PLCs). *Procedia Computer Science*, *200*, 710–717. DOI: 10.1016/j.procs.2022.01.269

Tian, D., Ke, C., Sun, B., Liang, H., Qian, Z., Wen, Q., Chen, X., Yang, C., Xu, M., & Zhao, Y. (2025). Quantum dot-based memristors for information processing and artificial intelligence applications. *Nanoscale*, *17*(17), 10485–10505. Advance online publication. DOI: 10.1039/D5NR00136F PMID: 40211906

Ventuneac, C., & Gaitan, V. G. (2024). Industrial Internet of Things Gateway with OPC UA Based on Sitara AM335X with ModbusE Acquisition Cycle Performance Analysis. *Sensors (Basel)*, *24*(7), 2072. DOI: 10.3390/s24072072 PMID: 38610283

Verdugo-Cedeño, M., Jaiswal, S., Ojanen, V., Hannola, L., & Mikkola, A. (2023). Simulation-Based Digital Twins Enabling Smart Services for Machine Operations: An Industry 5.0 Approach. *International Journal of Human-Computer Interaction*, *40*(20), 6327–6343. DOI: 10.1080/10447318.2023.2254607

Voss, R. F., & Clarke, J. (1976). Flicker (1f) Noise: Equilibrium Temperature and Resistance Fluctuations. *Physical Review. B, Solid State*, *13*(2), 556–573. DOI: 10.1103/PhysRevB.13.556

Warr, W. A. (2012). Scientific workflow systems: Pipeline Pilot and KNIME. *Journal of Computer-Aided Molecular Design*, *26*(7), 801–804. DOI: 10.1007/s10822-012-9577-7 PMID: 22644661

Werth, A. W., & Morris, T. H. (2023). A Digital Twin Internal to a PLC to Detect Malicious Commands and Ladder Logic that Potentially Cause Safety Violations. *Journal of Cyber Security Technology*, *7*(2), 53–82. DOI: 10.1080/23742917.2023.2171538

Wu, R., Zhang, K., Xu, Y., & Zhou, M. (2023). *Application of Optocoupler Isolation in Rotation Speed Measuring Circuit of Maglev Blower. Lecture Notes in Electrical Engineering*. Springer Nature.

Xu, H., Wu, J., Pan, Q., Guan, X., & Guizani, M. (2023). A Survey on Digital Twin for Industrial Internet of Things: Applications, Technologies and Tools. *IEEE Communications Surveys and Tutorials*, *25*(4), 2569–2598. DOI: 10.1109/COMST.2023.3297395

Yashika, A., Patra, N. S., & Das, (2024) Human-Machine Interaction in Industrial Automation: Gesture-Based PLC Control. *Proceeding of 15th International Conference on Computing Communication and Networking Technologies (ICCCNT)*, Kamand, India, 2024, pp. 1-7, DOI: 10.1109/ICCCNT61001.2024.10726180

Yu, Y., Xiao, M., Fieser, D., Zhou, W., & Hu, A. (2024). Nanoscale Memristor Devices: Materials, Fabrication, and Artificial Intelligence. *Journal of Materials Chemistry. C, Materials for Optical and Electronic Devices*, *12*(11), 3770–3810. DOI: 10.1039/D3TC04510B

Zafar, M. H., Langås, E. F., & Sanfilippo, F. (2024). Exploring the Synergies Between Collaborative Robotics, Digital Twins, Augmentation, and Industry 5.0 for Smart Manufacturing: A State-of-the-Art Review. *Robotics and Computer-integrated Manufacturing*, *89*(102769), 102769. DOI: 10.1016/j.rcim.2024.102769

Zhao, H., Chen, Z., Shu, X., Shen, J., Liu, Y., & Zhang, Y. (2023). Multi-Step Ahead Voltage Prediction and Voltage Fault Diagnosis Based on Gated Recurrent Unit Neural Network and Incremental Training. *Energy*, *266*, 126496. DOI: 10.1016/j.energy.2022.126496

Zhao, S., Blaabjerg, F., & Wang, H. (2021, April). An Overview of Artificial Intelligence Applications for Power Electronics. *IEEE Transactions on Power Electronics*, *36*(4), 4633–4658. DOI: 10.1109/TPEL.2020.3024914

KEY TERMS AND DEFINITIONS

Digital Twin (DT): model simulating an electronic part of the machine in industrial scenarios.

Electronic Circuit Models: designed circuits providing electrical simulated signals.

Artificial Intelligence (AI) Engine: AI algorithms integrated in DT models and predicting or classifying the electrical signal.

Industry 5.0: upgraded tools and facilitates of Industry 4.0.

Circuit Control Systems: circuits or part of circuits able to control industrial machines.

Sensors: circuits reading machine or production parameters.

Noises: intentional or unintentional industrial environmental disturb of the electronic signal.

Interferences: superposition of intentional or unintentional signals on the electronic signal.

Chapter 6
Integration of Bluetooth Technology and Artificial Neural Networks in High Accuracy Indoor Positioning

Atıl Emre Cosgun
https://orcid.org/0000-0002-4889-300X
Aksaray University, Turkey

ABSTRACT

Nowadays, with the increasing digitalization and widespread use of Internet of Things (IoT) applications, accurate and reliable indoor positioning system applications have become increasingly important. With the increasing need in the areas of use, researchers have proposed various indoor positioning system technologies and methods. In this context, Bluetooth Low Energy (BLE) technology offers a cost-effective, energy-efficient, and easy-to-integrate solution within existing infrastructures. The integration of BLE technology and ANNs for indoor positioning will be the subject of this section, which will cover the theoretical basis and practical applications. A comprehensive review of BLE-based positioning algorithms, performance evaluations, and the design and training of ANN models will be presented. In addition, the limitations of existing systems will be highlighted through the examination of the challenges. The aim of this study is to make a meaningful contribution to the field of indoor positioning and to provide applicable solutions in both academic and industrial environments.

DOI: 10.4018/979-8-3693-7994-3.ch006

1. INTRODUCTION TO INDOOR POSITIONING SYSTEMS (IPS)

This section researches the technologies and methods developed for indoor positioning systems (IPS), particularly emphasizing the use of Bluetooth technology and Artificial Neural Networks (ANN). These technologies are used in basic applications such as positioning individuals or objects in indoor environments where GPS system signals are weak, asset tracking, and smart building management. Also includes large-scale applications, including inventory management, equipment tracking, and product monitoring(Han et al., 2024). At the present time, there is no other system that is as dominant in the field of indoor positioning as GNSS. Deciding on an appropriate indoor positioning system necessitates a compromise between various perspectives. The effectiveness of IPS is determined by the technology used, the algorithms created, and the needs. In this respect, the technologies and methods used offer different levels of accuracy and applicability to the user. Location-based systems are often classified into three categories: (i) interior location systems, (ii) wide-area location systems utilizing cellular networks, and (iii) global location systems. Indoor location technologies are categorized into two groups according to the signal type: (i) radio-based signals and (ii) non-radio-based signals. Radio-based technologies encompass Wi-Fi, RFID, Bluetooth, Zigbee, and UWB, whilst non-radio-based technologies comprise ultrasonic, infrared, and geomagnetic fields (Hailu et al., 2024). The most important technologies include Ultra Wideband (UWB), Wi-Fi Round Trip Time (RTT), Bluetooth Low Energy (BLE), and complementary techniques such as Visible Light Communication (VLC) and Pedestrian Estimation (PDR), as well as AI-supported approaches using Received Signal Strength Indicator (RSSI) and Channel State Information (CSI). When comparing the technologies used for IPS, they have their own strengths and weaknesses in terms of accuracy and reliability. Factors such as the need for sensitivity according to the user's demand, cost constraints and environmental conditions determine which technology will be used to meet the need. However, at this point, the integration of artificial neural networks (ANN) into Bluetooth technology significantly increases the accuracy of indoor positioning systems. An artificial neural network (ANN) for indoor positioning generally has a multi-layer architecture, with the input layer receiving diverse signal properties, hidden layers processing these inputs, and the output layer delivering location predictions. The selection of inputs, quantity of neurons, and activation functions are essential for enhancing the performance of the ANN in indoor settings. RSSI comes signals from various Wi-Fi access points are frequently utilized as inputs. Signal-to-Noise Ratio (SNR) and N values are complementary signal components. They increase the sensitivity of the model by providing more detailed signal features to the developed model (Borenovic & Neskovic, 2011). In some models, Channel Impulse Response (CIR) is used as the input signal. This input parameter increases the accuracy of location estimates by encapsulating the multipath properties of the signal(Dao & Salman, 2022). The number of neurons in the input layer of the models is equal to the number of input features. A model using RSSI from several access points contains one neuron for each access point (Mehmood, 2010). However, the number of hidden layers and the number of neurons within each layer may vary. While some models use a single hidden layer for simplicity and efficiency (Puckdeevongs, 2021), others use multiple hidden layers to distinguish complex patterns. A model consisting of four hidden layers has shown great accuracy. The output layer usually has one or two neurons that specify the expected location coordinates (Malik et al., 2019; Mehmood, 2010).

The following sections provide a detailed comparison of these technologies based on literature, the techniques and technologies used, and the difficulties encountered in indoor positioning. Wireless networks such as Wi-Fi and Bluetooth are used in positioning in IPS as RF-based systems. Wi-Fi reduces investment costs by using existing infrastructure while providing improvements in accuracy with newer technologies such as Wi-Fi 6. Furthermore, Wi-Fi works well in densely populated areas due to its higher frequency, which minimizes interference from human energy absorption(Han et al., 2024). Algorithms such as K-Nearest Neighbor (KNN), Decision Tree, Naïve Bayes and Random Forest regressor are generally used to process Received Signal Strength Indicator (RSSI) data in positioning. In certain scenarios, the Naïve Bayes method has been shown to provide the highest precision accuracy(Prasanthi et al., 2024). Metamaterial antennas are often used to increase efficiency in RF technologies and can provide positioning accuracy of up to 2 cm In VLC-based IPS, where Light Emitting Diodes (LEDs) and photodiodes are used, techniques such as fingerprint recognition and triangulation are frequently used to reduce errors. This method shows high precision and offers positioning errors as low as 18 millimeters (Khafagy & Abdellatif, 2024; Kong et al., 2024). Pedestrian Estimation (PDR) method uses accelerometer, gyroscope and magnetometer sensors in smartphones to estimate users' walking distance and direction. The current position is calculated based on the previous position, walking length and direction. The PDR approach is effective for short-distance positioning, but it creates drift errors at longer distances(Zou et al., 2017). Another method used in IPS is UWB. Its accuracy is increased by using Kalman filter and Time of Flight (ToF) measurements. It works by calculating the transmission times of a signal between devices. It is less susceptible to interference compared to other methods. High bandwidth and precise positioning can be achieved with this method(Qiao et al., 2024). In the system tested in various scenarios, it was observed that the average errors decreased to 17.21 cm under line-of-sight conditions using this method (Warnakulasuriya et al., 2024). Similarly, in the Wi-Fi RTT method used in the field by industry leaders such as Google, signal transmission times were measured and positive results were obtained against precision and signal interference (Qiao et al., 2024). Environmental factors such as signal attenuation and multipath propagation affect the Received Signal Strength Indicator (RSSI) and Angle of Arrival (AoA) measurements used in BLE-based systems (Jianyong et al., 2014; Wu, Wang, et al., 2024). Nevertheless, BLE systems are also popular due to their low cost, low power consumption, and ease of installation. However, BLE location determination performance is generally lower than UWB, Wi-Fi RTT, RSSI and CSI, Visible Light Communication (VLC), Angle of Arrival (AoA) Positioning and Artificial Intelligence-Assisted Indoor Positioning Systems technologies. However, advanced algorithms such as K-Nearest Neighbors (KNN) and deep learning techniques are used to increase the performance in location determination(Syed Salman Rahman et al., 2024; Wu, Wong, et al., 2024). Furthermore, developments in AoA contribute to further expanding the potential of BLE (Alghananim et al., 2025). Adapting models such as Convolutional Neural Networks (CNNs), Recurrent Neural Networks (RNNs), and Autoencoders, which are the underlying models of Artificial Neural Networks (ANNs), to Bluetooth data significantly improves the performance of RSSI and AoA by reducing noise, and accuracies as high as 0.79 meters are achieved in double-branch CNN configurations (Han et al., 2024; Wu, Wong, et al., 2024). Moreover, compared to traditional methods, the use of RNNs in IPS has greatly reduced error rates in location detection (Sahnoun et al., 2025). CSI-based deep learning approaches such as CNNs have further improved positioning accuracy by effectively overcoming environmental dynamics and signal barriers with successful results in complex indoor environments (Bizon et al., 2024; Homma et al., 2024). Furthermore, Han et al. demonstrated that errors may be markedly decreased by integrating Artificial Neural Networks (ANN) with Bluetooth technology. Applying the Kalman filter into the Convolutional Autoencoder model position detection, an error has gotten of under one meter (Han et al., 2024). It shows that such techniques used for location detection are important in reducing signal fluctuations and thus obtaining more stable and reliable results in location estimations. In addition, successful results are obtained in technologies developed within the scope of deep learning, such as the LocNet model. According to Kotturi et al., LocNet achieved a remarkable positioning accuracy of up to 9 cm with a 90% success rate in simulated environments. In particular, this model exhibits strong generalization performance even when trained on incomplete datasets(Kotturi et al., 2024). These features highlight the potential of advanced deep learning algorithms to address long-standing challenges in location detection tasks, paving the way for their broader adoption in a variety of contexts. Furthermore, these studies demonstrate the increased efficiency of integrating ANNs and deep learning with Bluetooth technology. Reducing errors and improving stability are significant improvements for both research and practical applications. Bluetooth, especially Bluetooth Low Energy (BLE), provides a cost-effective and widely available platform for indoor localization, while ANNs offer advanced data processing capabilities to increase accuracy and reliability.

Table 1. Comparison of indoor positioning technology(Han et al., 2024; Huang et al., 2023; Li et al., 2020)

Positioning Method	Accuracy Range (m)	Infrastructure Requirements	Data Collection Effort	Computational Complexity	Network Bandwidth Demand
Ultra-Wideband (UWB)	<0.1	Large	None	High	High
Wi-Fi Round-Trip Time (RTT)	1	Large	High	Moderate	Moderate
Bluetooth Low Energy (BLE)	1-3	Large	High	Moderate	Low
RF-based (RSSI & CSI)	0.2 - 2	Large	High	High	Moderate
Visible Light Communication (VLC)	0.018	Large	None	High	Low
Pedestrian Dead Reckoning (PDR) & Map Matching	Varies	Small	More	High	Low
Angle of Arrival (AoA) Positioning	<1	Moderate	High	High	Moderate
AI-Enhanced Indoor Positioning Systems (IPS)	0.09 - 1	Large	Very High	Very High	Very High

1. INDOOR POSITIONING TECHNIQUES

Triangulation, Trilateration, Proximity, Scene Analysis, Fingerprinting, and Pedestrian Dead Reckoning are the primary techniques for indoor location, besides this there are techniques for integrating Bluetooth and ANNs as seen in Fig. 1. Such as Hybrid Data Fusion, Machine Learning Models, Deep Learning Models, Attentional Filtering and Spatial Regularization, Crowdsourcing and Model Updating(J et al., 2022).

Figure 1. **Techniques used in location finding in indoor environments**

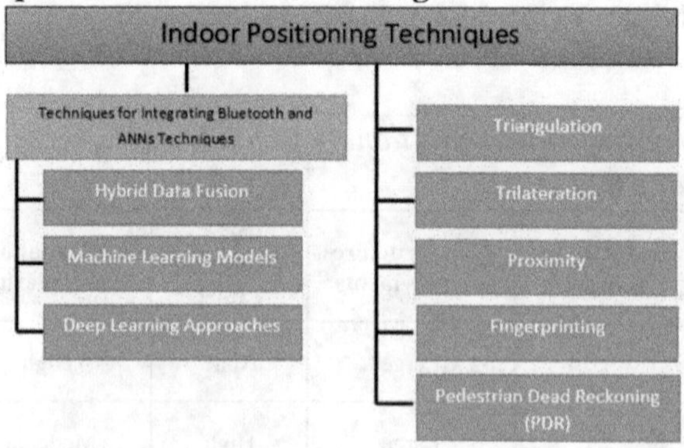

Triangulation

Triangulation calculates the location of objects using the angle-side-angle triangle congruence theorem, by looking at the angles of the object to be located from three signal emitting measuring devices and the distances of the measuring devices to each other.

Figure 2. **Triangulation method**

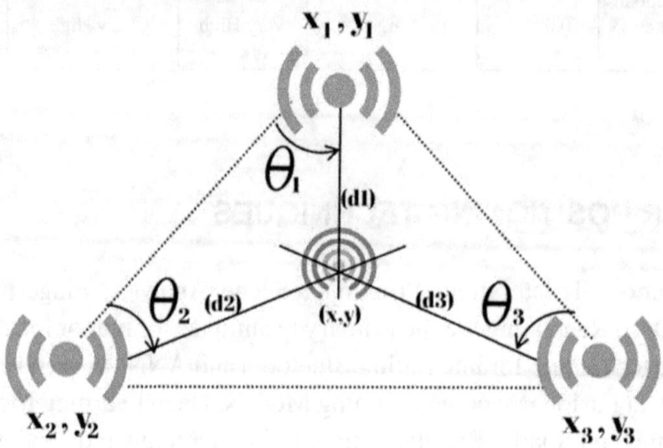

Trilateration

In the trilateration method, the position of the object is found by the distance measurement method based on the coordinates of at least three known reference points.

Figure 3. **Trilateration method**

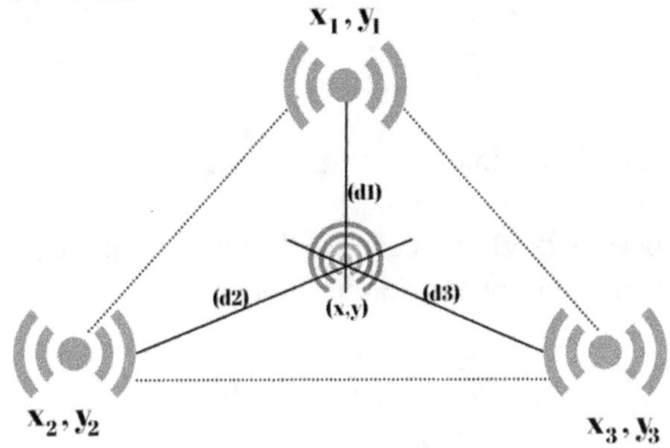

Proximity

Proximation-based localization uses certain reference points for determining the location of objects or people. This method does not determine a certain point; it just determines whether an object is within a certain area or which reference point it is closest to. To mathematically formalize proximity-based localization, it can model the method using distance metrics and set theory, which define regions of influence around reference points (e.g., Bluetooth beacons). The number derived from measuring the input RF signal at the receivers (expressed in dBm) is referred to as RSSI. The RSSI variation measurement exhibits minimal noise, which is contingent upon the propagation speed of RF signals inside the communication environment. Below is a rigorous framework:

a.Reference Points and Regions:

Let N reference points (e.g., Bluetooth anchors) be positioned at coordinates $\{(x_1, y_1),(x_2, y_2),\ldots,(x_N, y_N)\}$. Each reference point i defines a *proximity region*A_i, typically a circular area with radius r_i, determined by signal range or empirical thresholds (e.g., RSSI > -70 dBm).

Figure 4. **Proximity method**

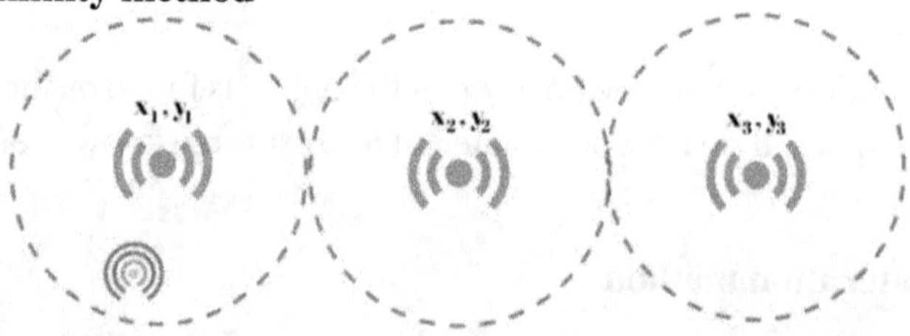

b. Binary Proximity Detection:

An object at location (x,y) is assigned to the reference point j if it lies within Aj . This is expressed as a binary decision function:

$$\text{Proximity } (i) = \begin{cases} 1 \; if \sqrt{(x-x_i)^2 + (y-y_i)^2} \; \leq \tau_{i \, (or \, RSSI_i \geq \tau)} \\ 0 \; otherwise \end{cases}$$

Here, τ is a signal strength threshold, and RSSI follows a log-distance path loss model:

$RSSI = -10nlog(d) + k$

Here, n=signal propagation constant,

d =distance between the two sensors,

k= signal strenght at a distance of 1 meter

c. Voronoi Partitioning:

In ideal geometric terms, the proximity regions form a Voronoi diagram, partitioning the space into cells where each cell Vi contains all points closer to the anchor i than to any other anchor:

$$V_i = \left\{ (x,y) \middle| \forall_{j \neq i} \sqrt{(x-x_i)^2+(y-y_i)^2} \leq \sqrt{(x-x_j)^2+(y-y_j)^2} \right\}$$

The object's location is then the centroid or identifier of the Voronoi cell it occupies.

Fingerprinting

In closed areas, the signal strengths of the signal emitters in the environment they are located are measured and recorded in the database, and the location is determined by matching the signal information at the user's location with the recorded signals.

Figure 5. **Fingerprinting method**

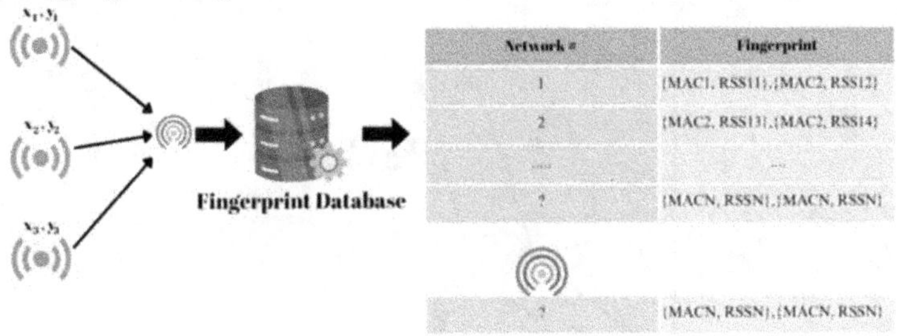

Pedestrian Dead Reckoning (PDR)

Global Navigation Satellite Systems (GNSS), including GPS, determine external location using trilateration using time-of-flight readings from satellite signals. However, GPS is ineffective indoors due to signal attenuation by walls and ceilings. Indoors, accelerometers measure linear acceleration using microelectromechanical systems (MEMS), facilitating step detection and displacement estimation through double integration; however, error accumulation (drift) limits its independent application. Gyroscopes measure angular velocity (degrees/second) via the Coriolis force, allowing orientation changes necessary for predictive calculation and orbit adjustment when integrated with accelerometer data. Magnetometers evaluate the Earth's magnetic field to determine absolute orientation, counteracting gyroscopic drift when affected by internal magnetic disturbances such as steel buildings. Collectively, these inertial sensors provide an inertial measurement unit (IMU) that facilitates pedestrian dead reckoning (PDR) by monitoring relative motion from a predetermined source. In indoor localization, sensor fusion such as Kalman filters integrates IMU data with infrastructure-based techniques (BLE, Wi-Fi) to reduce drift and increase precision. GPS can activate interior systems at building entrances, while IMUs provide seamless tracking in the absence of satellite signals, resulting in hybrid systems that are resilient to signal interference.

It is an inertial navigation (INS) technique that uses sensors such as accelerometer, gyroscope and magnetometer to determine location in closed areas where GPS is insufficient. It uses technologies such as Wi-Fi or BLE.

Figure 6. **Pedestrian Dead Reckoning (PDR) method**

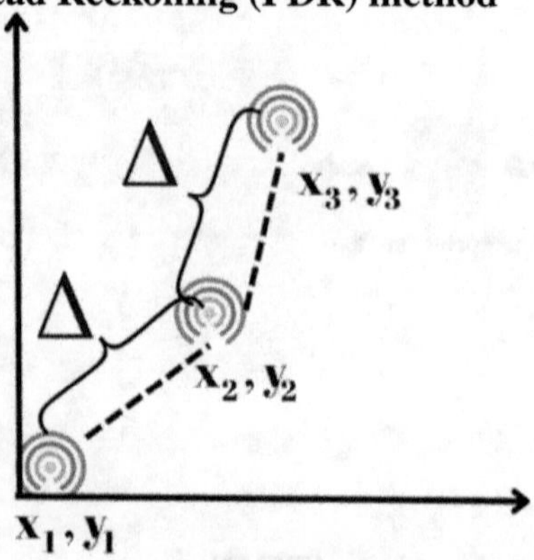

Hybrit Data Fusion

It is an approach that aims to provide more accurate, reliable and uninterrupted location detection in indoor positioning systems (IPS) by using multiple sensor data and technologies.

Figure 7. **Hybrit Data Fusion**

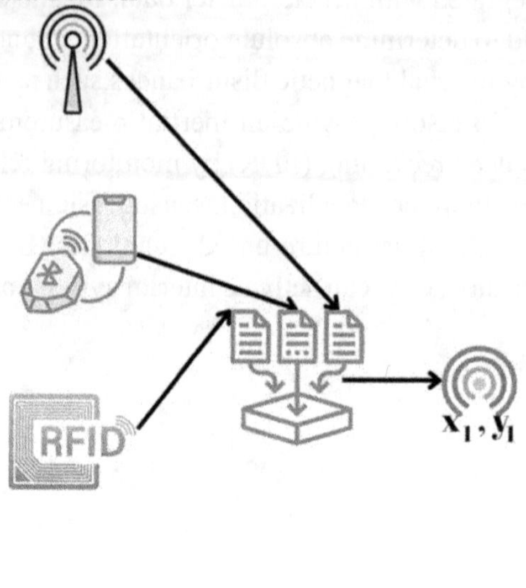

Machine Learning Model

Machine learning (ML) is a method used in indoor localization with hybrid data fusion, especially by combining different data sources such as Wi-Fi, BLE, magnetic field, PDR and image processing.

Figure 8. **Machine Learning Model**

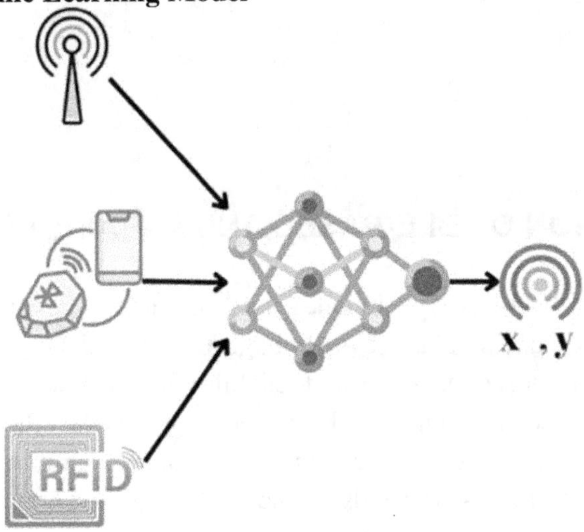

Deep Learning Approaches

Within the Deep Learning method, multilayer neural networks (MLP) can directly perform location detection by inferring raw RSSI or BLE signal intensities. Convolutional Neural Networks (CNN) interpret signals based on a grid and capture spatial patterns. The location shift problem can be eliminated with the help of LSTM/GRU based models with the help of IMU sensors. Sensor fusion techniques (CNN-LSTM hybrid architectures) compensate for the weaknesses of a single sensor by combining Wi-Fi, Bluetooth, IMU and camera data.

Figure 9. **Deep Learning Approaches**

2. COMPARISON OF BLE-BASED AND AI-BASED TECHNIQUES

BLE-based and AI-based techniques used for indoor positioning each have their own strengths and weaknesses. Although BLE-based systems have a higher margin of error in terms of accuracy and reliability in dynamic environments, their affordability and ease of installation are an advantage. On the other hand, AI-based methods offer higher accuracy and robustness, but require more computational work. Which method to use depends on the application requirements, and hybrid approaches usually provide the best results between accuracy, reliability, and cost. Table 2 was prepared by looking at the studies in the literature(Chen et al., 2022; Hu et al., 2023; Lyu et al., 2024; Thanasit Rithanasophon & Chatchai Wannaboon, 2023; Wang et al., 2023; Xiao et al., 2024).

Table 2. Comparison of BLE-Based and AI-Based Indoor Positioning Techniques

Characteristic	BLE-Based	AI-Based
Accuracy	1 to 3 meters	<1 m(to centimeter-level)
Reliability	Vulnerable to multipath effects and environmental disruptions	Resistant to signal distortions and attenuations
Complexity and Cost	Low-cost and easy to deploy	Higher computational requirements and high cost
Environmental Adaptability	Less adaptable to dynamic environments without additional sensors	Excels in dynamic environments due to learning and adaptation capabilities

3. INTEGRATION OF BLUETOOTH TECHNOLOGY AND ARTIFICIAL NEURAL NETWORKS IN HIGH ACCURACY INDOOR POSITIONING

The combination of Bluetooth technology with Artificial Neural Networks (ANN) in indoor navigation systems is possible due to the synergistic advantages of both technologies. Bluetooth technology, designed for short-range communication, can efficiently send data from anchor sensors in indoor navigation systems to mobile devices (Tuncer & Tuncer, 2015). It is low-cost, easy to use, and can be effectively used in indoor positioning due to its wide accessibility. ANNs, on the other hand, increase the precision and reliability of positioning systems thanks to their capabilities in complex data processing. Collectively, these technologies will provide more accurate and efficient navigation solutions by alleviating the problems of indoor navigation, including signal variations and environmental interferences. Combining artificial neural network (ANN) based algorithms with Bluetooth technology, especially Bluetooth Low Energy (BLE), has become easier with advanced algorithms using the techniques discussed in Part Two. Since these methods use wireless signals for location determination, many of the inherent problems are solved in software (the hardware is usually used to collect the signal). These problems include multipath interference, signal attenuation, and ambient noise. These problems usually result in less accurate Received Signal Strength Indicator (RSSI) or Angle of Arrival (AoA) readings. With the help of ANN architectures such as convolutional neural networks (CNNs) or recurrent neural networks (RNNs), these systems process spatially and temporally correlated signal data on the fly to model nonlinear relationships between signal parameters (such as RSSI and time of flight) and locations. It is also allows adaptive learning for external features such as changes in the indoor environment or how obstacles propagate. This increases locational accuracy in difficult, non-linear line-of-sight (NLOS) situations. Regression models based on Artificial Neural Networks (ANN) have a lot of potential to fix signal degradation at certain places by learning from real-world information gathered from anchor nodes or marker arrays. These models use ANNs' ability to find complicated patterns and connections in data, which makes them useful for tasks like figuring out where you are inside and predicting when a signal will lose power. A new ANN-based design for indoor localization, called the C2R model, transforms a classification model into a robust regressor. The model simplifies the task by using a classifier followed by a probability matching method that improves the output from the regression. This method has been shown to reduce the average localization error by 7.9% compared to previous methods such as weighted k-nearest neighbors (kNN) and 15.4% compared to dense neural networks (DNN)(Klus et al., 2024). Generalized Adaptive Regression and Radial Basis Function ANN Models are used in signal strength loss estimation. By using a vector Nonlinear Median Filter (NLMF), the models increase the quality of the datasets, which improves the learning and prediction capabilities of ANN models. In particular, the Generalized Adaptive Regression (GR)-ANN model has shown superior performance in solving function approximation problems compared to the Radial Basis Function (RBF)-ANN model, although the Radial Basis Function (RBF)-ANN model requires less training time (Ebhota et al., 2021).

Bluetooth Low Energy (BLE) technology is extensively utilized for assessing Round-Trip Time (RTT), Received Signal Strength Indicator (RSSI), and BLE signal intensities, which are essential for distance calculation and indoor location. A multitude of software and algorithms have been created to properly exploit these metrics. The selection of software and the formulae utilized for distance computation are contingent upon the particular methodology applied, such as RSSI-based techniques or Time-of-Flight (ToF) approaches. Although RSSI-based systems are favored for their simplicity and cost-effectiveness, they are vulnerable to environmental interference and noise, which can compromise accuracy. Advanced filtering techniques and machine learning models are utilized to improve precision and alleviate these difficulties. Conversely, ToF-based approaches, while theoretically more precise, need further hardware and software development to successfully extract and utilize the data. The selection of these approaches is contingent upon the particular application needs and environmental circumstances.

4. CHALLENGES FACED IN INDOOR ENVIRONMENTS FOR INDOOR POSITIONING SYSTEMS (IPS)

In today's world where asset tracking, navigation and location-based services are gaining importance, factors such as complex structuring of spaces, signal interference, presence of animate and inanimate objects, dynamic environmental changes and non-line-of-sight (NLOS) conditions prevent the healthy operation of indoor location systems (IPS). In order to obtain healthy and reliable results, advanced methodologies are needed. However, their implementation faces several obstacles, including security and privacy issues, system scalability, user experience, and emerging trends. The problems are made harder by the complexity of the indoor environment and the need for high accuracy and flexibility. Data collection and system maintenance is a major issue in creating and maintaining radio maps for fingerprinting methods. Innovative methodologies such as cyber-physical systems and digital twins are being investigated to address these challenges (Karakusak et al., 2023). Moreover, Indoor positioning systems, especially those employing location fingerprinting, present considerable security and privacy issues due to their capacity to reveal sensitive user location information to service providers and adversaries, thereby creating risks of unauthorized tracking or exploitation (Fathalizadeh et al., 2024). Decentralized authentication protocols, particularly those utilizing zero-knowledge proofs, are emerging as privacy-enhancing solutions that facilitate secure verification of user credentials without dependence on centralized entities, thus reducing data exposure and safeguarding user anonymity (Casanova-Marqués et al., 2023). Although there are still issues associated with the implementation of indoor positioning systems, research and emerging techniques are providing solutions. The combination of AI, IoT, and advanced sensors is facilitating the development of more precise, scalable, and user-centric systems. However, addressing privacy issues is an important area for future work and development. In addition, the scalability of indoor positioning systems (IPS) encounters obstacles due to dependence on extensive infrastructure and substantial data processing requirements, which may restrict their adaptability in dynamic settings; nonetheless, AI-driven solutions are being developed to enhance resource allocation and augment system scalability by facilitating real-time adjustments to variable conditions (Farahsari et al., 2022)(Pešić, 2024). Integrating IPS with IoT ecosystems simultaneously offers a complementary approach to enhancing scalability, as utilizing interconnected IoT devices and existing network infrastructure diminishes reliance on dedicated hardware while facilitating cost-effective, large-scale implementations (Farahsari et al., 2022). Also, basic error sources and their effects are available. First of all, one of the important problems is the interference of signals and multipath effects. In indoor areas, there are usually electronic devices and structural components that produce interference. This negatively affects the Received Signal Strength Indicator (RSSI) values and causes incorrect location estimates (Arnob et al., 2025; Castro-Arvizu et al., 2017). Additionally, multipath propagation causes signals to reflect off walls, floors, and other surfaces, creating multiple paths to the receiver, which can lead to errors in the time of arrival (ToA) and angle of arrival (AoA) measurement. This makes it hard to find the real signal line, which leads to mistakes (Bizon et al., 2024). The error margin of IPS is further increased in non-line-of-sight (NLOS) situations. Some of the factors that obstruct line-of-sight are walls, furniture, or dynamics environmenral. These prevent signals from traveling directly from the transmitter to the receiver, making it difficult to determine how far away the object to be located is. Furthermore, if signals travel via longer, indirect paths, inaccurate results will occur. To deal with NLOS errors, researchers have tried a number of methods. Hybrid positioning method is one of the methods that combines several techniques, automatically selecting the best positioning method(He et al., 2024; Liu et al., 2024). Moreover, as mentioned above, environmental dynamics affect the accuracy of IPS. Because signal paths and intensity can change with the presence of moving objects or individuals. This can eventually cause fluctuations in positioning accuracy in high-traffic areas (Arnob et al., 2025; Zhang et al., 2023). Adaptive algorithms have been developed by integrating real-time data into the system and taking into account environmental changes. In addition, techniques such as Kalman filters and machine learning have been used to provide dynamic improvement (Cao & Yang, 2023; Zhang et al., 2023). On the other hands the level of technological development also limits IPS. The low error rates of IPS are related to the quality of the devices used, including sensor sensitivity and port density (Castro-Arvizu et al., 2017). In addition, unlike GPS, which is used universally, IPS has no standard (Hailu et al., 2024). It is challenging to determine the most suitable method due to the variety of circumstances and conditions in which the systems have been tested. It is known that when a new IPS is proposed, it is necessary to compare various alternative methods in various dimensions, including accuracy, precision, robustness, and computational cost. The final proposed method should be selected based on pre-defined criteria, such as the best positioning error, the lowest execution time, or a trade-off between the two(Pascacio et al., 2021). Despite these challenges, researchers continue to seek solutions to improve the accuracy and reliability of indoor positioning systems. In particular, data fusion, machine learning, and hybrid positioning models have been intensively studied recently. However, the inherent complexity of indoor environments remains a significant obstacle, and IPS technologies need to be continuously innovated and adapted to meet the demands for precise and reliable indoor localization.

5. CONCLUSION

Notwithstanding considerable progress in indoor positioning systems, stable, scalable, and globally applicable solutions are deficient. Current techniques, like RSSI-based trilateration and ANN-assisted fingerprinting, are constrained by dynamic environmental variables such as non-line-of-sight (NLOS) propagation, multipath interference, and temporal obstructions, which diminish accuracy and necessitate frequent recalibration. Furthermore, device heterogeneity results in systemic inconsistencies, since variations in hardware specifications (e.g., antenna gain, transmit power) diminish the generalizability of models across platforms. The trade-offs between energy and accuracy hamper deployments, particularly in IoT ecosystems where resource limitations clash with the computing requirements of high-precision algorithms like UWB or deep neural networks. Emerging developments like UWB-5G hybrid systems and quantum inertial sensors present exciting opportunities for subcentimeter accuracy and drift mitigation; nevertheless, these advances are constrained by cost, scalability, and theoretical validation. The emergence of semantic localization, utilizing vision-language models to derive contextual significance from location data, presents ethical challenges about user privacy and monitoring. The absence of universal standards, infrastructure-light scalability, and real-time latency for AR applications, together with other examined concerns, underscores the necessity for multidisciplinary study. Future initiatives must emphasize self-calibrating algorithms, cross-modal sensor fusion, and ethical frameworks to synchronize technology progress with social needs, hence enhancing indoor localization towards resilient, adaptive, and privacy-conscious systems. This chapter has illustrated that the applicability between Bluetooth Low Energy (BLE) and Artificial Neural Networks (ANNs) and high-precision indoor positioning systems (IPS). Signal attenuation, multipath effects, and fluctuating environmental conditions negatively affect conventional BLE-based approaches. However, advanced ANN models such as Convolutional Neural Networks (CNNs), Recurrent Neural Networks (RNNs), and hybrid models significantly reduce these problems. Due to the capability of Artificial Neural Networks (ANN) to process complex data inputs, it effectively eliminates inaccuracies and signal variability such as Received Signal Strength Indicator (RSSI), Angle of Arrival (AoA), Channel State Information (CSI) and non-line of sight (NLOS). In addition, the comparative evaluations given in Table 1 show that although Bluetooth Low Energy (BLE) technology performs worse compared to Ultra Wide Band (UWB), Wi-Fi Round Trip Time (RTT), RF-based (RSSI & CSI), Visible Light Communication (VLC), Angle of Arrival (AoA) Positioning and AI-Enhanced Indoor Positioning Systems (IPS), the cost-effectiveness, energy efficiency and scalability of BLE make this technology attractive. Also, the use of hybrid methodologies with sensor fusion makes BLE stand out in terms of system reliability for indoor use, too. The combination of BLE and ANN is important in many areas, especially in sectors that depend on accurate indoor localization, such as logistics. With these hybrid technologies, asset tracking and positioning in complex environments can be facilitated.As a result of, many studies are being conducted by researchers to create effective indoor localization systems. In this regard, this study analyzes different AI-BLE based indoor localization methods presented in the literature and their comparative analysis. Last but not least, if it is also touched upon upcoming trends, sensor technologies like 5G, LiDAR, and edge computing are likely to drive the next generation of indoor positioning systems (IPS). These technologies will make systems smarter, more responsive in real time, and more reliable in complex settings (El-Sheimy & Li, 2021). The combination of AI and machine learning is expected to transform IPS by enhancing localization precision, automating maintenance procedures, and facilitating innovative applications in smart infrastructure and customized services (Qi et al., 2024)Simultaneously, as these innovations enhance system capabilities, increasing apprehensions regarding data security are driving the creation of privacy-preserving mechanisms, such as decentralized protocols and sophisticated encryption techniques, to protect sensitive user information while preserving functionality (Fathalizadeh et al., 2024). These developments collectively indicate a progression towards more intelligent, scalable, and secure IPS systems that reconcile technical innovation with ethical and privacy concerns.

REFERENCES

Alghananim, M. S., Feng, Y., & Ochieng, W. Y. (2025). *The first calibration model for bluetooth angle of arrival: Enhancing positioning accuracy in indoor environments* (arXiv:2501.08805). arXiv. https://doi.org//arXiv.2501.08805DOI: 10.48550

Arnob, S. S., AlQahtani, A. A. S., Alshayeb, T., & Alamleh, A. H. (2025). *Precision in Motion: An Approach to Indoor Localization and Navigation for Complex Spaces.* DOI: 10.20944/preprints202501.0356.v1

Bizon, I., Nimr, A., Fettweis, G., & Chafii, M. (2024). Indoor Positioning Using Correlation Based Signal Analysis and Convolutional Neural Networks. *2024 19th International Symposium on Wireless Communication Systems (ISWCS)*, 1–6. DOI: 10.1109/ISWCS61526.2024.10639139

Borenovic, M., & Neskovic, A. (2011, November). ANN based models for positioning in indoor WLAN environments. *2011 19thTelecommunications Forum (TELFOR) Proceedings of Papers.* 2011 19th Telecommunications Forum Telfor (TELFOR), Belgrade, Serbia. DOI: 10.1109/TELFOR.2011.6143551

Cao, F., & Yang, X. (2023). UWB-Based Intelligent Positioning Method for Indoor Complex Environments. *2023 3rd International Conference on Mobile Networks and Wireless Communications (ICMNWC)*, 1–5. DOI: 10.1109/ICMNWC60182.2023.10435759

Casanova-Marqués, R., Torres-Sospedra, J., Hajny, J., & Gould, M. (2023). Maximizing privacy and security of collaborative indoor positioning using zero-knowledge proofs. *Internet of Things : Engineering Cyber Physical Human Systems*, *22*, 100801. DOI: 10.1016/j.iot.2023.100801

Castro-Arvizu, J. M., Fernández Rubio, J. A., & Closas Gómez, P. (2017). *Robust indoor positioning in WLAN networks* [Universitat Politècnica de Catalunya]. DOI: 10.5821/dissertation-2117-107945

Chen, M.-C., Cheng, Y.-T., & Chen, R.-W. (2022). A Novel Indoor Positioning Framework. *Computer Modeling in Engineering & Sciences*, *130*(3), 1459–1477. DOI: 10.32604/cmes.2022.015636

Dao, V.-L., & Salman, S. M. (2022). Deep Neural Network for Indoor Positioning Based on Channel Impulse Response. *2022 IEEE 27th International Conference on Emerging Technologies and Factory Automation (ETFA)*, 1–8. DOI: 10.1109/ETFA52439.2022.9921735

El-Sheimy, N., & Li, Y. (2021). Indoor navigation: State of the art and future trends. *Satellite Navigation*, *2*(1), 7. Advance online publication. DOI: 10.1186/s43020-021-00041-3

Farahsari, P. S., Farahzadi, A., Rezazadeh, J., & Bagheri, A. (2022). A Survey on Indoor Positioning Systems for IoT-Based Applications. *IEEE Internet of Things Journal*, *9*(10), 7680–7699. DOI: 10.1109/JIOT.2022.3149048

Fathalizadeh, A., Moghtadaiee, V., & Alishahi, M. (2024). *Indoor Location Fingerprinting Privacy: A Comprehensive Survey* (arXiv:2404.07345). arXiv. https://doi.org//arXiv.2404.07345DOI: 10.48550

Hailu, T. G., Guo, X., Si, H., Li, L., & Zhang, Y. (2024). Theories and Methods for Indoor Positioning Systems: A Comparative Analysis, Challenges, and Prospective Measures. *Sensors (Basel)*, *24*(21), 6876. DOI: 10.3390/s24216876 PMID: 39517773

Han, T.-T., Nguyen Dinh, P., Nguyen Duc, T., Nguyen Long, V., & Tan, H. D. (2024). An Efficient Method for BLE Indoor Localization Using Signal Fingerprint. *EAI Endorsed Transactions on Industrial Networks and Intelligent Systems*, *12*(1). Advance online publication. DOI: 10.4108/eetinis.v12i1.6571

He, J., Deng, Z., Hu, E., Huang, Y., & Pan, T. (2024). Hybrid Positioning Framework with Innovation Based Judgment in Complex Indoor Environments. *2024 IEEE 14th International Conference on Electronics Information and Emergency Communication (ICEIEC)*, 1–4. DOI: 10.1109/ICEIEC61773.2024.10561735

Homma, S., Ida, Y., Ohira, Y., Kuroda, S., & Matsumoto, T. (2024). Study on Effective CSI Feature Extraction and Deep Learning Based Indoor Positioning Method. *2024 IEEE VTS Asia Pacific Wireless Communications Symposium (APWCS)*, 1–5. DOI: 10.1109/APWCS61586.2024.10679303

Hu, X., Chen, L., Lian, X., Wang, T., & Cai, J. (2023). Crowdsourcing-based high-precision Bluetooth indoor location method for adapting to environmental dynamics. *2023 IEEE 98th Vehicular Technology Conference (VTC2023-Fall)*, 1–7. DOI: 10.1109/VTC2023-Fall60731.2023.10333616

Huang, J., Junginger, S., Liu, H., & Thurow, K. (2023). Indoor Positioning Systems of Mobile Robots: A Review. *Robotics (Basel, Switzerland)*, *12*(2), 47. DOI: 10.3390/robotics12020047

Jianyong, Z., Haiyong, L., Zili, C., & Zhaohui, L. (2014). RSSI based Bluetooth low energy indoor positioning. *2014 International Conference on Indoor Positioning and Indoor Navigation (IPIN)*, 526–533. DOI: 10.1109/IPIN.2014.7275525

Karakusak, M. Z., Kivrak, H., Watson, S., & Ozdemir, M. K. (2023). Cyber-WISE: A Cyber-Physical Deep Wireless Indoor Positioning System and Digital Twin Approach. *Sensors (Basel)*, *23*(24), 9903. DOI: 10.3390/s23249903 PMID: 38139747

Khafagy, A. A., & Abdellatif, M. M. (2024). VLC-IPS: A Visible Light Communications Based Indoor Positioning System. *2024 International Telecommunications Conference (ITC-Egypt)*, 172–177. DOI: 10.1109/ITC-Egypt61547.2024.10620517

Klus, R., Talvitie, J., Torres-Sospedra, J., Quezada Gaibor, D. P., Casteleyn, S., Cabric, D., & Valkama, M. (2024). C2R: A Novel ANN Architecture for Boosting Indoor Positioning With Scarce Data. *IEEE Internet of Things Journal*, *11*(20), 32868–32882. DOI: 10.1109/JIOT.2024.3420122

Kong, R., Li, R., Ma, Z., & Yu, Y. (2024). Indoor positioning systems based on visible light communication. *Applied and Computational Engineering*, *78*(1), 162–168. DOI: 10.54254/2755-2721/78/20240460

Kotturi, S. P., Yerrapragada, A. K., Prasad, S., & Ganti, R. K. (2024). *On the Application of Deep Learning for Precise Indoor Positioning in 6G* (arXiv:2410.19436). arXiv. https://doi.org//arXiv.2410.19436DOI: 10.48550

Li, X., Yan, Z., Huang, L., Chen, S., & Liu, M. (2020). High-Accuracy and Real-Time Indoor Positioning System Based on Visible Light Communication and Mobile Robot. *International Journal of Optics*, *2020*, 1–11. DOI: 10.1155/2020/3124970

Liu, J., Deng, Z., & Hu, E. (2024). An NLOS Ranging Error Mitigation Method for 5G Positioning in Indoor Environments. *Applied Sciences (Basel, Switzerland)*, *14*(9), 3830. DOI: 10.3390/app14093830

Lyu, Z., Chan, T. T.-L., Hung, T. Y.-T., Ji, H., Leung, G., & Lun, D. P.-K. (2024). Deep Learning-Based Bluetooth Low-Energy 5.1 Multianchor Indoor Positioning with Attentional Data Filtering. *Advanced Intelligent Systems*, *6*(1), 2300292. DOI: 10.1002/aisy.202300292

Malik, R. F., Gustifa, R., Farissi, A., Stiawan, D., Ubaya, H., Ahmad, M. R., & Khirbeet, A. S. (2019). The Indoor Positioning System Using Fingerprint Method Based Deep Neural Network. *IOP Conference Series. Earth and Environmental Science*, *248*, 012077. DOI: 10.1088/1755-1315/248/1/012077

Mehmood. (2010). Indoor Positioning System Using Artificial Neural Network. *Journal of Computer Science*, *6*(10), 1219–1225. DOI: 10.3844/jcssp.2010.1219.1225

Pascacio, P., Casteleyn, S., Torres-Sospedra, J., Lohan, E. S., & Nurmi, J. (2021). Collaborative Indoor Positioning Systems: A Systematic Review. *Sensors (Basel)*, *21*(3), 1002. DOI: 10.3390/s21031002 PMID: 33540703

Pešić, S. (2024). *Modern Challenges in Indoor Positioning Systems: AI to the Rescue*. IntechOpen., DOI: 10.5772/intechopen.1005354

Prasanthi, T. S., Rayavarapu, S. M., Rao, G. S., Goswami, R. K., & Kumar, G. S. (2024). RADIO FREQUENCY BASED INPAINTING FOR INDOOR LOCALIZATION USING MEMORYLESS TECHNIQUES AND WIRELESS TECHNOLOGY. *Informatyka, Automatyka. Pomiary w Gospodarce i Ochronie Środowiska*, *14*(4), 10–15. DOI: 10.35784/iapgos.6236

Puckdeevongs, A. (2021). Indoor Localization using RSSI and Artificial Neural Network. *2021 9th International Electrical Engineering Congress (iEECON)*, 479–482. DOI: 10.1109/iEECON51072.2021.9440373

Qi, L., Liu, Y., Yu, Y., Chen, L., & Chen, R. (2024). Current Status and Future Trends of Meter-Level Indoor Positioning Technology: A Review. *Remote Sensing (Basel)*, *16*(2), 398. DOI: 10.3390/rs16020398

Qiao, J., Yang, F., Liu, J., Huang, G., Zhang, W., & Li, M. (2024). Advancements in Indoor Precision Positioning: A Comprehensive Survey of UWB and Wi-Fi RTT Positioning Technologies. *Network (Bristol, England)*, *4*(4), 545–566. DOI: 10.3390/network4040027

Sahnoun, S., Souissi, R., Chiboub, S., Chabchoub, A., Baazaoui, M. K., Fakhfakh, A., & Derbel, F. (2025). Enhancing Localization Accuracy and Reducing Processing Time in Indoor Positioning Systems: A Comparative Analysis of AI Models. *Sensors (Basel)*, *25*(2), 475. DOI: 10.3390/s25020475 PMID: 39860845

Thanasit Rithanasophon & Chatchai Wannaboon. (2023). *Integration_of_Machine_Learning_and_Kalman_Filter_Approach_for_Fingerprint_Indoor_Positioning.pdf*. DOI: 10.1109/ECTI-CON58255.2023.10153162

Tuncer, S., & Tuncer, T. (2015). Indoor localization with Bluetooth technology using Artificial Neural Networks. *2015 IEEE 19th International Conference on Intelligent Engineering Systems (INES)*, 213–217. DOI: 10.1109/INES.2015.7329709

Wang, Y., Zou, M., Peng, L., Chen, Z., & Zhang, L. (2023). Adaptive High-Precision Positioning Method Based On BLE RSSI/INS/ODO Tightly-Coupled Integrated Navigation System. *2023 China Automation Congress (CAC)*, 1881–1886. DOI: 10.1109/CAC59555.2023.10452079

Warnakulasuriya, H. N., Dorji, T., & Horanont, T. (2024). Indoor Positioning Based on Time of Flight and Kalman Filtering Using Ultra-Wideband Sensors. *2024 IEEE International Conference on Big Data and Smart Computing (BigComp)*, 117–123. DOI: 10.1109/BigComp60711.2024.00027

Wu, C., Wang, Y., Ke, W., & Yang, X. (2024). A Dual-Branch Convolutional Neural Network-Based Bluetooth Low Energy Indoor Positioning Algorithm by Fusing Received Signal Strength with Angle of Arrival. *Mathematics*, *12*(17), 2658. DOI: 10.3390/math12172658

Wu, C., Wong, I.-C., Wang, Y., Ke, W., & Yang, X. (2024). Experimental Study of Bluetooth Indoor Positioning Using RSS and Deep Learning Algorithms. *Mathematics*, *12*(9), 1386. DOI: 10.3390/math12091386

Xiao, K., Hao, F., Zhang, W., Li, N., & Wang, Y. (2024). Research and Implementation of Indoor Positioning Algorithm Based on Bluetooth 5.1 AOA and AOD. *Sensors (Basel)*, *24*(14), 4579. DOI: 10.3390/s24144579 PMID: 39065976

Zhang, H., Wang, Q., Li, Z., Mi, J., & Zhang, K. (2023). Research on High Precision Positioning Method for Pedestrians in Indoor Complex Environments Based on UWB/IMU. *Remote Sensing (Basel)*, *15*(14), 3555. DOI: 10.3390/rs15143555

Zou, H., Chen, Z., Jiang, H., Xie, L., & Spanos, C. (2017). Accurate indoor localization and tracking using mobile phone inertial sensors, WiFi and iBeacon. *2017 IEEE International Symposium on Inertial Sensors and Systems (INERTIAL)*, 1–4. DOI: 10.1109/ISISS.2017.7935650

Chapter 7
Multiscopic Topological Twin for Cyber-physical Systems

Naoyuki Kubota
https://orcid.org/0000-0001-8829-037X
Tokyo Metropolitan University, Hino, Japan

Chyan Zheng Siow
https://orcid.org/0009-0003-6132-1457
Tokyo Metropolitan University, Tokyo, Japan

Takenori Obo
https://orcid.org/0009-0009-0318-470X
Tokyo Metropolitan University, Tokyo, Japan

ABSTRACT

Recently, digital twin has been discussed with the integration of information, communication, intelligence, and robot technologies. In order to conduct multiscale and multiphysics simulations on a real-world phenomenon in the cyber world based on digital twin, topological features and structures should be extracted from given or measured big data. Therefore, the concept of topological twin was proposed to complementarily use digital twin. The physical dynamics is dealt with at the microscopic level, while the spatiotemporal qualitative relationships are used between objects, people, culture, and knowledge at the macroscopic level. Furthermore, a mesoscopic integration method is required to connect microscopic and macroscopic topological features. This chapter explains the multiscopic concept of topological twin and shows an application example to path planning problems and navigation tasks of mobility support robots. Finally, this chapter discusses the applicability and

DOI: 10.4018/979-8-3693-7994-3.ch007

future direction of topological intelligence to Industry 5.0 and Society 5.0.

1. INTRODUCTION

Recently, digital twin has been discussed with the integration of information, communication, intelligence, and robot technologies. The digital twin is one of the most important technologies in cyber-physical systems (CPS) and Industry 4.0 (Greer, Burns, Wollman, & Griffor, 2019; Banyai, Petrillo, & De Felice, 2020). Furthermore, the fundamental idea of CPS has been expanded to cyber-physical-social systems (CPSS) and cyber-physical-human systems (CPHS), since we explicitly have to deal with human factors in CPS (Wang, 2010; Wang et al., 2018). This transition is similar to the update to Industry 5.0 from 4.0 (Publications Office of the European Union, 2021). The most important factor of CPSS, CPHS, and Industry 5.0 lies in the human-centered and personalized services in customer experience. Especially, human-robot interaction is one of the most important applications in the human-centered and personalized services.

In fact, we can perform the real time recognition and tracking of people and moving objects in a wide area, and represent them as objects in digital twin of the cyber space. However, it is difficult to simulate and analyze the interaction of people and moving objects in cyber space, because we have to clarify the relationship between objects. Therefore, relationships and features must be extracted from the sensing data in order to perform computer simulations. Such a relationship in sensing data is often represented as topological information.

To provide more context on real-world challenges, it is important to note that digital twins primarily focus on replicating physical entities and their behaviors, whereas topological twins emphasize the underlying relational structures and connectivity patterns within data. This distinction is of paramount importance for applications necessitating multi-scale analysis, wherein both physical dynamics and abstract relationships must be taken into account.

In order to conduct multiscale and multiphysics simulations (Saputra, Toda, Botzheim, & Kubota, 2018; Saputra, Botzheim, Sulistijono, & Kubota, 2016) on a real-world phenomenon based on digital twin, we often need to extract topological features and structures from given or measured big data. Therefore, we proposed the concept of *topological twin (Kubota, 2021)*. The aim of topological twin is to (1) extract topological structures hidden implicitly in the real world, (2) reproduce them explicitly in the cyber world, and (3) simulate and analyze phenomena of the real world in the cyber world. The digital twin is used as a platform for simulations, while the topological twin is used to represent the relationship between data or variables for multi-scale multi-physics simulation.

Actually, while we need to deal with the physical dynamics in the microscopic level, we need to deal with spatiotemporal qualitative relationships between objects, people, culture, expand knowledge in the macroscopic level. Furthermore, we need a mesoscopic integration method connecting microscopic and macroscopic topological features. Thus, the topological twin plays the important role in extracting and connecting structures hidden in real world from the multiscopic point of view. Furthermore, we need a multiscopic approach to deal with inference, learning, search, and prediction based on topological and graphical data as the methodology of *topological intelligence (Kubota, 2021)*.

In this paper, we explain how to deal with topological intelligence for mobile robots as an example of human-robot co-existing environments in CPHS (Oshio, Kaneko, & Kubota, 2021; Ardilla et al., 2022; Doteguchi & Kubota, 2022; Watanabe, Obo, & Kubota, 2025). First, we explain the role of topological intelligence in detail, and how to use topological twin in robot applications. Next, we show the map topological building method. Furthermore, we show several examples of topological tracking and topological collision avoidance. Finally, we discuss the applicability and future direction of topological intelligence to Industrie 5.0 and Society 5.0.

Figure 1. Digital twin and topological twin

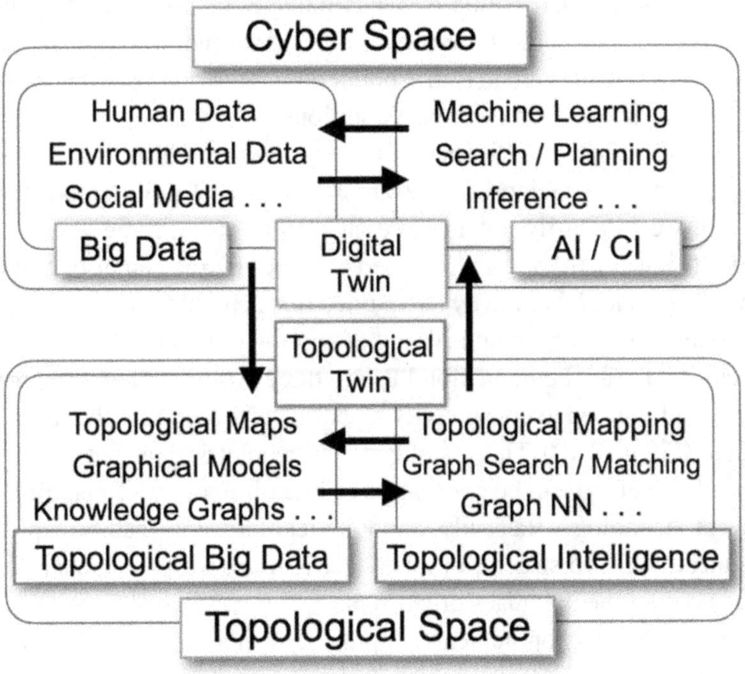

2. TOPOLOGICAL INTELLIGENCE

2.1. Topological Mapping

Digital twin and topological twin are used complementarily, not exclusively. Figure 1 shows the concept of topological twin (Kubota, 2021). We can extract topological features from big data, e.g., human data, environmental data, and social media. Topological mapping methods are used to obtain the topological features composed of nodes and edges. Various types of unsupervised learning methods such as k-means clustering algorithms, k-nearest neighbors algorithms (kNN), self-organizing maps (SOM), growing cell structures (GCS), and growing neural gas (GNG), have been used for data analysis (Eppstein, Paterson, & Yao, 1997; Kohonen, 2001; Martinetz & Schulten, 1991; Fritzke, 1991a; Fritzke, 1994; Fritzke, 1997).

Topological intelligence plays a crucial role in cyber-physical systems (CPS) by enhancing adaptability, resilience, and efficiency in data-driven decision-making. The integration of topological intelligence with CPS allows for real-time analysis of network structures, enabling predictive maintenance, anomaly detection, and optimization of distributed systems (Jurczenia & Rak, 2022; Deka et al., 2024). For example, in smart grids, topological mapping methods facilitate the monitoring and adjustment of energy distribution by recognizing node connectivity patterns and detecting irregularities in the system. Similarly, in autonomous transportation systems, topological features aid in traffic prediction, route optimization, and dynamic resource allocation based on evolving conditions.

First, we discuss the difference between standard clustering problems and topological mapping problems. For example, in case of point cloud A shown in Figure 2 (a), we can easily obtain three clusters by using three line segments or three nodes (Figure 2 (b)). A node is used to represent the center of a cluster, and we can make Voronoi Diagram by using three nodes in this example. Next, we just add two nodes to the point cloud A, and obtain the point cloud B in Figure 2 (c). It is difficult to divide the point cloud B into three clusters. Therefore, we can use three nodes and their edge connectivity to extract and represent the feature of data distributions (Figure 2 (d)). This is similar to the geometric dual, but it is incomplete. In general, this method can be considered as one of topological mapping methods in intelligent technology. Basically, a topological mapping method represents the order, connectivity, and structure in data distribution, while a standard clustering method can obtain the boundary of clusters.

We can use kNN for topological mapping, known as kNN graph. However, the computational cost of kNN is very high, because this method need to calculate the distance between all data. SOM is often used for the topological mapping from high-dimensional data to low-dimensional space, e.g., 2D space for the visualization. The number of nodes and connectivity in the original SOM is predefined. On the other hand, the number of nodes in GCS is increasing while preserving the topology. Both SOM and GCS does not delete nodes and edges, but GNG can delete nodes and edges based on the concept of age. Figure 3 shows the learning results of topological mapping methods. These three methods can almost cover the data distribution, but only GNG can conduct clustering and topological mapping. Therefore, GNG is called a topological clustering method. In this way, GNG has been applied to various topological mapping and clustering tasks. Table 1 summarizes the feature of topological mapping methods.

Figure 2. Clustering problems: (a) point cloud A, (b) clustering result A, (c) point cloud B, (d) clustering result B

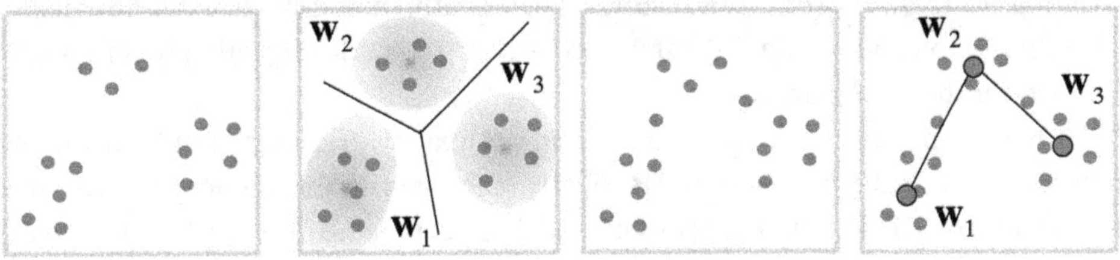

Figure 3. Examples of learning results by topological mapping methods: (a) data distribution, (b) SOM (c) GCS, (d) GNG

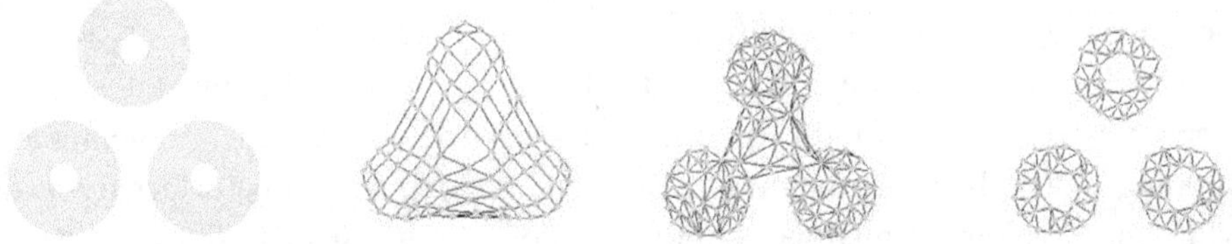

Table 1. Features of topological mapping methods

	SOM	GCS	GNG
Topology Preservation	✓	✓	
Incremental Learning		✓	✓
Topological Clustering			✓

2.2. Map Building and Path Planning

The methodology of digital twin and topological twin has an analogy with map building in robotics (Paul, 1981; Fu et al., 1987; Fukuda & Kubota, 1999; Thrun, Burgard, & Fox, 2005). In a traditional approach, a real environment is represented by approximate polygonal map (Figure 4 (a)). A visibility graph is generated by connecting two vertices of polygonal objects. In the visibility graph, the shortest path between two points, e.g., start and goal, can be generated easily by selecting edges (see Figure 4 (b)). However, it is dangerous for a robot to move along the generated path, because the path is adjacent to the vertices of the polygonal objects. To overcome this problem, a Voronoi diagram can be used to generate a safe path far away from any vertices (see Figure 4 (c)), but its trajectory becomes much long than that in Figure 4 (b). As a result, the best path should be generated by using visibility graph and Voronoi diagram according to a multi-objective function pre-defined for the path planning.

A metric map is often represented as an occupancy grid map built by using the measurement data (Figure 5 (a)). Basically, each cell in the occupancy grid map has a value of {0, 1} where 0 is empty and 1 is occupied. If the least size of a grid is larger than the size of the robot, we don't need to take into account the size of the robot for path planning in the grid search space. The path planning can be done on a discrete grid map by using A* algorithm and evolutionary computation, and rapidly exploring random tree (Russell et al., 1995; Fogel, 1995; LaValle & Kuffner, 2001) (Figure 5 (b)). We can use these search methods, but it takes more computational time, as the grid size is larger. Therefore, topological map is often used to reduce the computational memory and time.

Topological maps are categorized into topological environmental map and topological roadmap (Figure 6). Basically, the topological environmental map building is done by using the measurement data of depth sensors (external sensor), while the topological roadmap building is done by using IMU sensors (internal sensors) as a result of movement. The topological environmental map is useful to reduce the computational memory and to represent the 2D or 3D surface features of rough terrains. The roadmap is used to easily represent a path or trajectory of mobile robots. The above Visibility graph and Voronoi diagram are also considered as topological maps. In this way, topological features are very useful in robotics in the map building and path planning.

Figure 4. An example of path planning in a polygonal map: (a) polygonal map, (b) visibility graph, (c) Voronoi diagram

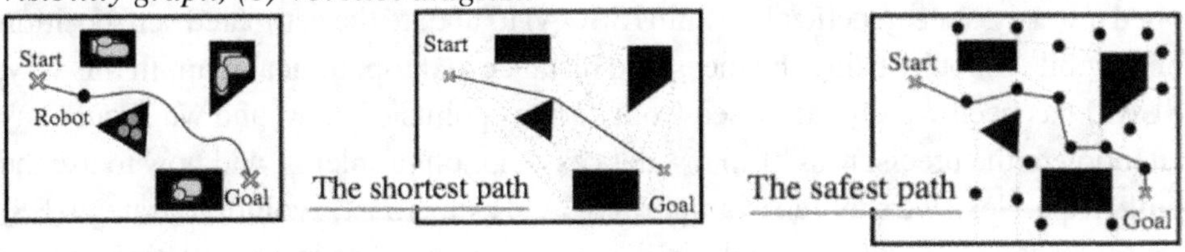

Figure 5. Map building and path planning by an occupancy grid map: (a) occupancy grid map, (b) grid search

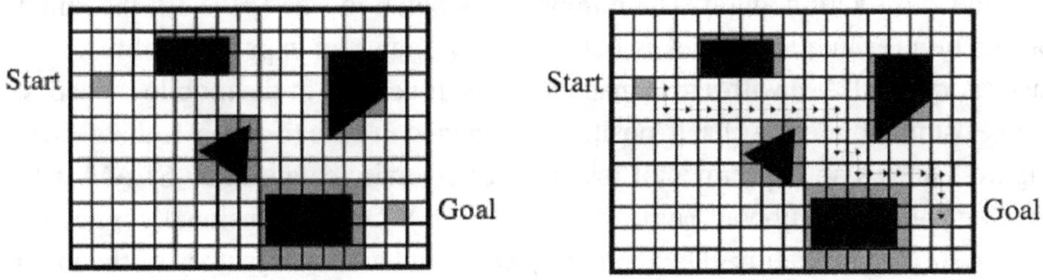

(a) Occupancy grid map

(b) Grid search

Figure 6. Topological map building: (a) environmental map, (b) roadmap

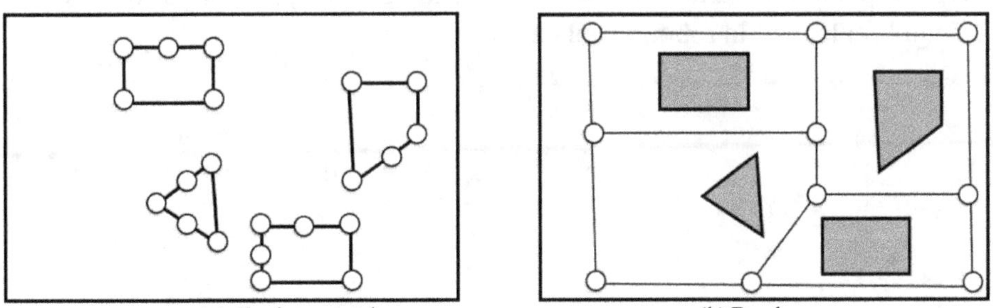

(a) Environmental map

(b) Roadmap

Simultaneous localization and mapping (SLAM) is a fundamental task given to a robot in unknown environments (Thrun, Burgard, & Fox, 2005). Various types of methods for SLAM have been proposed such as Extended Kalman Filter (EKF) SLAM, Graph SLAM, visual SLAM, and cooperative SLAM. We also proposed a topological map building method based on GNG (Fukuda & Kubota, 1999; Thrun, Burgard, & Fox, 2005; Sasaki, Kubota, & Taniguchi, 2008). Furthermore, we applied a steady-state genetic algorithm (SSGA) to update the estimated self-position of a mobile robot by using the measured distance and topological map. In this way, SLAM has been widely discussed from various points of view, and we should pay attention to the preciseness of maps, data association problems, and how to use the built maps. Next, we show an example of SLAM using (μ+λ) evolution strategy (ES) for self-localization where μ and λ indicate the number of parent population and the number of offspring population produced in a single generation, respectively (Toda & Kubota, 2013). We use (μ+1)-ES to enhance the local hill-climbing search as a continuous model of generations, which eliminates and generates one individual in a generation. The above SSGA is considered as one of (μ+1)-ES methods. A candidate solution is composed of numerical parameters of the update to the current position and rotation angle. The number of population size (μ) is 30; the number of iteration (evaluation) times is 600. Figure 7 shows an experimental result of a laboratory by SLAM with a grid map where the trajectory of the mobile robot based on the estimation result of self-position is depicted in blue (Toda & Kubota, 2013). Figure 8 shows an experimental result of a university cafeteria by SLAM with a topological map (Kubota, Yuki, & Baba, 2009b). One-dimensional connectivity is used for simplification. The target trajectory and actual trajectory of the mobile robot are depicted in red. The experimental results shows that topological mapping method can be also useful for SLAM. In general, the computational memory of the topological map is much smaller than the grid map, but the computational time of topological map may be longer than the grid map, because the nearest node should be detected in the topological map after the distance calculation between the measurement point and all nodes. Furthermore, the grid mapping algorithm is much easier to implement in robotic application.

Figure 7. An experimental result of SLAM using a grid map: (a) target environment, (b) map building and localization

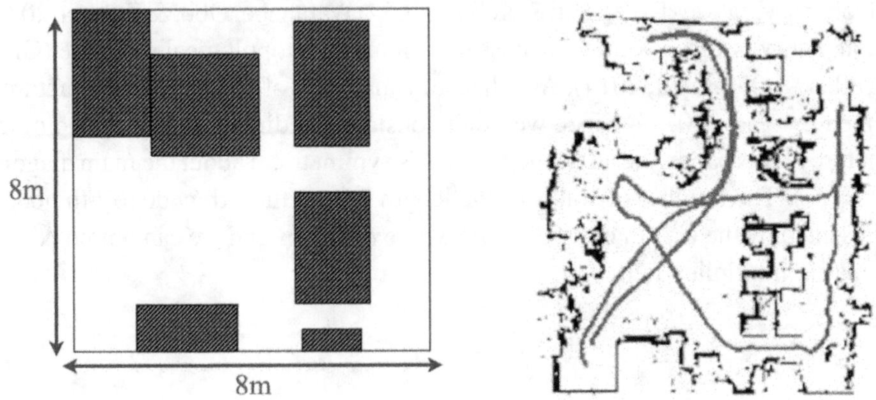

(a) Target environment (b) Map building and Localization

Figure 8. An experimental result of SLAM using a topological map: (a) target environment, (b) map building and localization

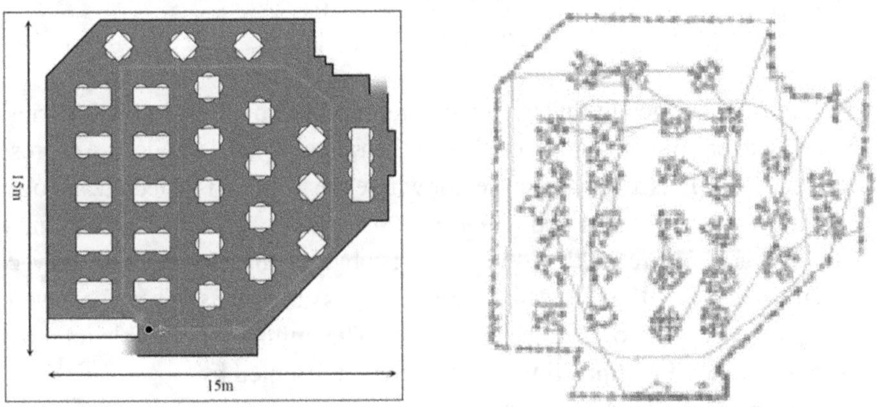

(a) Target environment (b) Map building and Localization

2.3. Topological Analysis

This section explains a method of topological analysis for people flow simulations and robot navigation (Doteguchi & Kubota, 2022; Watanabe, Obo, & Kubota, 2025). Graph theory is very useful to discuss the feature of a topological structure (Gross & Yellen, 2005; Bapat, 2014). An adjacency matrix **A** of a topological structure is defined by $c_{i,j} = \{0, 1\}$. Since we don't consider the direction in the topological structure in this paper, the adjacency matrix is symmetrical about the main diagonal ($c_{i,j} = c_{j,i}$). The number of walks of the length k from the i-th node to j-th node is represented by the (i,j)-th entry $c^k_{i,j}$ of $\mathbf{A}^k$ where $c^1_{i,j} = c_{i,j}$ and we can obtain $\mathbf{A}^{k+1}$ by the matrix multiplication,

$$c^{k+1}_{i,j} = \sum_{h=1}^{n} c^k_{i,h} \cdot c^1_{h,j} \tag{1}$$

where the number of nodes is n. Figure 9 shows a simple example of the number of walks where the number of nodes is 6. The adjacency matrix **A** in Figure 9 and its matrix multiplication is represented by,

$$A = \begin{pmatrix} 0 & 1 & 0 & 1 & 0 & 0 \\ 1 & 0 & 1 & 1 & 0 & 0 \\ 0 & 1 & 0 & 1 & 1 & 0 \\ 1 & 1 & 1 & 0 & 1 & 0 \\ 0 & 0 & 1 & 1 & 0 & 1 \\ 0 & 0 & 0 & 0 & 1 & 0 \end{pmatrix}, A^2 = \begin{pmatrix} 2 & 1 & 2 & 1 & 1 & 0 \\ 1 & 3 & 1 & 2 & 2 & 0 \\ 2 & 1 & 3 & 2 & 1 & 1 \\ 1 & 2 & 2 & 4 & 1 & 1 \\ 1 & 2 & 1 & 1 & 3 & 0 \\ 0 & 0 & 1 & 1 & 0 & 1 \end{pmatrix}$$

The number of edges in this example is represented by the diagonal entries of $\mathbf{A}^2$. When an agent (a person or robot) is in the range of the 1st node, we represent $h = (1 \ \ 0 \ \ \cdots \ \ 0)^T$ as an existence vector where the 1st node is selected and obtain $A \bullet h = (0 \ \ 1 \ \ 0 \ \ 1 \ \ 0 \ \ 0)^T$ and $A^2 \bullet h = (2 \ \ 1 \ \ 2 \ \ 1 \ \ 1 \ \ 0)^T$. If the i-th entry is positive, the agent can reach the i-th node. It is obvious that the value of the vector calculated by $A^k \cdot h$ is the degree of variety of possible paths. Therefore, we use this idea for the tracking of persons and robots. The number of possible total paths from i-th to j-th node within k times of move is represented by

$$S_k = \sum_{i=1}^{k} A^i$$

. (2)

In general, since we can approximately estimate the possible moving direction from the previous path, we use the following time-series of tracking data of the b-th agent,

$$h_b = (h_{b,1} \; h_{b,2} \; \cdots \; h_{b,n})^T$$

,

$$h_{b,i} \leftarrow \begin{cases} 1.0 & if\, i\text{th node is selected as } s_1 \text{ or } s_2 \\ \gamma h_{b,i} & otherwise \end{cases} \tag{3}$$

where γ is the temporal discount rate, $s_1$and s_2 are the first and second nearest nodes to b-th agent. By taking the negative possibility of the visit to the node visited previously in account (Figure 9 (d)), we represent the time-series of tracking data in the following,

$$u_b = (u_{b,1} \; u_{b,2} \; \cdots \; u_{b,n})^T$$

,

$$u_{b,i} \leftarrow \begin{cases} 1.0 & if\, i\text{th node is selected as } s_1 \text{ or } s_2 \\ -h_{b,i} & otherwise \end{cases} \tag{4}$$

As a result, we can estimate the current situation in spatiotemporal tracking,

$$v_b = S_k \cdot u_b$$

where

$$v_{b,i} \leftarrow \max\left(v_{b,i}, 0\right)$$

. (5)

Next, we consider the collision avoidance with other people, robots, and objects by using the positions of other agents in the following,

$$\tilde{u} = (\tilde{u}_1 \; \tilde{u}_2 \; \cdots \; \tilde{u}_n)^T$$

,

$$\tilde{u}_i \leftarrow \begin{cases} 1.0 & if\, i\text{th node is selected as } s_1 \text{ or } s_2 \text{ of others} \\ 0 & otherwise \end{cases} \tag{6}$$

As a result, we can estimate the current situation in spatiotemporal tracking based on the possibility of colliding with other agents,

$$v_b \leftarrow v_b - \tilde{v}$$

,

$$\tilde{v} = S_k \cdot \tilde{u} \tag{7}$$

Furthermore, we can estimate the approximate future state if we use larger number of *k* for the multiplication. However, the reliability of the future estimation may be low, because we don't use the geometric information in topological maps and graphs. In this way, we can use the simple matrix calculation to estimate the current situation of each agent. The essence of this discussion is how to use adjacency matrix and how to design the existence vector.

Figure 9. Possible moves based on walks: (a) an example of graph, (b) walks from the 1st node by **A**, *(c) walks from the 1st node by* $\mathbf{A}^2$, *(d) move from the 2nd node, (e) collision avoidance*

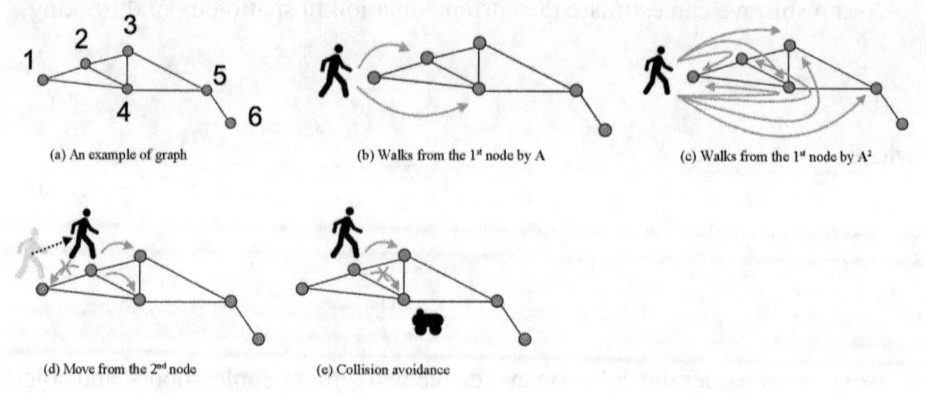

3. MULTISCOPIC TOPOLOGICAL TWIN

3.1. Basic Learning Process of Growing Neural Gas

Various methods based on growing neural gas have been proposed so far (Satomi, Masuta, & Kubota, 2009); Toda & Kubota, 2013; Toda, Matsuno, & Minami, 2021; Toda & Kubota, 2018; Toda, Chin, & Kubota, 2017; Saputra et al., 2019; Iwasa, Kubota, & Toda, 2021; Ardilla, Saputra, & Kubota, 2023; Obo, Kubota, Toda, & Masuyama, 2024). In this subsection, we explain the basic learning process of GNG in brief (Fritzke, 1994). In the initialization, two or three nodes and their edge connections are generated by directly selecting sample data (Figure 10 (a)). Each node and edge have the accumulated error and age information, respectively. When a sample data (v) is given to the network, the nearest and second nearest nodes (s_1 and s_2) are selected (Figure 10 (b)). The accumulated error of the selected nodes is updated by adding the distance. After several iterations, a new node (r) is added between the node (q_1) with the maximal accumulated error and the node (q_2) with the maximal accumulated error in its connected nodes (Figure 10 (c)). If there is no edge between these the nearest and second nearest nodes, a new edge is added to the network. Otherwise, the age of its edge is reset to zero. Furthermore, the edge age from the nearest node to the rest of connected nodes is incremented. If the edge age is the predefined threshold, its edge is removed. The dotted edge is removed from the network in Figure 10 (d). As a result, if there is no edge from the connected node, the node is also removed. In this way, the topological learning is done incrementally.

The most important advantage of GNG is to realize topological mapping and clustering simultaneously. However, the learning convergence is not so good, because the original GNG uses an incremental learning. Furthermore, the computational cost to find the nearest node among all nodes is expensive, as the number of nodes is large. Therefore, we proposed a fast multi-scale batch-learning GNG (Fast GNG in short) and multi-layer GNG (ML-GNG) to improve the performance related with the above problems, respectively.

Figure 10. Learning procedure of GNG: (a) initialization, (b) node selection, (c) edge addition, (d) edge removal

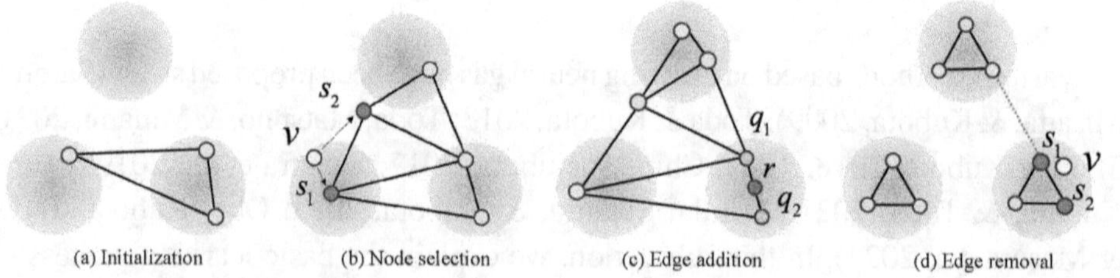

3.2. Fast Multi-Scale Batch-Learning Growing Neural Gas

In order to reduce the computational cost and improve the convergence property, we proposed a method of Fast Multi-scale Batch-Learning Growing Neural Gas (Fast GNG), which is applied to relatively large temporal data in real-time (Kubota, 2021; Iwasa, Kubota, & Toda, 2021; Ardilla, Saputra, & Kubota, 2023; Obo, Kubota, Toda, & Masuyama, 2024). We explain how to deal with dynamic data over time by using Fast GNG. Figure 11 shows the learning phases of Fast GNG from the viewpoint of multi-scale batch-learning. When a data set D is given to the current network, the first mini-batch learning is done at Phase 1 (Figure 11). If the current data set is almost overlapping with the previous one, most nodes are selected as the first nearest nodes. As a result, the transition to the next phase (Phase 2) is done. On the other hand, if the current data set is rarely overlapping with the previous one, very few nodes are selected as the nearest nodes. Therefore, we restart the learning as the second initialization for a current different data set (Node Initialization) by the second mini-batch data set as a result of initial screening. After that, the phase transition is done one by one shown in Figure 11. Finally, the full batch learning will be done at least once in order to update reference vectors and its edge connectivity. Based on the above phase transition, we need less than or equal to 2D of data sampling,

$$D_{\text{sum}} = \sum_{i=1}^{L_{\max}} \frac{1}{2^{i-1}} D < 2D \tag{8}$$

If the restart is done in the Phase 1, Fast GNG needs $2D$ where $L_{\max}$ is the maximal phase level.

The most important difference between the original GNG and Fast GNG is in the aging process. When the edge connection is updated by the mini-batch dataset, the non-selected edge connection and node without having any edge are automatically deleted. Therefore, Fast GNG does not need to use the age parameters in the batch learning procedure. The next important feature is to use as-if-silent strategy to enhance the increase of nodes in the topological structure. If the sample data is far away from the current network, it is added to the network as a node. Figure 12 shows an example of topological adaptation to dynamic data stream by Fast GNG (Iwasa, Kubota, & Toda, 2021). Each donut sample data takes circular motion with different speed. So, several donut sample data are overlapping over time, but Fast GNG can trace the circular motion of four donut sample data.

Figure 11. Learning phase in MS-BL-GNG (= {D/16, D/8, D/4, D/2, D}, = {N/10, N/3, N/2, 4N/5, N} where D and N are the number of data and maximal number of nodes, respectively, in the above example)

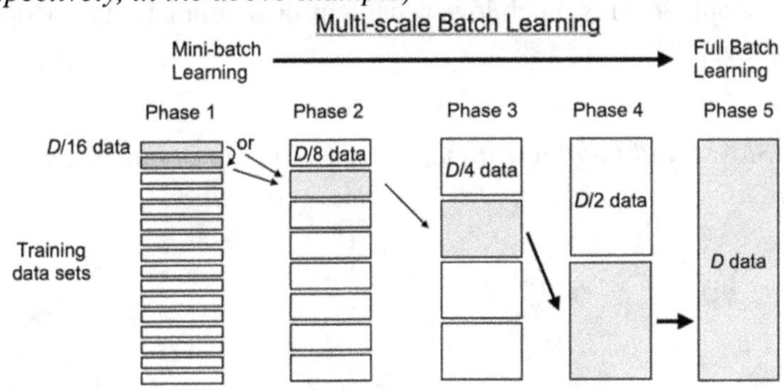

Figure 12. Topological adaptation to dynamic data stream

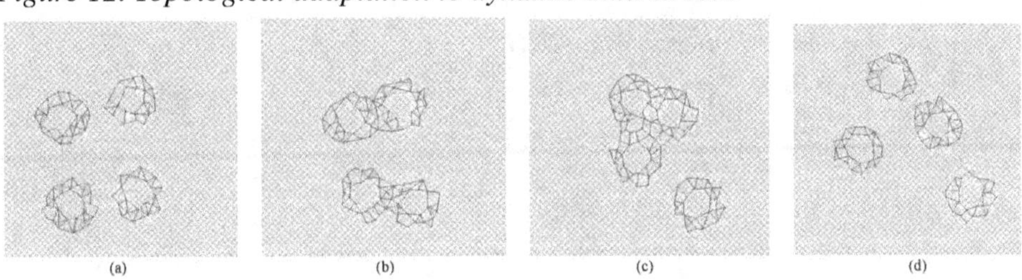

Figure 13 shows an example of people flow simulation using the topological map obtained by Fast GNG (Doteguchi & Kubota, 2022). SLAM is done by 2D LiDAR equipped with the mobile robot using a CAD data of a rough image of the open space at the microscopic level (Figure 13 (a)). At this microscopic level, it generates a detailed environment map. Next, the digital model for the simulation is generated in the mesoscopic level (Figure 13 (b)). At this mesoscopic level, it discretizes the environment into navigable areas. The topological map is generated according to the measurement data of human trajectories obtained by simple AI simulation of people flow at the macroscopic level (Figure 13 (c)). Finally, Figure 13 (d) show people flow simulation based on the above topological analysis by using the built topological map at the mesoscopic level. In Figure 13 (d), the adjacent nodes in $\mathbf{A}^2$ are depicted in red, the selected nodes (s_1 and s_2) in a large red sphere; and the nodes with negative value in yellow. Each agent (person) moves and avoids collisions with others according to the moving direction calculated by the above topological analysis. In this way, we can build an environmental map and digital model as the digital twin and topological roadmap as a topological twin. Furthermore, the computational cost of the topological analysis for people flow simulations performed at the macroscopic level is much lower than physical simulations performed at the mesoscopic level.

Figure 13. SLAM and topological mapping: (a) a map built, (b) an approximate digital map, (c) topological map, (d) behavioral simulations

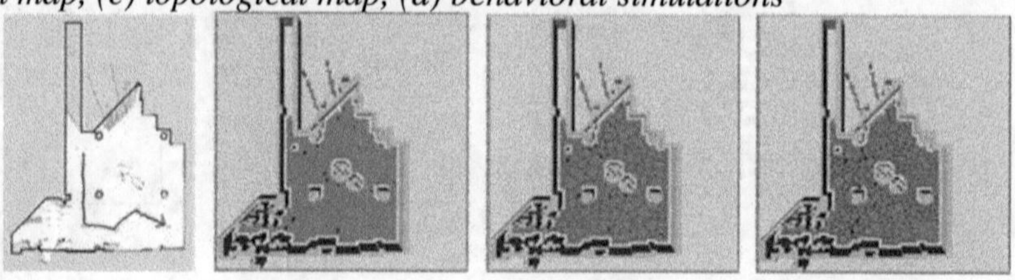

Figure 14. Node inheritance in ML-GNG

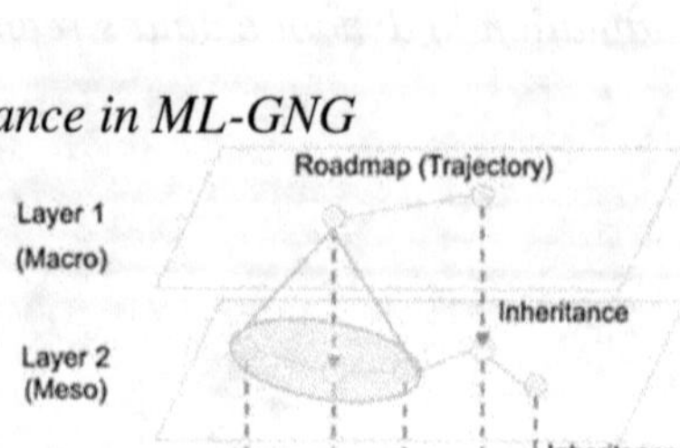

3.3. Multi-Layer Growing Neural Gas

The multiscopic topological twin is a hierarchical digital representation of a system that captures its structure (topology) and behavior across different scales—microscopic (local), mesoscopic (intermediate), and macroscopic (global). This approach facilitates more precise modeling, analysis, and decision-making in complex systems. Furthermore, it enables robots to efficiently navigate dynamic spaces by leveraging topological intelligence at different resolutions. At the macroscopic level, it delineates the highest-level navigable space, thereby functioning as a foundational map for global navigation. The utilization of this system is effective for the planning of long-range missions, as demonstrated by the following example: "Go from Point A to Point B." At the mesoscopic level, it captures environmental geometry for the purpose of feasible path planning, taking into account the robot's size and kinematics. Robots have the capacity to modify their trajectories without necessitating global replanning, a process referred to as "local optimization." At the microscopic level, the system under consideration is capable of detecting fine-grained, dynamic obstacles (e.g., humans, moving objects). It functions in conjunction with SLAM to facilitate dynamic obstacle mapping.

Basically, hierarchical or multilayer approach for growing neural gas can be used to extract hierarchical structure of topological information. We have proposed several different types of hierarchical GNG (H-GNG) and applied H-GNG to extract different scale of topological feature (Satomi, Masuta, & Kubota, 2009; Toda & Kubota, 2013; Toda, Matsuno, & Minami, 2021). Furthermore, in order to reduce the computational cost and deal with multiscopic topological intelligence, we proposed a new method of multi-layer Growing Neural Gas (ML-GNG) (Watanabe, Obo, & Kubota, 2025).

We explain the main difference between H-GNG and ML-GNG. We use the node inheritance from the upper layer to lower layer in order to keep the relationship between two layers in ML-GNG, and we don't update the reference vector of the inherited nodes (Figure 14). The nearest node selection is performed in the local nodes subset surrounding the inherited node (depicted in the green circle of Figure 14). As a result, ML-GNG can reduce the computational cost. Furthermore, in the learning process of ML-GNG, we use the sequential learning of each layer from the top to bottom (top-down learning), and then, the sequential update from the bottom to top (bottom-up learning). Since some unnecessary edges may still exist in the learning at the higher layer, we use the bottom-up learning to cut such edges.

Next, we show an application example of ML-GNG to robot navigation tasks (Watanabe, Obo, & Kubota, 2025). First, we perform SLAM by a mobile robot based on GLIM framework including a front-end SLAM system and a back-end Smoothing and Mapping. Point cloud data was collected using a Livox Mid-360 sensor mounted on a mobility support robot (MSR). In this experiment, we used three-layers of ML-GNG. The 1st layer is used to learn the trajectory of the MSR as a topological structure at the macroscopic level. The topological structure plays the role in the minimal constrain of movable range. The second 2nd layer is used to learn the topological structure of environmental map for path planning according to the size of MSR at the mesoscopic level. The 3rd layers is used to extract the topological structure of objects in the environment at the microscopic level. In fact, topological structures in the 3rd layer are used to extract dynamic human movement and static objects for collision avoidance. The maximal nodes in the 1st, 2nd, and 3rd layers are 50, 3,000, and 10,000, respectively. Figure 15 shows the learning result of topological maps. The point cloud data consisting of 400,000 points was reduced to 10,000 nodes in the 3rd layer. Figure 16 shows an experimental result based on the topological analysis for robot navigation tasks. The MSR traced the preplanned target path (depicted in green) in the beginning (Figure 16 (a-1) and (b-1)), but the MSR tried to search for a path (in blue) in the 2nd layer to avoid the collision with the wall at the left side and in Figure 16 (a-2) and (b-2). Next, the MSR encountered a walking person in Figure 16 (a-3) and (b-3) and avoided the collision in Figure 16 (a-4) and (b-4).

Figure 15. Topological mapping results of ML-GNG based on trajectory and point cloud data: (a) point cloud and trajectory, (b) 1st layer (50 nodes), (c) 2nd layer (3,000 nodes), (d) 3rd layer (10,000 nodes)

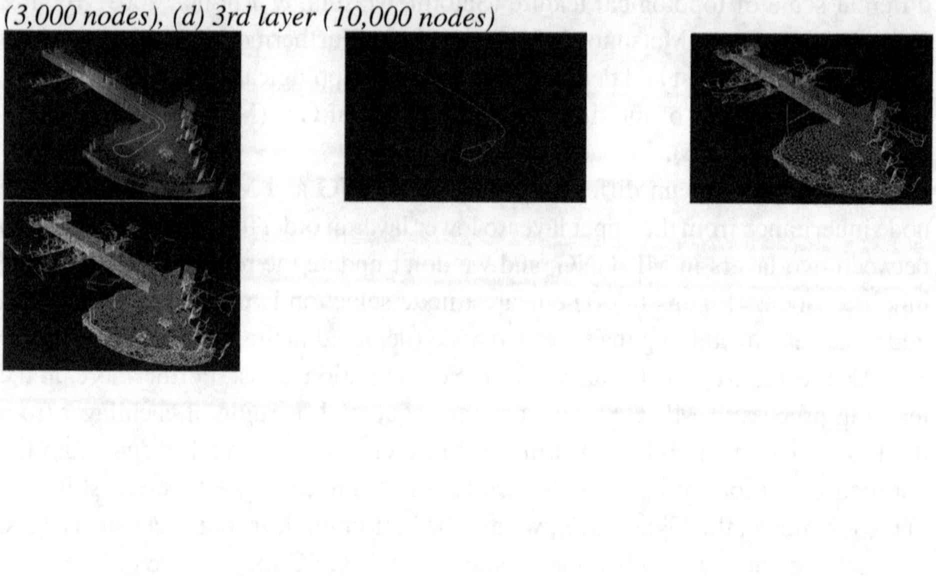

Figure 16. Collision avoidance and navigation of MSR based on topological analysis

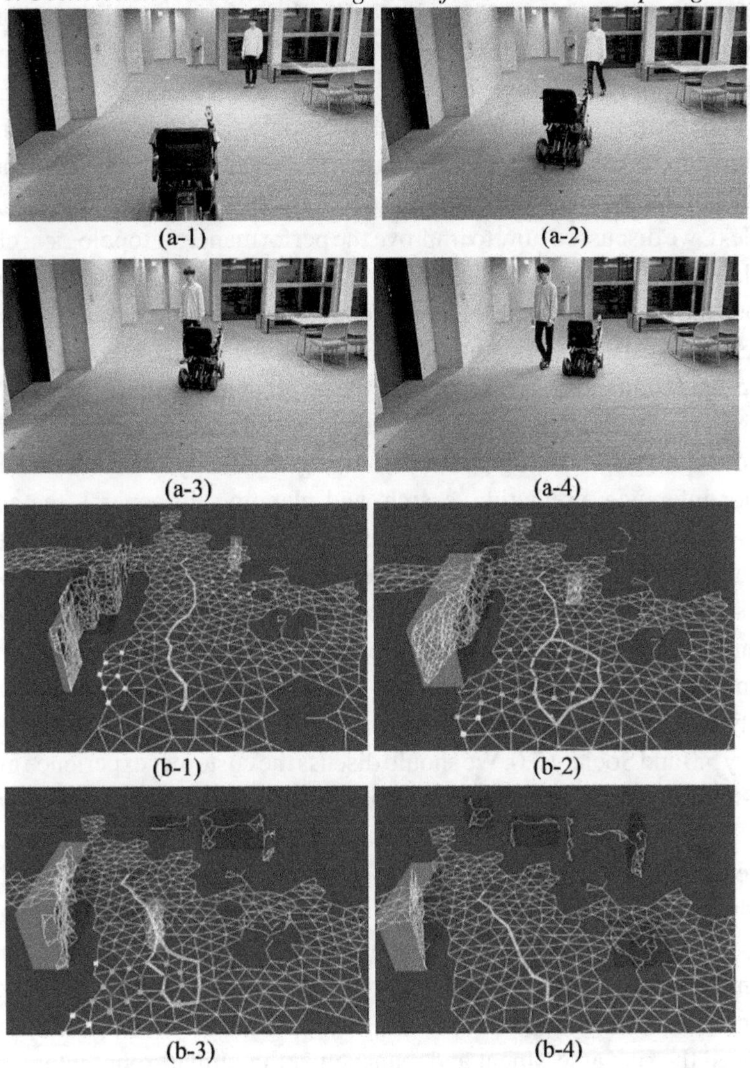

4. SUMMARY

This paper discussed the methodology and applicability of multiscopic topological twin in cyber-physical-human systems and Industry 5.0. The most important factor lies in the human-centered and personalized services in customer experience. Especially, human-robot interaction is one of the most important applications, and we should discuss the safety of robots. In this paper, we focused on mobility support robots and discussed multiscopic topological intelligence for mobility support robots. First, we explained the role of topological twin and the topological intelligence using topological mapping, topological clustering, and graph theory. The digital twin is used as a platform for simulations, while the topological twin is used to represent the relationship between data or variables for multi-scale multi-physics simulations. Next, we explained the methodology of topological mapping and topological clustering. Next, we discussed how to improve the performance of topological clustering and explained the methods of fast multi-scale batch-learning growing neural gas to improve the performance of dynamic adaptation and multi-layer growing neural gas to deal with different tasks in different layers. Furthermore, both methods can reduce computational costs comparing to the original growing neural gas. Next, we showed several application examples of agent-based topological intelligence. The essence of topological intelligence leis in using adjacency matrices and existence vectors for inference, prediction, search, and planning. In general, the adjacency matrix is often used to discuss the connectivity of nodes, but we can predict the future state of agents by use the existence vector and adjacency matrix with both positive and negative values. Next, we apply multiscopic topological twin to people flow simulations and navigation tasks of mobility support robots. The simulation results and experimental results show the effectiveness of the proposed method.

Finally, we discuss the applicability and future direction of topological intelligence to Industry 5.0 and Society 5.0. We should discuss the customer experience in human-robot co-existing situations, and we need large scale of simulations of human-robot interactions before the social implementation. However, it takes huge computational cost if we use physics simulation. Therefore, we need simplified intelligent simulations to predict the future situations. The multiscopic topological intelligence plays the important role to reduce computational costs, while the multiscopic topological intelligence can connect the relationship between micro-, meso-, and macro-scopic topological structures. We should know that human behaviors are unpredictable and we need the risk assessment and management in human-robot interactions. As a future work, we intend to apply multiscopic topological intelligence to the risk assessment and management. We can create a topological structure of humans and robots and use it to predict the intersection point when they meet, as well as their future movements. Furthermore, we develop a cognitive model based on topological mapping methods to deal with various types of spatio-temporal data streams.

ACKNOWLEDGMENT

This work was partially supported by JST [Moonshot RnD][Grant Number JP-MJMS2034].

REFERENCES

Ardilla, F., Saputra, A. A., Besari, A. R. A., Naoki, D., Kohei, O., Obo, T., & Kubota, N. (2022, October). Topological Twin for Mobility Support Robots. In *2022 IEEE 8th World Forum on Internet of Things (WF-IoT)* (pp. 1-6). IEEE. DOI: 10.1109/WF-IoT54382.2022.10152223

Ardilla, F., Saputra, A. A., & Kubota, N. (2023). Multi-Scale Batch-Learning growing neural gas efficiently for dynamic data distributions. *International Journal of Automotive Technology*, *17*(3), 206–216. DOI: 10.20965/ijat.2023.p0206

Bapat, R. B. (2014). *Graphs and matrices*. Hindustan Book Agency

Deka, D., Kekatos, V., & Cavraro, G. (2023). Learning Distribution Grid Topologies: A tutorial. *IEEE Transactions on Smart Grid*, *15*(1), 999–1013. DOI: 10.1109/TSG.2023.3271902

Doteguchi, N., & Kubota, N. (2022, May). Topological Tracking for Mobility Support Robots Based on Multi-scale Batch Learning Growing Neural Gas. In *Mobile Wireless Middleware, Operating Systems and Applications:10th International Conference on Mobile Wireless Middleware, Operating Systems and Applications (MOBILWARE 2021)* (pp. 17-31). Cham: Springer International Publishing. DOI: 10.1007/978-3-030-98671-1_2

Eppstein, D., Paterson, M. S., & Yao, F. F. (1997). On nearest-neighbor graphs. *Discrete & Computational Geometry*, *17*(3), 263–282. DOI: 10.1007/PL00009293

Fogel, D. B. (1995). *Evolutionary Computation: Towards a New Philosophy of Machine Intelligence*. IEEE Press.

Fritzke, B. (1991, July). Unsupervised clustering with growing cell structures. In *IJCNN-91-seattle international joint conference on neural networks* (Vol. 2, pp. 531–536). IEEE. DOI: 10.1109/IJCNN.1991.155390

Fritzke, B. (1994). A growing neural gas network learns topologies. *Advances in neural information processing systems, 7*. Vol. 7, 1995, pp. 625-632.

Fritzke, B. (1997). A self-organizing network that can follow non-stationary distributions. In *Artificial Neural Networks—ICANN'97:7th International Conference Lausanne,Switzerland,October 8–10, 1997Proceeedings 7* (pp. 613-618). Springer Berlin Heidelberg. DOI: 10.1007/BFb0020222

Fu, K. S., Gonzalez, R. C., Lee, C. G., & Freeman, H. (1987). *Robotics: control, sensing, vision, and intelligence*. McGraw-Hill.

Fukuda, T., & Kubota, N. (1999). An intelligent robotic system based on a fuzzy approach. *Proceedings of the IEEE*, *87*(9), 1448–1470. DOI: 10.1109/5.784220

Greer, C., Burns, M., Wollman, D., & Griffor, E. (2019). *Cyber-physical systems and internet of things*. DOI: 10.6028/NIST.SP.1900-202

Gross, J. L., & Yellen, J. (2005). Graph Theory and its applications. In *Chapman and Hall/CRC eBooks*. DOI: 10.1201/9781420057140

Industry 5.0 - Towards a sustainable, human-centric and resilient European industry. (n.d.). Research and Innovation. https://research-and-innovation.ec.europa.eu/knowledge-publications-tools-and-data/publications/all-publications/industry-50-towards-sustainable-human-centric-and-resilient-european-industry_en

Iwasa, M., Kubota, N., & Toda, Y. (2021, October). Multi-scale batch-learning growing neural gas for topological feature extraction in navigation of mobility support robots. In *The 7th International Workshop on Advanced Computational Intelligence and Intelligent Informatics (IWACIII 2021)*.

Jurczenia, K., & Rak, J. (2022). A survey of Vehicular Network Systems for Road traffic management. *IEEE Access : Practical Innovations, Open Solutions*, *10*, 42365–42385. DOI: 10.1109/ACCESS.2022.3168354

Kohonen, T. (2001). Self-Organizing Maps. In *Springer series in information sciences*. DOI: 10.1007/978-3-642-56927-2

Kubota, N. (2021, December). Multiscopic topological twin in robotics. In The 28th International Conference on Neural Information Processing (ICONIP2021), Bali, Indonesia, Online.

Kubota, N., Yuki, K., & Baba, N. (2009). Integration of intelligent technologies for simultaneous localization and mapping. *2009 ICCAS-SICE*, 4981–4986. https://ieeexplore.ieee.org/iel5/5305883/5332438/05334643.pdf

[] LaValle, S. M., & Kuffner, J. J. (2001). Rapidly-exploring random trees: Progress and prospects. *Algorithmic and computational robotics*, 303-307.

Martinetz, T., & Schulten, K. (1991). A" neural-gas" network learns topologies. *Artificial Neural Networks, 1*, 397–402.

Obo, T., Kubota, N., Toda, Y., & Masuyama, N. (2024). Fast multi-scale Batch-Learning growing neural gas. In *Topics in intelligent engineering and informatics* (pp. 13–33). DOI: 10.1007/978-3-031-58257-8_2

Oshio, K., Kaneko, K., & Kubota, N. (2021, October). Multi-scopic simulation for human-robot interactions based on multi-objective behavior coordination. In *International Workshop on Advanced Computational Intelligence and Intelligent Informatics* (pp. 3-8).

Paul, R. P. (1981). *Robot manipulators: mathematics, programming, and control.* The MIT Press.

Russell, S., Norvig, P., & Intelligence, A. (1995). A modern approach. *Artificial Intelligence. Prentice-Hall. Egnlewood Cliffs, 25*(27), 79–80.

Saputra, A. A., Botzheim, J., Sulistijono, I. A., & Kubota, N. (2015). Biologically inspired control system for 3-D locomotion of a humanoid biped robot. *IEEE Transactions on Systems, Man, and Cybernetics. Systems, 46*(7), 898–911. DOI: 10.1109/TSMC.2015.2497250

Saputra, A. A., Chin, W. H., Toda, Y., Takesue, N., & Kubota, N. (2019). Dynamic density Topological structure generation for Real-Time ladder affordance detection. *2021 IEEE/RSJ International Conference on Intelligent Robots and Systems (IROS)*, 3439–3444. DOI: 10.1109/IROS40897.2019.8968003

Saputra, A. A., Toda, Y., Botzheim, J., & Kubota, N. (2017). Neuro-Activity-Based Dynamic Path Planner for 3-D rough terrain. *IEEE Transactions on Cognitive and Developmental Systems, 10*(2), 138–150. DOI: 10.1109/TCDS.2017.2711013

Sasaki, H., Kubota, N., & Taniguchi, K. (2008). Topological map and cell space map for SLAM of a mobile robot. *GESTS International Transactions on Computer and Engineering, 45*(3).

Satomi, M., Masuta, H., & Kubota, N. (2009, March). Hierarchical growing neural gas for information structured space. In *2009* IEEE Symposium Series on Computational Intelligence 2009 (pp. 54-59). IEEE. DOI: 10.1109/RIISS.2009.4937906

Thrun, S., Burgard, W., & Fox, D. (2005). *Probabilistic Robotics*. The MIT Press.

Toda, Y., Chin, W., & Kubota, N. (2017, December). Unsupervised neural network based topological learning from point clouds for map building. In *2017 International Symposium on Micro-NanoMechatronics and Human Science (MHS)* (pp. 1-6). IEEE. DOI: 10.1109/MHS.2017.8305188

Toda, Y., & Kubota, N. (2013). Self-Localization based on multiresolution map for remote control of multiple mobile robots. *IEEE Transactions on Industrial Informatics, 9*(3), 1772–1781. DOI: 10.1109/TII.2013.2261306

Toda, Y., & Kubota, N. (2013, August). Feature extraction based on hierarchical growing neural gas for informationally structured space. In *The 2013 International Joint Conference on Neural Networks (IJCNN)* (pp. 849-855). IEEE. DOI: 10.1109/IJCNN.2013.6706825

Toda, Y., & Kubota, N. (2018). Topological structure learning based enclosing formation behavior for monitoring system. *2022 IEEE International Conference on Systems, Man, and Cybernetics (SMC)*, 831–836. DOI: 10.1109/SMC.2018.00149

Toda, Y., Matsuno, T., & Minami, M. (2021). Multilayer batch learning growing neural gas for learning multiscale topologies. *Journal of Advanced Computational Intelligence and Intelligent Informatics, 25*(6), 1011–1023. DOI: 10.20965/jaciii.2021.p1011

Wang, F. (2010). The emergence of intelligent Enterprises: From CPS to CPSS. *IEEE Intelligent Systems, 25*(4), 85–88. DOI: 10.1109/MIS.2010.104

Wang, P., Yang, L. T., Li, J., Chen, J., & Hu, S. (2018). Data fusion in cyber-physical-social systems: State-of-the-art and perspectives. *Information Fusion, 51*, 42–57. DOI: 10.1016/j.inffus.2018.11.002

Watanabe, M., Obo, T., & Kubota, N. (2025, July). Topological mapping based on multi-layer growing neural gas for topological twin. In *2025 international joint conference on neural networks (IJCNN)* (pp. 1-8). IEEE.

Woo, J., Kubota, N., & Lee, B. (2010). Steady-State genetic algorithms for growing topological mapping and localization. In *Pacific Rim International Conference on Artificial Intelligence (PRICAI 2010),* (pp. 558–569). DOI: 10.1007/978-3-642-15246-7_51

Chapter 8
Predictive Maintenance Using Sensor Data Processing

Manoj Himmatrao Devare
https://orcid.org/0000-0002-9530-3914
Amity University Mumbai, India

Anita Manoj Devare
https://orcid.org/0009-0009-4159-6289
Amity University Mumbai, India

ABSTRACT

The chapter explores the various types of equipment failure, techniques to discover the root cause, and different approaches to find the solution to predict the event of the failure. Moreover, this work addresses the equipment failure prediction on the assembly line of truck wheel manufacturing plant. The step-by-step process of the wheel production includes one after another usage of the machines. Any workstation failure leads to the idle time of the subsequent machines and reduction in the working efficiency of the labor. The sensor-based data of the machine is collected using the process automation. Historical data is used to predict the remaining useful life. K-Nearest Neighborhood technique has given the accuracy of 96.40%. The Decision Tree with Gini Index and Entropy have has achieved the 96% accuracy in the machine state prediction. The multi-regression based model developed to calculate the values of coefficient and intercept as learning parameters to be useful in the further deployment.

DOI: 10.4018/979-8-3693-7994-3.ch008

1. INTRODUCTION

There is a wide range of industries from assembly lines, manufacturing industry, oil and gas, construction industry, and the vessels industry, where the Predictive Maintenance (PM) is high in demand. Whereas, due to the unavailability of the right data, ineffective data preparation techniques, and lack of expertise in the domain of the expertise in Machine Learning, the success rate is not high. Predictive maintenance projects need effective tools and techniques for a high amount of success rate. The manufacturing industry expects enhanced outcomes in terms of high availability of the machines, and a reduction in unexpected failures. The successful manufacturing industry needs continuously running equipment during the active labor shifts. If the machinery is not working, then the labor goes into an idle state; this waste of the resources leads to serious consequences such as financial loss, time and reputation loss. The delay in the small parts delivery may lead to further cancellation of the contracts with the clients.

The automation industry requires smart solutions based on amalgamation Machine Learning, AI based tools, range of sensors, and intelligent data processing techniques. The intentional or unintentional wrong handling of the equipment can be avoided by monitoring the labor activities with the help of video recording. However, automation in the process of monitoring and advanced intimation is important to understand the activities. The crucial logic of identifying reasons for failure is collecting sensor-based data of the machine using the process automation. Historical data is used to predict the remaining useful life. The maintenance of the equipment in the assembly requires the domain knowledge of Electrical, Mechanical and Electronics Engineering with specialization of usage of parts in the high temperature. The knowledge of voltage fluctuation, current, temperature on the various sub parts of the machine also requires to effectively design the predictive algorithm in such an arrangement. Sensors are installed by the process automation engineer, with systematic analysis can research.

The equipment such as hydraulic press failure, Centrifugal Pump, may have several reasons including electrical system, inlet pipe leakage, seal leakage, lubrication failure, motor failure, bearing failure, mechanical failure, or operator error. The fault-tree analysis calculates the minimal cut sets i.e. smallest combination of the events responsible for the failure. This can lead to the root cause analysis of failure. These failures can be classified into basic, intermediate, and top-level failures.

A comprehensive predictive maintenance management program uses the most cost-effective tools such as vibration monitoring, thermography, and tribology to obtain the actual operating condition of critical plant systems and based on this actual data schedules all maintenance activities on as-needed basis. PdM should be a part of an overall equipment strategy that begins with the most critical processes and systems. The problems can be certain mechanical or electrical problems or whether the condition of the equipment is causing process problems. The outcome of the predictive analytics can decrease downtime, optimize asset performance, and increase the lifespan of machines. The other targets are double-digit increases in asset utilization. Smart machines that flag performance issues before they happen. Huge bumps in Overall Equipment Effectiveness (OEE) such as availability, effectiveness and quality. The Total Effective Equipment Performance (TEEP) which takes account of both Equipment Losses, as measured by OEE and Schedule Losses as measured by Utilization. It takes Overall Production Effectiveness (OPE) parameters into consideration. PdM enables manufacturers to perform maintenance only when necessary. Asset performance metrics like Mean Time to Repair (MTTR) which include the repair time and testing period. The MTTR is also calculated as the total maintenance time divided by the total number of repairs. The Mean Time Between Failures (MTBF), and MTTF.

There are various events that occur around equipment under observation. All such events can be listed to make it contribution to the fault of that machinery. There is dependency on the one event that can cause the reason for the failure of the machine or some accident. The events can be categorized into the basic event. For example, simple component failure or the man-made mistake can be a basic event. There are certain probabilities associated with each event. One event can occur when two other events can happen at the same time. It is also possible that the failure of the machine is considered a top-level event. The dependency of the one event as a failure can be a combination of the logical condition or the or logical condition. The tree structure with the top node as the top-level failure event. This problem is also known as system failure diagnosis, prognostics and health management. The Remaining Useful Lifetime (RUL) is another technique to find the hidden state of the components of the system. The approach of failure prediction can be based on feature extraction and status classification. The other approach is based on time series modeling. One more option is to find the anomaly using statistical or some other techniques.

Vision computing based solutions can identify abnormal activities by the labor such as wrong approach of handling equipment, and specific spark generation by the Butt Weld machine identifies the possibility of failure. The manufacturing industry seeks machine learning based solutions to predict the sudden failure of the machines, equipment, and high demand manufacturing units. It has been observed that the record of replacement of the faulty parts, dates, times, and shift of work also shows the clues towards the identification of the root cause of the machine failure.

Tools like thermography imaging tools or ultrasonic lubricators are useful in predictive maintenance. The conditions under which an individual machine operates, the differential wear-and-tear of different machine parts, or other factors that might predict failure. Find meaningful statistical associations between problems and root causes. Spindle speed, hours running, temperature, vibration, humidity, usage–these are just a few of the parameters that interact in unique ways and, in the aggregate, have variable impact on machine life. Understanding how machines are used. Some of the more common predictive areas include calculating machine lifetime before failure; identifying a window in which a failure is likely to occur; identifying the most common types of failures; and detecting anomalous machine behavior.

Vibration analysis is useful for finding the stability of the rotating machines. Vibration data analysis is another technique combined with the sensors collecting the data in the frequency or time domain for vibration data. The acceleration, velocity, and displacement data can be captured using the accelerometer, velocity and displacement sensors. The data related to the bearing, gears, conveyers are useful to find the condition of the machine. Fast Fourier transformation is useful to convert from the time to the frequency domain. Shocks, transient waves, misalignment, gear wear, imbalance in the rotational motion.

The change in the vibration patterns indicates the change in the condition of the machine. The abnormal condition is proportional to the amount of variation detected. The LabVIEW software is useful for experimental purposes, with the setup made up using spur gears, piezoelectric sensors and accelerometers. Acoustic emissions and ultrasonic sensors are useful in early-stage fault detection. The piezoelectric, MEMS, Capacitive accelerometers are good for the finding faults, monitoring the vibrations in the rotating machinery like motors, pumps, and fans. The electromagnetic velocity and displacement sensors are good for the low frequency vibrations and slow-speed machinery such as large bearing, and turbines. The turbines can be monitored using Piezoelectric velocity sensors. The acoustic emission sensors are good for bearing and gearbox fault detection. The structural health monitoring of the bridges and buildings can be cone using accelerometers and laser displacement sensors. The relative motion between two surfaces can be detected using the eddy current proximity probes such as shaft vibration in the rotating equipment. Laser displacement sensors are also useful in the fault detection. The lubrication and cracks detection can be done for early fault detection such as cavitation, leak detection, and high-speed machinery faults using ultrasonic, and microphone sensors.

The Principal Component Analysis (PCA) and Linear Discriminant Analysis (LDA) are useful to predict failure of the manufacturing device. The data is collected during the manufacturing of the medical device. The same technique is also used to predict the failure of the grinding tool. A tool like oscilloscope is used to measure the readings. The data is recorded from the different wheels of the grinding machine. The wheels burning or non-burning are recorded as predicted. The model is trained offline using historical data, and the prediction of the failure is done at real time.

The book chapter addresses the prediction of equipment failure on the assembly line of Truck wheel Manufacturing plant. This chapter provides data collection, data processing techniques, and challenges in the overall process of Predictive Maintenance (PM). A few unique data processing techniques are presented for data exploration and a set of results on the ML models applied on the sensor dataset collected from IoT enabled working machine. The step-by-step process of the wheel rim production includes one after another usage of the machines towards the final finishing product. The metal sheets of iron are used from the first assembly line machine called as Butt Weld Machine to the final finished rim as the finished object. The quality of the rim production is important, as per the specified standard of the time taken by the labor. Machine number N is dependent on machine number N-1 and so on. All machines are operated by the machine operator, in semi-automatic mode. There are certain interventions required at each workstation. Any workstation failure leads to the idle time of the subsequent machines. For example, if the machine number P fails, then all subsequent machine P+1, P+2 remains idle till the fault is identified and rectified. Prediction of machine failure well-in advance is highly required to efficiently utilize the labor hours and equipment.

The lazy learner approach of K-Nearest Neighbor technique is applied with the accuracy of 96.40%. The KNN approach is moderately only finds the distance between the current activity on the assembly line and the past incidences of machine failure. This approach is not giving any importance to the specific attribute. Hence, it is important to assign weight to certain attributes with their descending order of contribution to the failure of the machine. The real-time monitoring and subsequent indicating the wrong doings can avoid the failures. The computations of algorithms at real-time require effective and timely indicators. The data collected through sensor reading is available to test the machine learning algorithm. The training and testing phases on the recorded datasets are conducted using the Decision Tree algorithm having Gini Index and Entropy has 96% accuracy.

2. BUTT WELD MACHINE

A butt weld machine is used to join two pieces of metal end-to-end (butt joint) by heating and applying pressure. It is commonly used in industries for welding pipes, rods, rails, and frames where strong, continuous joints are required. Two metal components such as rods or pipes are aligned end-to-end. The ends are cleaned and clamped into the machine's fixtures. Then during the heating phase such as by flash or resistance, the ends are pushed together with low pressure. The Electric current in resistance butt welding or flashing i.e. arcing between ends is applied. The metal at the joint heats up to a plastic melting state due to resistance to current or arcing.

During the upset phase also called Forge Phase, once sufficient heat is reached, the machine increases pressure and forces the two parts together. This causes plastic deformation and the molten or softened metal fuses at the joint. The excess material, i.e. flash, is squeezed out. Then during the cooling phase the current is stopped, the welded joint is allowed to cool under pressure to ensure a strong, solid bond. During the finishing phase, i.e. excess metal is trimmed off. The joint may be cleaned or machined further depending on the application.

Figure 1 shows the Butt Weld machine which is used to do the welding to the plate used to create rim in the truck wheel. Figure 2 shows the Butt Weld machine in operation of Rim welding. The failure of the machine may have the reason that, chiller temperature is low, jaw temperature is not in the range, there is wire break etc.

Figure 1. Butt Weld Machine

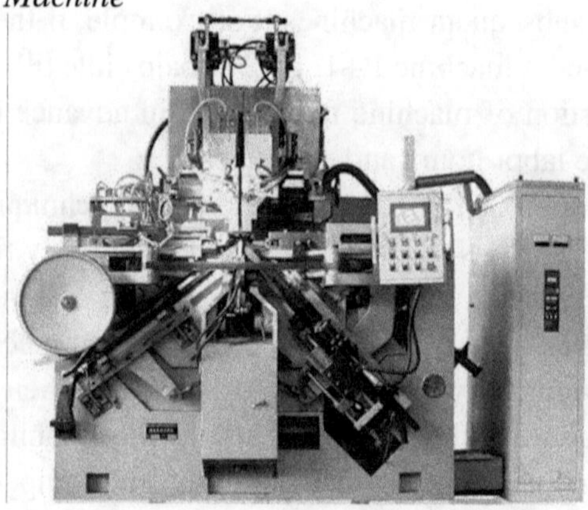

Figure 2. Close view of Butt Weld Machine during Rim Welding Operation

There are three types of Butt Welding i.e. Flash, Resistance and Upset Butt Welding. In case of flash butt welding which uses the arcing i.e. sparks to heat the ends before joining. I resistance Butt Welding, the electrical resistance is used for heating. The upset welding uses the pressure throughout the heating cycle. The butt welding has strong and uniform joints, and no filler material is needed. It is automated and suitable for mass production. It is cost-effective for joining rods, pipes and rails. The butt welding is not suitable for thick and dissimilar metals. The proper alignment and surface preparation are crucial. It requires precise current and pressure control.

Figure 3 shows different parts of the Butt-Welding machine such as two electrodes required for clamping the workpieces and conducting the welding current. Each electrode holds one of the two metal pieces that are to be joined. They ensure proper alignment and firm contact between the faces of the two workpieces. Mechanical or hydraulic force may be applied through the electrodes to press the parts together, especially during the forging stage. The electrodes are connected to the power source and conduct high current through the workpieces. The electrical resistance at the joint where the pieces touch causes localized heating. This heating brings the metal to a plastic or molten state, allowing them to fuse together under pressure. A Current is passed through the joint area. Resistance causes heat generation at the interface. When sufficiently heated, pressure is applied and increased, and the joint is formed by forging or melting and solidifying.

Figure 3. Different parts of Butt-Welding machine.

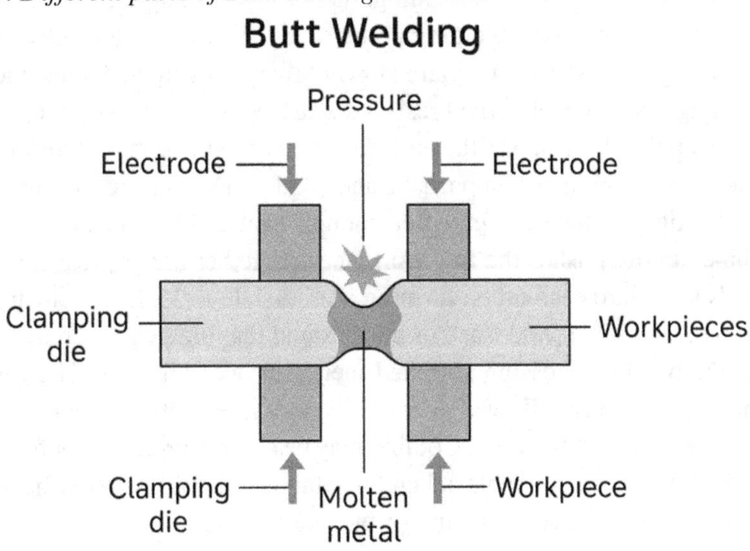

Clamping die refers to the mechanical part that grips and holds the workpieces (typically rods, wires, or bars) in place during the welding process. The clamping die firmly holds the workpieces. It prevents movement during current flow and forging. Clamping die ensures accurate alignment of the ends to be welded. Applies mechanical or hydraulic pressure to push the workpieces together during the forging phase of welding. Clamping die ensures good electrical contact Maintains stable contact so that welding current can pass efficiently through the joint. In some butt-welding machines (especially resistance butt welders), electrodes and clamping dies can be part of the same component, i.e., the electrode clamps the workpiece and carries current. In other machines (especially flash butt welders), separate clamping dies hold the parts, and the current may be introduced through integrated conductors.

Figure 4. Different stages of Butt Weld machine like flash, upset, and cooling

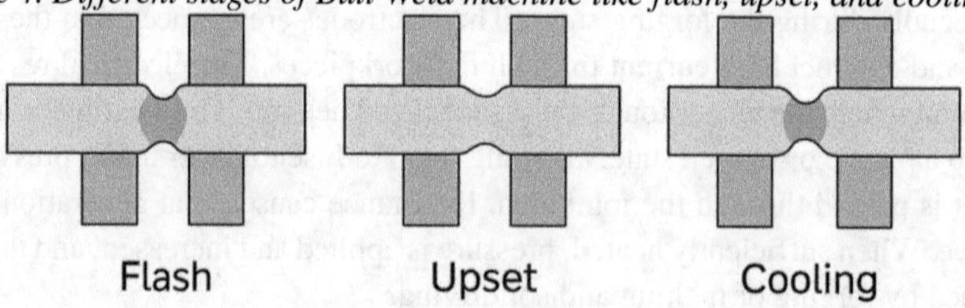

Figure 4 shows the flash butt welding, the terms flash, upset, and cooling refer to three distinct stages in the welding process. In the flash stage electrical current is passed through the parts as they are slowly brought together. As the ends of the workpieces lightly touch, electrical resistance causes arcing and sparking. The heat generated from the flash causes the ends to soften or melt slightly. This also cleans the surfaces by burning off impurities and oxides. During the upsetting phase, an actual welding is happening. After enough heat is built up during flashing, the machine rapidly pushes the two workpieces together under pressure. The hot, plastic ends forge into each other, forming a solid-state weld. Excess metal may be squeezed out, forming a weld flash or burr around the joint. Upset creates a strong metallurgical bond by pressing softened metal surfaces together. The final stage where the welded joint is allowed to cool. The pressure is often maintained during cooling to ensure proper bonding. Cooling may be natural air-cooled or forced water or fan-cooled depending on material and machine. The cooling solidifies the joint and helps stabilize the microstructure of the weld.

Failure of a flash butt welding machine can result from various mechanical, electrical, or operational issues. Mechanical failures can be misalignment of workpieces leading to poor contact, improper heating and weak joints. The worn or damaged clamping dies cause the slipping or uneven pressure. The insufficient force during the upset phase results in poor bonding. The improper guide rails or movement affects smooth motion during flashing and upsetting. Another reason can be electrical failures such as poor electrical contact cause the arcing at the clamps instead of at the joint. The faulty power supply or transformer such as inadequate current delivery leads to incomplete heating or flashing. Another reason can be Inconsistent current flow or voltage drops, and results into the weak welds. Control System Issues such as Malfunctioning timers or sensors, results into incorrect timing of flash, upset, or cooling stages. Sometimes the software or PLC faults in automated machines, which causes erratic operation or improper sequencing. Another reason is that inaccurate parameter settings, such as wrong values for current, time, or pressure can compromise weld quality. Operational Errors such as improper surface preparation, dirty, oxidized, or uneven surfaces prevent proper welding. The incorrect material pairing such as incompatible metals or sizes can cause weld defects. The operator's inexperience, poor judgment during setup or manual overrides can damage the machine or ruin the weld. The thermal issues such as overheating or underheating can weaken the weld or cause cracking. The inadequate cooling after the weld leads to residual stress or metallurgical defects. All these kinds of problems in Butt Welding machine leads to incomplete fusion or weak joints, excessive flash or no flash. There can be burnt or oxidized welds, machine stalling during upset, and Irregular or noisy operation.

3. BACKGROUND

Component failures, capacitor failure, faulty diodes, broken connections, power fluctuations can damage the expensive equipment such as voltage regulators. The improper ventilation, high loads, transformer overheating, harmonic, and electrical noise can be a reason for the failure. The servo controller, motor driven controller, incorrect voltage adjustment, corrosion and loose connections can lead to sparking and fire. The excessive vibration in the rotational motion of the wheels, spindles, shafts can cause damage to the electrical connections. Accordingly, these factors play an important role while deciding the predictive maintenance. Cleaning, dust removal, firmware updates, thermal imaging, component replacement, and regular maintenance can lead to less damage to the equipment. Damon et al. (2008) have performed a study on reliability centered maintenance (RCM) and predictive modeling on voltage regulators at an electric utility. The regression equation is developed based on the operating variables by analyzing the number and types of failures occurring on various classifications of voltage regulators under normal operations.

Identifying the dependent and independent variables requires the domain knowledge about the various parts of the equipment. The unsupervised machine learning algorithms can apply the features to cluster the similarity of the data available. The anomaly detection and threshold of the various factors such as current, voltage, pressure, temperature and heat are required to be decided through the statistical analysis and clustering methods such as DBScan or K-Means. The failure of the equipment cannot be higher records, as compared to the running state of the machine. Otherwise, such frequently failed equipment is already replaced with better stable and reliable equipment. Techniques such an isolation forests, and autoencoders can be used to find the pattern different from the normal. The anomaly does not directly imply the reason for the failure but indicates some deep analysis to find the root cause analysis. The unlabeled dataset does not have the indicators of precision and recall. In any case, the expert domain knowledge is required to understand the dataset. Liang et al. (2007) discusses unlabeled data and its importance in predictive modeling. It emphasizes Probability based, Regression based and the Bayesian analysis on such a dataset. In such a situation, past records of the failure of the equipment are not available.

Turbofan is used in Jet engines of aircraft as an alternative to turbojets or turboprops useful in the Boeing, Airbus and Passenger Jets. It has two types such as a low bypass used in military fighter jets and a high bypass used in Cargo Planes. Turbofans consist of the gas combustion unit to generate power and fan system to generate the thrust. Thakkar & Chaoui, (2022) performed the Deep Neural Network (DNN) method to find the Turbofan Engine Damage propagation modeling based on data stemming from a run-to-failure simulation considering the Remaining Useful Lifetime (RUL). The real-time work and the noisy data handling is the key characteristics of the Kalman Filter. The Kalman Filter (KF) method is applied for linear and non-linear methods as well, using the Matlab Toolbox. The noisy and uncertain situation is handled by KF. The algorithm of KF first predicts the state of the turbofan and then corrects the prediction if there is any noise. It also works like the weight updates in the neural network, to predict the next state of the system. KF is used in the GPS and navigation and RADAR system to track vehicles, missiles, and aircraft. Robots and self-driving vehicles can also be tracked using Kalman Filter.

The early failure detection of sensor parts based on feature extraction and status classification; time series modeling approach based on anomaly detection using autoencoders i.e. the feed forward neural network is used. The autoregressive integrated moving average (ARIMA) model is tested using the confusion matrix and the receiver operating characteristics curve (Kuzin & Borovicka, 2016).

The milling machine is one of the important pieces of equipment in the manufacturing industry where the metal work is done. The metal objects can be reshaped using milling machines. Now the milling machines come in various facilities like turret, vertical, horizontal, manual or CNC operating machines. The plastic, metal or wooden pieces of objects can be reshaped using power driven milling machine. Spindles, tables, columns, knee, saddle, and cutting tool are the main components of the milling machine. Such a machine can have slow deterioration and hidden micro-pitting. Techniques such as ultrasound, infrared thermography, and X-ray inspections can reveal internal wear and hidden cracks. The efforts have been made to Apply find out the hidden degradation process, and decision support method for condition-based optimal predictive maintenance on the milling machine (Wei et al., 2013). Wavelet transformation and the fast Fourier transformation is useful in finding such a hidden degradation. The micro cracks over time cause major damage to the equipment. The temperature and acoustic signals along with the vibration data can be useful for finding material fatigue in rotating equipment due to repeated stress. Similarly, material insulation wear can be detected so that further breakdowns can be avoided. The non-destructive technique of the infrared thermography detects the heat and temperature patterns and variations in the objects. The thermogram is generated using the infrared rays which are further useful in the building inspections, medical diagnostics, and security inspections. Active and passive thermography are the two types. An active method is using the external heat to the object and the thermal response is analyzed. Similarly, in the passive method the natural temperature difference is analyzed without external heating. Thermography is a non-contact; non-invasive method works in the limited lighting conditions.

The study by Baradaran (2025), addresses the challenge of accurately diagnosing the health of electric motors to prevent unexpected failures. Operational data from electric motors, including various sensor readings is used. The research compares multiple supervised learning algorithms Naive Bayes, Support Vector Machines (SVM), Regression models, Random Forest, K-Nearest Neighbors (KNN), and Gradient Boosting to classify motor conditions into "Healthy," "Needs Preventive Maintenance," or "Broken" stages. Among the models tested, one demonstrated superior accuracy in classifying motor health states, highlighting the effectiveness of supervised learning in predictive maintenance. The study underscores the practicality of using supervised learning for electric motor diagnostics, providing a foundation for efficient maintenance scheduling and minimizing unplanned downtimes in industrial applications.

The research by Nascimento et al. (2022) aims to enhance fault diagnosis in rotating machinery by leveraging advanced deep learning architectures. MaFaulDa and Case Western Reserve University (CWRU) bearing datasets is used for the diagnosis of the fault. Study introduces T4PdM, a modified Transformer-based deep neural network, to classify multiple fault types in rotating machinery. T4PdM achieved accuracy of 99.98% on the MaFaulDa dataset and 98% on the CWRU dataset, outperforming existing models. The Transformer-based model significantly improves machinery fault analysis and diagnostic processes in Industry 4.0 applications.

Zheng et al. (2020) explores the evolution from traditional predictive maintenance to intelligent maintenance systems integrating AI and Industrial Internet of Things (IIoT). Turbofan Engine Degradation Simulation Dataset is used along with a novel framework combining probabilistic deep learning, real-time data collection via smart sensors, and big data technologies for intelligent maintenance. The framework demonstrated improved reliability modelling and maintenance decision-making capabilities. It concludes that integrating AI and IIoT facilitates the transition to intelligent maintenance, enhancing system reliability and efficiency.

The study by Juodelyte et al. (2022) focuses on predicting degradation stages of rolling-element bearings to enhance maintenance in pharmaceutical manufacturing. FEMTO Bearing dataset with high-frequency vibration signals is used and employed k-means clustering for bearing lifetime segmentation and utilized autoencoders to embed vibration signals into a low-dimensional space, facilitating supervised classification of degradation stages. The framework reliably predicted degradation stages across various bearings, demonstrating scalability and effectiveness. The proposed method offers actionable insights for maintenance scheduling, ensuring compliance and efficiency in pharmaceutical production.

The research work by Ali et al (2021) addresses the need for effective predictive maintenance strategies for rotational machinery to prevent unexpected downtimes. Operational data including vibration and current parameters collected via Bluetooth Low Energy (BLE) networks. An unsupervised deep learning model, specifically a Long Short-Term Memory (LSTM) autoencoder, was utilized to detect anomalies and estimate the remaining useful life (RUL) of machinery. The system accurately identified faults and predicted failures, and enhanced machinery reliability.

Shah et al. (2024), worked towards the predictive maintenance of the wind turbines are often located in remote areas, making timely maintenance challenging. Traditional predictive maintenance methods struggle to provide reliable advance notice for necessary interventions. The study utilized operational data from wind turbines, focusing on seven distinct failure types. The researchers introduced two deep learning models, ForeNet-2d and ForeNet-3d, employing a multi-parametric attention-based approach. This methodology bypasses manual feature engineering, reducing potential human errors. The models achieved accuracy, with the most precise forecast deviating by only ten minutes from the actual Remaining Useful Life (RUL) and the least accurate prediction off by 1.8 days. Most predictions were within a few hours of the actual RUL. The proposed deep learning approach offers a substantial time frame for accessing remote wind turbines and performing necessary maintenance, enhancing the practicality of predictive maintenance in such settings.

Industrial machinery such as centrifuges, pumps, and compressors require effective predictive maintenance strategies to minimize downtime and maintenance costs. The study by Levin (2024) analyzed a dataset reflective of centrifuge, pump, and compressor operations, capturing various operational parameters. The research evaluated the efficacy of machine learning models Random Forest, Gradient Boosting Machine (GBM), and Deep Neural Networks (DNNs) using metrics like accuracy, precision, recall, F1 score, and ROC AUC. Feature importance analysis was conducted, identifying vibration levels and operational hours as key predictors. DNNs demonstrated superior performance in predictive accuracy and operational efficiency. However, challenges such as computational demands and the need for specialized expertise were noted. The study provides a comprehensive framework for applying machine learning to predictive maintenance, offering valuable insights into developing more efficient and reliable industrial maintenance strategies.

Autran et al. (2024) used the existing publicly available datasets for predictive maintenance often lack the complexities present in real-world industrial settings, such as missing data and irregular data acquisition. The study introduced AI4I-PMDI, derived from the AI4I 2020 predictive maintenance dataset, enhanced with irregularities common in real maintenance data. Machine learning algorithms were applied to both the original and modified datasets for binary and multi-class classification tasks. The impact of the introduced irregularities on model performance was assessed. The modifications significantly affected the performance of predictive maintenance models, emphasizing the importance of using datasets that reflect real-world complexities. AI4I-PMDI provides a more realistic benchmark for evaluating predictive maintenance algorithms, highlighting the need for datasets that capture the irregularities of complex industrial environments.

Al Mamlook et al. (2024), mentions that timely detection of potential machine failures is crucial in industrial settings to prevent unexpected downtimes and associated costs. The study utilized datasets with class imbalance issues, which were addressed using the Synthetic Minority Over-sampling Technique (SMOTE). Seven ML classifiers, including Gradient Boosting, Support Vector Machine (SVM), Random Forest, XGBoost, LightGBM, CatBoost, AdaBoost, and Artificial Neural Networks (ANN), were assessed using various performance metrics. Gradient Boosting achieved the highest accuracy of 98.19%. Key indicators for predicting machinery failure were identified, enhancing the understanding of failure patterns. Integrating machine learning into predictive maintenance provides accurate early warnings of machine failures, enabling proactive maintenance actions and reducing unexpected downtimes.

Belim et al. (2024) selecting appropriate forecasting models is essential for effective predictive maintenance, especially in industrial environments with varying data patterns. The study analyzed sensor data from BA Glass equipment, including temperature and pressure readings, collected via SCADA systems and OPC Router. Various forecasting models, including Linear and Polynomial Regression, Simple and Double Exponential Smoothing, ARIMA, and Prophet, were evaluated using blocked cross-validation and rolling window methodologies. Performance metrics such as MSE, RMSE, and MAE were calculated. Double Exponential Smoothing with an additive damped trend and linear models performed best for datasets with daily seasonality and gradual oscillations. ARIMA and Prophet models were more accurate for datasets with stable trends and higher frequency oscillations. The choice of forecasting models significantly impacts the effectiveness of predictive maintenance systems. Tailoring model selection to the specific characteristics of the dataset enhances prediction accuracy and maintenance decision-making.

Ohalete et al. (2023) mentioned that the oil and gas industry face challenges in equipment maintenance due to the complexity and scale of operations. Traditional maintenance approaches often lead to unplanned downtimes and increased operational costs. This study is a comprehensive literature review and does not utilize a specific dataset. The authors conducted a systematic literature review focusing on the integration of Artificial Intelligence (AI) and Data Science in predictive maintenance within the oil and gas sector. Databases such as IEEE Xplore, ScienceDirect, SpringerLink, and Web of Science were utilized to gather relevant literature from 2010 onwards. The review highlights that AI and data science significantly enhance predictive maintenance strategies by enabling accurate failure predictions and optimized maintenance scheduling. These advancements lead to reduced downtime and operational costs. However, challenges such as data management complexity and the need for high-quality, real-time data are identified. The study concludes that the continuous evolution of AI will play a crucial role in shaping future maintenance strategies in the oil and gas industry.

Sengupta et al. (2023), Armored vehicles operate in high-stress environments, making timely maintenance crucial to ensure operational readiness and safety. Traditional maintenance schedules may not effectively predict failures, leading to unexpected downtimes. Sensor data collected from armored vehicles, including parameters such as temperature, pressure, and vibration readings. The study proposes an ensemble machine learning model incorporating Light Gradient Boosting, Random Forest, Decision Tree, Extra Tree Classifier, and Gradient Boosting algorithms. K-fold cross-validation and Technique for Order Preference by Similarity to Ideal Solution (TOPSIS) analysis were employed to evaluate model stability and performance. The ensemble model achieved an accuracy of 98.93%, precision of 99.80%, and recall of 99.03%. These results indicate the model's effectiveness in accurately predicting maintenance needs, thereby reducing vehicle downtime. The study demonstrates the potential of machine learning-based predictive maintenance solutions in enhancing the reliability and operational efficiency of armored vehicles. Implementing such models can lead to proactive maintenance strategies, minimizing unexpected failures.

Kamariotis et al. (2023), while data-driven prognostic algorithms are widely used for predictive maintenance, there is a lack of standardized metrics to assess their impact on maintenance decision-making and cost optimization. Simulated turbofan engine degradation data, designed to mimic real-world operational conditions and failure modes. The authors propose a novel metric that evaluates prognostic algorithms based on their influence on maintenance policies and associated costs. The study examines two predictive maintenance decision settings, i.e. component ordering and replacement planning. Four data-driven prognostic algorithms were assessed using this metric to determine their effectiveness in optimizing maintenance strategies. The proposed metric effectively quantifies the impact of prognostic algorithms on maintenance decisions, enabling the identification of algorithms that minimize long-term maintenance costs. The study demonstrates that integrating this metric into the evaluation process leads to more informed and cost-effective maintenance planning. The research introduces a decision-oriented performance assessment tool that bridges the gap between algorithm accuracy and practical maintenance outcomes. This approach facilitates the selection and optimization of prognostic algorithms that align with organizational maintenance objectives.

Siddique et al. (2023), focus on the predictive maintenance systems leveraging deep learning and IoT technologies are susceptible to adversarial attacks, which can compromise the accuracy of failure predictions and lead to unexpected equipment downtimes. NASA's turbofan engine degradation dataset, providing multivariate time series data representative of real-world engine performance and failures. The study introduces novel adversarial attack methods tailored for multivariate time series data, including MTS Projected Gradient Descent (PGD) and MTS PGD with random restarts. The impact of these attacks on various deep learning models such as Long Short-Term Memory (LSTM), Gated Recurrent Unit (GRU), Convolutional Neural Network (CNN), and Bi-directional LSTM was evaluated. Additionally, an approximate adversarial training method was proposed to enhance model robustness. The adversarial attacks significantly degraded the performance of predictive maintenance models, with RUL prediction errors increasing up to 11 times. The proposed adversarial training method improved model robustness up to 54 times, outperforming existing defense strategies. The research highlights the vulnerability of predictive maintenance systems to adversarial attacks and underscores the importance of incorporating robust defense mechanisms. The proposed methodologies offer a pathway to developing more secure and reliable predictive maintenance frameworks.

Latrach (2023), worked on the oilfield equipment that operates under harsh conditions, making predictive maintenance essential to prevent unexpected failures and optimize operational efficiency. Traditional maintenance approaches may not effectively capture the complex patterns leading to equipment degradation. Publicly available datasets, including NASA's C-MAPSS, Case Western Reserve University Bearings, and FEMTO Bearings datasets, encompassing various sensor readings and failure modes. The study explores the application of various deep learning architectures such as fully connected networks, CNNs, and RNNs for equipment health diagnostics and RUL prediction. Data preprocessing techniques, including Principal Component Analysis (PCA) and signal processing methods like Fourier and Wavelet analyses, were employed to enhance feature extraction. The deep learning models demonstrated strong capabilities in accurately diagnosing equipment health states and predicting RUL. The integration of advanced feature extraction techniques contributed to the robustness and reliability of the predictive models. The research confirms the efficacy of deep learning approaches in predictive maintenance for oilfield equipment. Implementing such models can lead to significant reductions in unplanned downtimes and maintenance costs, enhancing overall operational performance.

Bao et al. (2022) introduce the Maintenance Inspection Anomaly Detection (MIAD) dataset, aimed at advancing research in unsupervised anomaly detection within predictive maintenance systems. The dataset includes over 100,000 high-resolution images across six industry-relevant scenarios such as railways, bridges, and power stations. MIAD is notable for its synthetic generation of anomalies using generative techniques, which mitigates the typical scarcity of defect data. The dataset is designed for vision-based anomaly detection systems and offers pixel-level annotations, although it primarily supports unsupervised approaches. It deals with data diversity, providing a standardized platform for evaluating anomaly detection algorithms without reliance on labeled data. The unsupervised models like Convolutional Autoencoders and Isolation Forest illustrate the practical utility of the dataset. In case of lack of comprehensive datasets for unsupervised predictive maintenance, especially in visually intensive domains, this work addresses the issue. MIAD is synthetically generated; the domain transferability and real-world applicability remains a question. A critical gap is addressed in the predictive maintenance research landscape by facilitating scalable and generalizable model training in the absence of labeled failure data.

Carrasco et al. (2021) proposed a novel evaluation framework for unsupervised anomaly detection in time-series data, tailored specifically for predictive maintenance applications. Traditional approaches to evaluating anomaly detection models often rely on point-based metrics that fail to capture the temporal nature of real-world industrial anomalies. This paper redefines anomalies as intervals rather than isolated points, allowing for a more accurate and realistic assessment of model performance. The framework integrates multiple evaluation metrics such as precision, recall, and F1-score, adapted to the temporal context, and supports a variety of unsupervised algorithms. The validation of the framework is performed through extensive experiments using synthetic and real-world industrial datasets. A Python-based toolkit for researchers and practitioners to apply the framework directly in their workflows. The paper makes a compelling case for rethinking how unsupervised anomaly detection is assessed in time-series data, especially when labeled data is unavailable or inconsistent. While the work is highly valuable, the reliance on synthetic benchmarks in some cases limits the assessment of real-world variability.

Malawade et al. (2021) explore the intersection of neuroscience and predictive maintenance by proposing algorithms inspired by brain function, specifically Hebbian learning and spike-timing-dependent plasticity (STDP). The authors argue that these biologically inspired methods provide a strong foundation for unsupervised learning in systems where labeled data is rare or nonexistent. They develop a framework to model the temporal evolution of machine states based on sensory input patterns in manufacturing systems. The proposed neural algorithms autonomously learn system degradation patterns without supervision, detecting early signs of failure. Experiments on benchmark manufacturing datasets, including simulated fault scenarios, demonstrate that their method is competitive with, and in some cases surpasses, traditional unsupervised techniques such as k-means and autoencoders. The authors emphasize the adaptability and low computational complexity of their approach, making it well-suited for embedded industrial applications.

Patil et al. (2018) present an unsupervised methodology for analyzing time series data to extract error patterns useful for predictive maintenance. Their work addresses the common challenge in industrial applications where labeled failure data is unavailable or insufficient. The proposed technique involves a preprocessing pipeline that standardizes time series data and applies feature extraction using discrete wavelet transform (DWT). Following this, unsupervised clustering techniques specifically hierarchical and k-means clustering are employed to group similar behavior patterns, identifying deviations indicative of potential faults. The validation is done for this approach using synthetic and real-world sensor datasets from industrial machinery. The results demonstrate that their method can effectively identify anomalies without prior labeling, making it ideal for early-stage failure detection. A pattern extraction and interpretability, allowing maintenance engineers to understand the nature of the anomalies detected. The approach offers a practical, low-cost solution for many industries looking to implement predictive maintenance with minimal labeled data. It serves as a foundation for further exploration into unsupervised time series analysis in complex operational environments.

Vassiliadis et al. (2022) propose a comprehensive unsupervised predictive maintenance framework specifically for fleet management. The system integrates time-series analysis with context-aware modeling, enabling anomaly detection without labeled data. It uses vehicle telemetry such as engine temperature, speed, and GPS data to model normal and abnormal behavior patterns. The methodology combines clustering, statistical deviation modeling, and a decision module that flags anomalies considering contextual attributes like terrain and weather. A large-scale vehicular dataset is used and compared with baseline unsupervised algorithms such as isolation forest and DBSCAN. Their approach demonstrates superior accuracy in identifying mechanical faults early, reducing downtime and operational costs. The incorporation of contextual metadata, which enhances anomaly detection precision and reduces false positives. The model relies on domain-specific contextual inputs that may pose challenges for deployment in different industries without prior customization. A unsupervised predictive maintenance strategies illustrating the value of context-aware analytics in large and distributed vehicular systems.

Devare (2023) has applied the Multinomial Naïve Bayes machine learning algorithm with accuracy of 83.97%. The discrete Markov chain, and the Weibull process is also applied on the Butt Welding machine to predict the failure of the machine. The dataset used is the sensor data collected over the specific interval of the time.

4. SENSOR CONNECTIONS AND INDUSTIRAL EQUIPMENT

The sensors are connected to the various components of the equipment including the high heat producing element. The protection of the sensor is done through the insulation material or cooling provisions. The air or water cooling is used to maintain the operating temperature of the sensors. Stainless steel or ceramic insulation is used on the cables. Polytetrafluoroethylene (PTEF) is the high-performance polymer used in electrical cables and connectors. PTEF material stands from -200°C to +260°C of temperature. PTEF has dielectric strength, chemical resistance, and durable in harsh environments. The non-stick and low-friction properties are useful if they are connected to the rotating component of equipment. To avoid the short circuit and oxidation such as avoid moisture, and corrosion resistance property makes it not absorb water. These properties make the PTEF use in between center conductor and outer shield. The high temperature wires and cables are useful in aerospace, military and industrial applications. The optical transmission of the data is the best choice when there is very high, or very low temperature conditions are appearing.

Vacuum insulation is also one of the important tricks to make the sensors useful. The air molecules are responsible for the heat transfer. The conduction and convection are prevented through the vacuum to avoid the heat transfer. This trick is useful to connect the sensitive material sensors it the extreme equipment where the vibration readings are required to be collected. The radiation can be avoided by coating the reflective material. Cryogenic fluids like liquid nitrogen, are used along with the vacuum for the insulations.

The sensor connections to the various components of the equipment can be wireless or wired. Dependent on the environment, the connection types can be chosen. The wired connections such as fieldbus, ethernet, Profinet, MODBUS, CAN bus, HART, Wireless HART etc. are the choices. ISA100.11a, Zigbee, LORA are also useful along with BLE. There are pneumatic and hydraulic sensors useful in industrial automation.

5. INITIAL DATA ANALYSIS

The unlabeled dataset used in this research work consists of the sensor readings along with the machine production timing where the wheel rim production is continued or halted is indicated. This is converted into the meaningful states of the machine such as running, halted, or stopped. Figure 5 shows the number of failure points on six weekdays as from Wednesday to Monday. In some industrial areas in India, one day is weekly off such as Tuesday.

Figure 5. The number of failure points on subsequent days.

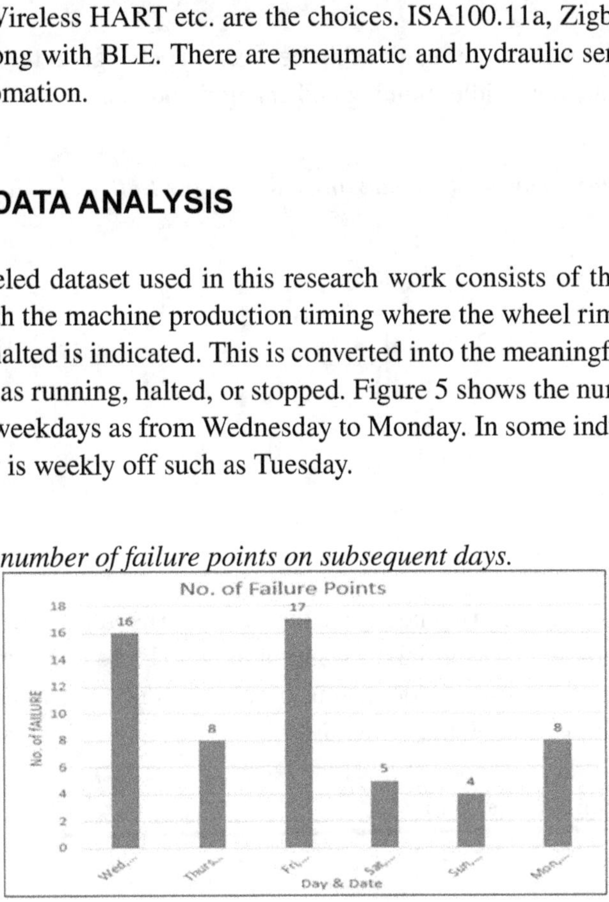

Figure 6 shows the time line of the machine state as change point T_n. Finding out the abnormal conditions at the prior points T_{n-1}, T_{n-2} can provide what went wrong due to which the state is changed. On the other side, the obserations and the statistical analysis of some sensors readings at the time T_{n+1}, T_{n+2} can probabiliy provide some irregulariteis as well, which can imply the failure as well.

Figure 6. Machine State Change point and finding the abnormal conditions

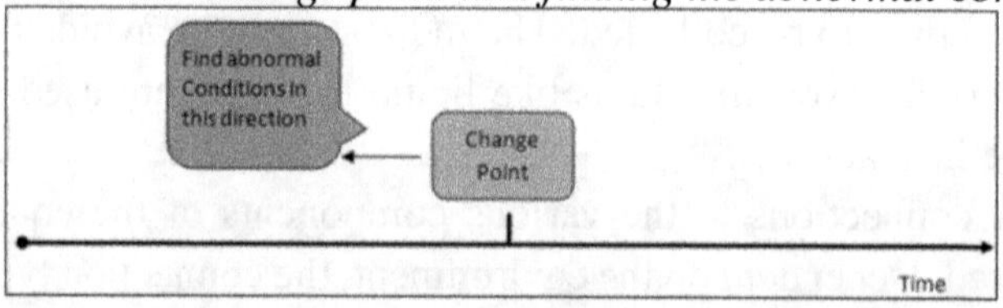

Figure 7 shows that, the relationship between the change in the three different states of the machine as idle, running and changing points.

Figure 7. Relation between the changing states of the Butt Weld Machine

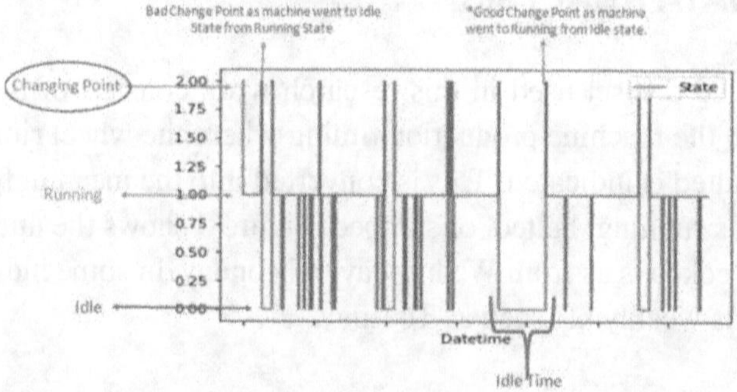

The analysis is required to understand the data, such as whether it is following the Gaussian distribution i.e. normal distribution or not. Hence the histogram plotting having bin size as 20 is taken for all-sensor values is done. One histogram for the chiller temperature is shown in figure 8.

```
import seaborn as sns

import matplotlib.pyplot as plt

machine_data = pd.read_csv('C:/datewise.csv')

sns.distplot(machine_data['G2-Chiller Temp'], bins=20)
```

```
plt.title("Histogram plot of Chiller Temp")
```

Figure 8. The histogram for the chiller temperature

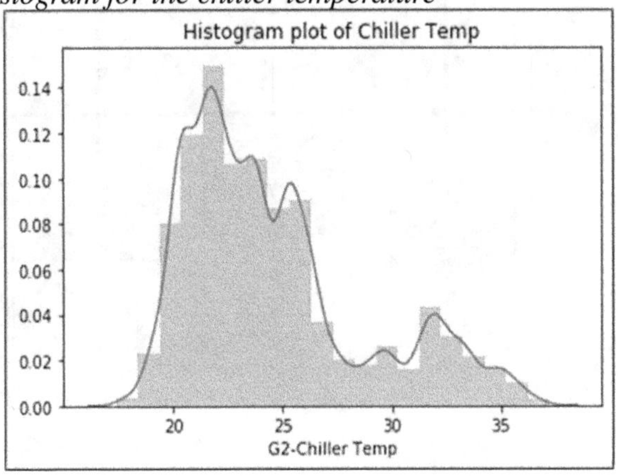

The box and whisker plot are the most suitable plots to understand the nature of the data series available. The state of the machine can be clearly observed when the machine was running, idle and changing point. The state values are Zero for idle or stopped machine, 1 for running machine and 2 for the changed state. Similarly, all the sensor values can be plotted using the box and whisker plot as shown in figure 9. The work carried out in the manufacturing industry has successfully deployed the industrial internet of things successfully deployed. Subsequently, the reports are generated by having the production count and the downtime. The sensor readings are available in the constant time interval. Not always but based on the longer down-time also plays role to indicate there may be a fault in the running condition. The labelling is done accordingly.

```
machine_data = pd.read_csv('datewise.csv')

plt.subplots(figsize=(6, 6))

sns.boxplot('State', 'G2-Chiller Temp', data=machine_data,
dodge=True, fliersize=2, whis=1.5, width=0.8, saturation=2,
linewidth=2)
```

Figure 9. Box Plot of the Machine State and the Chiller Temperature in degree Celsius

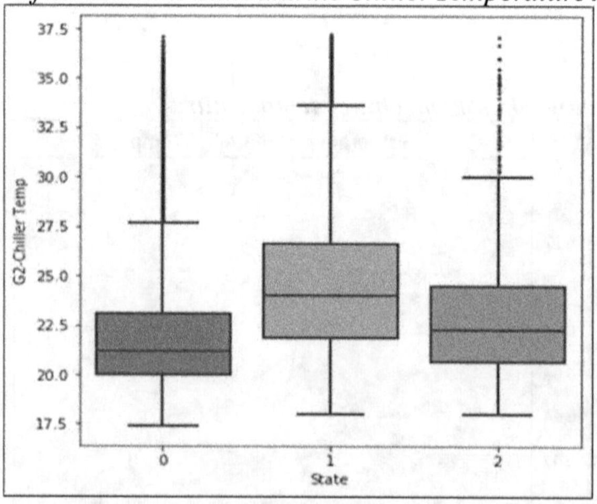

Unlabeled Dataset is one of the major challenges in finding the solution to predictive maintenance (PM). The problem of the PM needs enough data in numbers. The dataset can be created through historical documents, the experience from the assembly line engineer, and the purchase history of the parts. The dataset is an unlabeled dataset; in case the dataset is prepared without any IoT integration. Hence, the choice is an unsupervised machine learning algorithm for such a dataset. During the process of labeling the data, there is a possibility of wrong labeling. The reasons can be wrong understanding about the overall process happening on the shop floor. The logical error happened during the labelling, deficiency of the domain knowledge, lack of expertise, overall understanding of the constituent parts of the bigger assembly, and/or the data is not gathered as the failure events.

Another option is to create the labeled dataset manually by finding the idle points of the machine or equipment. It is a time-consuming and hard-working task, as the person must spend hours and hours making it labeled. The sensors may provide with the data thousands of times in a second. The system may record a few of the readings, out of the thousands. Although manual labeling is a time-consuming task, it is useful to understand the nature of the dataset.

The nature of the dataset can be understood by applying the initial queries on it, to find if anything is wrong inside the dataset. Such as finding null values, missing values, outliers and away points. Boxplot, five-point summary, and Z-Score are few of them to find the anomaly from the data series and datasets. The "failure" and "working" are mainly two classes for any predictive maintenance problem. However, the dataset labeled with these two labels may not have equal proportion. As the failure state of the machine is rare case, whereas the working state of the machine has more record sets. The labeling of the records may be text or numerical values. Such as zero for failure and one for working. The normal is the value in the range of the accuracy of the sensor deployed on the machine component. The anomaly may come if the sensor is malfunctioning, or the components are not getting the correct power supply, voltage, pressure or the temperature is not maintained due to certain weird conditions happening around the assembly line.

Creating the labeled dataset is one of the key aspects, to try and test the supervised learning algorithms. The reason behind creation of the labeled dataset is to detect the previous failure cases of the machine and the constituent unit. Another reason is the accuracy calculation can be easier. Here the data is divided into three states of the machine. Such as Running, Idle and Change Point. The change point is the point where either before that point there is something happening due to which machine went to the idle or stopped state.

6. RESULTS

The sensor readings from the seventeen different sensors are available. The data is cleaned, where a few entries were incomplete at the end of the dataset. There are 10 sensors having their temperature values of different parts of the machine. There are 3 current values of the sensors, and 3 voltage values. One more column indicates the date and time values which indicate the sensor recording. There are a total of 56982 rows available after cleaning. The null values were checked in the dataset. The additional numerical values column called "state" is added, which is the label column. The state values are Zero for idle or stopped machine, 1 for running machine and 2 for the changed state. The changed state can be either from the idle to the running or from the running to the idle state.

6.1 THE K-NEAREST NEIGHBORHOOD

The K-NN lazy learner is applied using the Euclidean distance method. Before applying the method, the date and time column is removed.

$$d(p,q)=\sqrt{\sum_{i=1}^{n}(q_i-p_i)^2} \tag{1}$$

p, q are two points in Euclidean n-space.

p_i, q_i are Euclidean Vectors, starting from the origin of the space.

Before applying the method, the date and time column is removed. The data is divided into 80% as the training dataset and 20% as the testing dataset. The value of K is decided as 3, and stratified sampling method is considered. The performance score is calculated as 96.4%.

In the next step we separated the data set of outliers detected using the IQR. Then we separated the data of "Changed State" which is outlier. Logically & statistically, these are the more anomaly data points which are the points responsible for the failure. In the future, also such a points need to be chosen carefully. However, in case of failure prediction, the line engineer needs the suitable time window beforehand. Otherwise, it is none of us to simply predict the data points of the basis of the past records. The live data must be aligned with the past data available.

```
df = pd.read_csv('Datewise2.csv')

X = df.drop(columns=['State'])

y = df['State'].values

from sklearn.model_selection import train_test_split

X_train, X_test, y_train, y_test = train_test_split(X, y,
test_size=0.2, random_state=1, stratify=y)

from sklearn.neighbors import KNeighborsClassifier

knn = KNeighborsClassifier(n_neighbors = 3)

knn.fit(X_train,y_train)

knn.predict(X_test)[0:5]
```

```
knn.score(X_test, y_test)

df=knn.predict(X_test)

df.to_csv('knnpredicted.csv')

df1 = pd.DataFrame(df)

df1.to_csv('knnpredicted.csv')

df2=pd.DataFrame(y_train)

df2.to_csv('knnoriginaloutput.csv')
```

The data is divided into 80% as the training dataset and 20% as the testing dataset. The value of K varies from 1 to 5, and the stratified sampling method is considered. The various readings are taken for different dataset divide and varying values of K. Similarly, Table 1 shows that there is a division of the data as 60 for training and 40 for testing, and the accuracy and precision is calculated. The division of the data is done with the labels assigned to each state of the machine as 70-30 and 80-20. There is 96% accuracy achieved in the K-NN model. Table 1 has the value of K as three, which means that only the nearest absolute distance of the 3 is taken.

Table 1. K-NN Evaluation with 3 Nearest Neighborhood & Dataset Division

	60-40	**70-30**	**80-20**
Accuracy	0.9643	0.9648	0.9657
Precision	0.9741	0.9735	0.9746
Recall	0.9835	0.9847	0.9846
F1 Score	0.9788	0.9791	0.9796
ROC-AUC-Score	0.9522	0.9506	0.9509

Table 2 shows that the data division for the training and testing is constant as 70-30, however the value of K is changed as from 1, 2, 4, and 5. The value of the K is considered as the 5, then the accuracy is achieved better. Figure 8 shows that the division of the dataset as 60-40 and the accuracy is achieved.

Table 2. K-NN 70-30 Split Evaluation with changing 'K'

	K=1	K=2	K=4	K=5
Accuracy	0.9616	0.9508	0.9631	0.9642
Precision	0.9754	0.9814	0.9778	0.9730
Recall	0.9788	0.9594	0.9781	0.9845
F1 Score	0.9771	0.9703	0.9779	0.9787
ROC-AUC-Score	0.9261	0.9460	0.9549	0.9567

Figure 10. K-NN Confusion Matrix with 60-40 Data Division

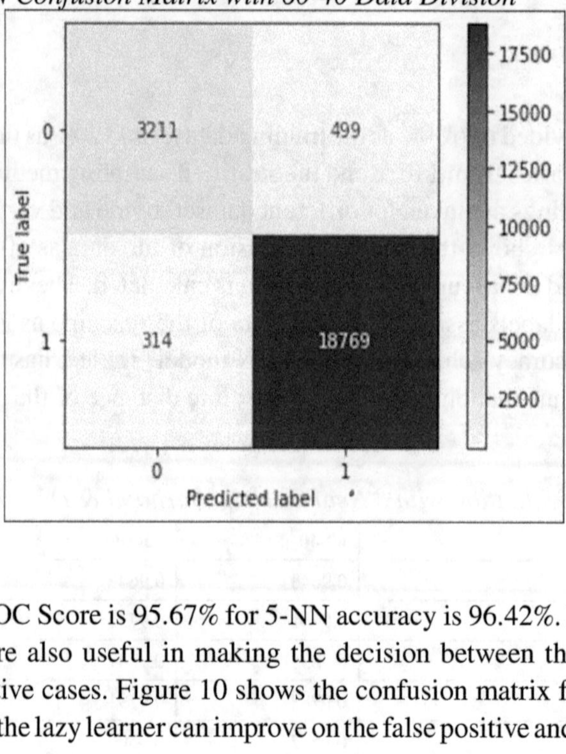

The AUC-ROC Score is 95.67% for 5-NN accuracy is 96.42%. Precision, recall and F1 Score are also useful in making the decision between the False Positive and False Negative cases. Figure 10 shows the confusion matrix for the nearest K neighbor where the lazy learner can improve on the false positive and false negatives.

6.2 DECISION TREE CLASSIFIER

Decision Tree (DT) is a non-parametric model that creates the IF-THEN-ELSE condition rules and visual representation of it. The Machine learns the simple rules generated by the Decision Tree. The classification tree takes the target variables as the discrete set of values. The leaves of the tree are class labels 0 for Stopped Machine and 1 for the Running Machine. The DT creates a subset of trees with variable length depending on the dataset. For the efficient and smallest DT generation the depth of the tree is controlled by the pruning process.

Entropy as a measure of uncertainty represented as

$$H(p) = H(p1,..,pn) := -\sum_{i=1}^{n} p_i \log 2 p_i \tag{2}$$

Gini index or G-measures impurity that is the degree of probability of a particular variable being wrongly classified when it is randomly chosen.

$$Gini = 1 - \sum_{i=1}^{n} (X_i)^2 \tag{3}$$

Information Gain is difference between the Entropy of the parent and sum of entropy of the children of a Tree.

$$IG = H(S) - \sum_{i=1}^{n} P(x) * H(x) \tag{4}$$

First, the Decision Tree is plotted with continuous values. The known threshold for the low set value for Temperature sensors is 18, and the high set value is 35. Hence, the importance of these values in the State of the Machine is higher. Another important criterion is decided for voltage. The voltage is 240V, and the lowest voltage allowed is 216V, and the highest voltage acceptable is 264V.

The data preparation for the categorical conversion as per the above-given criteria can be applied to the given dataset. For the Temperature sensor values, the criteria applied as

=IF(B2<18,"LOW",IF(B2>35,"HIGH",IF(AND(B2>18,B2<35),"NORMAL")))

Here, B2 is the cell containing the Temperature value.

For the voltage sensor values the criteria applied as

=IF(X2<216,"L",IF(X2>264,"H",IF(AND(X2>216,X2<264),"N")))

Here, X2 the cell containing the Voltage value.

Figure 11. Confusion Matrix for Decision Tree Classifier Using Gini Index Criteria and 70-30 Data Divide

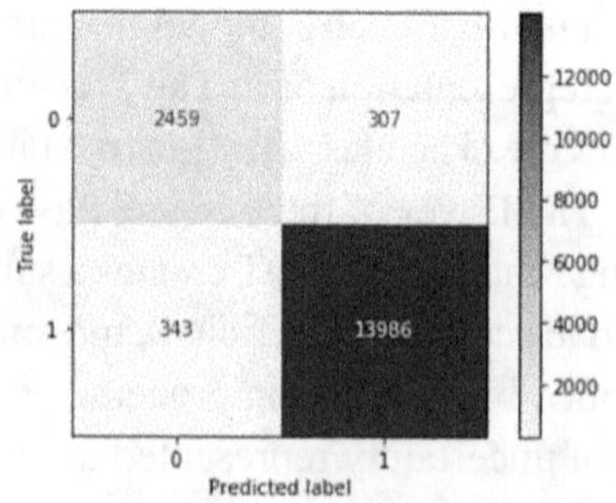

Figure 11 shows the confusion matrix for the decision tree classifier. For the data divide of 80-20, the accuracy is achieved 96.51%. As shown in table 3, the precision is 97.96% and Recall is 97.89%. The F1 score is 97.92%. The ROC-AUC score occupied 93.58% of the curve area.

Table 3. Decision Tree Evaluation with Gini Index Criteria & Data Divide

	60-40	70-30	80-20
Accuracy	0.9610	0.9619	0.9651
Precision	0.9787	0.9785	0.9796
Recall	0.9745	0.9760	0.9789
F1 Score	0.9766	0.9772	0.9792
ROC-AUC-Score	0.9329	0.9325	0.9358

Figure 12. AUC-ROC Curve for Decision Tree Classifier with Gini Index Criteria

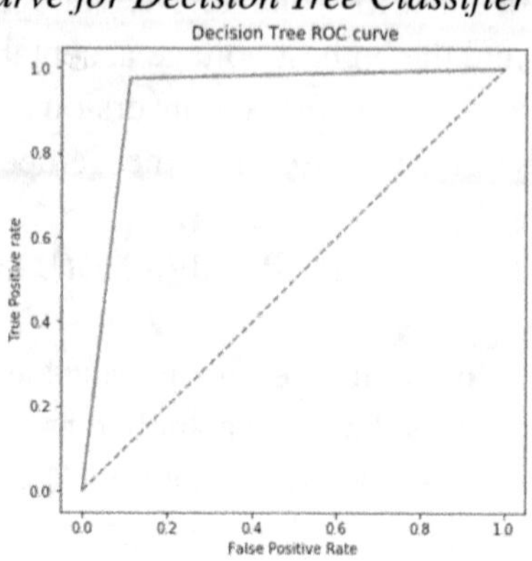

Figure 12 shows that the AUC-ROC curve for the decision tree classifier with Gini Index 93.58% of the area covered. It is expected that the model will perform stably during production. Table 4 shows that the data divided of 80-20 gives the better performance for the decision tree model with use of entropy. It is also clear that the difference in the division of the data is not having much difference in the performance of the decision tree model with entropy formula.

Table 4. Decision Tree (Entropy) Model Evaluation & Data Divide

	60-40	70-30	80-20
Accuracy	0.9616	0.9615	0.9656
Precision	0.9789	0.9789	0.9808
Recall	0.9752	0.9750	0.9782
F1 Score	0.9770	0.9770	0.9795
ROC-AUC-Score	0.9335	0.9333	0.9388

The graphical output is generated with the dimension 32767 by 2752 pixels. The diagram of the tree is having large number of nodes. The depth of the tree for the entropy criteria is 30. The Decision Tree applied with Gini Index, and Entropy produced 96.18% and 96.25% accuracy, respectively. Figure 13 shows the small portion of the bigger decision tree.

Figure 13. A Small Section of Decision Tree for the failure prediction.

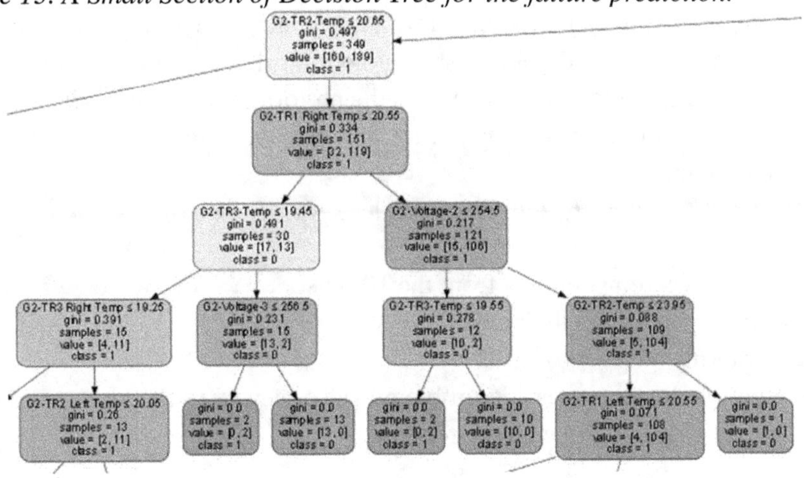

Figure 14. Prediction of failure and actual failure.

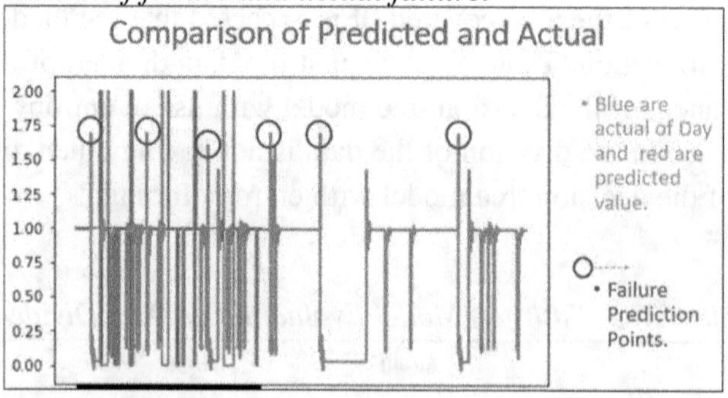

Figure 14 shows the state of the machine with actual failure and the predicted failure states over time. The states are annotated as 0, 1 and 2. The prediction of the failure points prior to the actual failure by the model is success to take preventive measures.

6.3 USE OF MULTI-LINEAR REGRESSION

After looking into the correlation matrix, it has been observed that there is a positive correlation among the temperature parameters. Hence the ten temperature sensor values are considered as the independent variables to calculate the state of the machine. For this, the problem is converted into the binary classification problem. Hence, all the changing states "2" are converted into the running states "1" and then the linear regression model is evaluated with the training and testing dataset division as 70% as to 30% division. Table 5 shows the multi-linear regression model with coefficient values, interceptive values and the performance of the model evaluated as the R^2 . These values can be useful for deployment of this model as a logic in the programming.

$$Failure\,prediction = \beta_0 + \beta_1 T_1 + \beta_2 T_2 + \ldots + \beta_{10} T_{10} + \varepsilon \quad (5)$$

Here,

T1,T2,…,T10: Temperatures from the 10 sensors

βi: Weights learned by the model

є: Error term

The multiple linear regression problem is also viewed as the target variable as dependent variable and three class problem.

Table 5. Multi-Linear regression with 10 Coefficients and one intercept.

Linear Regression	Coefficient Values	Intercept value	R^2 Score
When all the changing states "2" are converted into the running states "1"	[0.0682, -0.1902, 0.0447, -0.0082, 0.0304, 0.0245, 0.2172, 0.0700, -0.1145, -0.1582]	0.8942	0.4986
When all the changing states "2" are converted into the running states "0"	[0.0642, -0.1929, 0.0457, -0.01, 0.0285, 0.0255, 0.2361, 0.0783, -0.1180, -0.1738]	0.8871	0.5055
When considered as the target label or dependent variable as three class problem	[0.0652, -0.1880, 0.0452, -0.0107, 0.0274, 0.0213, 0.2124, 0.0758, -0.1180, -0.1466]	0.8973	0.4536

```
from sklearn import linear_model

df=pd.read_csv('Datewise-Heatmap-Positive-Corr.csv')

reg=linear_model.LinearRegression()

X = df.drop(columns=['State'])

y = df['State'].values

from sklearn.model_selection import train_test_split

X_train, X_test, y_train, y_test = train_test_split(X, y,
test_size=0.3, random_state=1, stratify=y)

reg.fit(X_train,y_train)

reg.coef_
```

The prediction of the state of the machine as a failure or not can be calculated based on the ten coefficients values and intercept values generated as the learning parameters from the linear regression. To deploy regression model based on the current learnt parameters, is possible to include in the suitable software code. These parameters can also change based on the future data generated during the various shifts of equipment. In that scenario this is evident to see the real-time indication of the prediction on the failure of the machine. The failure prediction in early stage can avoid major damage to the expensive equipment. The R^2 values show that there is moderate fit of the dataset in the regression.

7. CONCLUSION

Prediction of the State of the Butt Weld Wheel Rim Welding Machines is performed using sensor data and statistical calculations and machine learning models. The prediction of the machine's failure on the timeline gives the idea to prepare in advance. The learning parameters of the ML Models from K-NN, Decision Tree, and Multi-linear regression applied on the IoT based Sensor dataset. The time-to-failure and probability associated is calculated. The AUC and ROC curve assures the state prediction of the welding machine 93% for Entropy & Gini Indexed based Decision Tree, and 97.67% for 5-Nearest Neighborhood. The muti-linear regression model showed training and interpretations in the testing phase with moderate fit and showing some variability. Further improvements in multi-linear regression are possible by checking the other sensor readings.

All the models used in this case are highly data dependent, if the learnt parameters are used to deploy the model. However, the adaptive model of KNN, Decision Tree, and multi-regression will lead to more accurate predictions. Applying these models on the dataset generated over time can produce better results, if the window slides over time. The image and vision processing techniques will be the future scope of the work to find the role of harsh operating habits of the equipment operator. In this direction, the vision computing techniques will be applied be monitored to find the contribution of the human factor in failure.

REFERENCES

Al Mamlook, R. E., Obeidat, M. A., Elgeberi, N., & Nasayreh, A. (2024). Developing predictive maintenance for early warnings of machine failures using machine learning techniques in the fourth industrial era. In *Proceedings of the 2024 4th Interdisciplinary Conference on Electrics and Computer (INTCEC)*. DOI: 10.1109/INTCEC61833.2024.10602994

Ali, M. I., Lai, N. S., & Abdulla, R. (2021). Predictive Maintenance of Rotational Machinery Using Deep Learning. *International Journal of Electrical and Computer Engineering*, *14*(1), 1112–1121. . DOI: 10.11591/ijece.v14i1.pp1112-1121

Autran, J. V., Kuhn, V., Diguet, J.-P., Dubois, M., & Buche, C. (2024). AI4I-PMDI: Predictive maintenance datasets with complex industrial settings' irregularities. *Procedia Computer Science*, *246*, 1201–1209. DOI: 10.1016/j.procs.2024.09.546

Bao, T., Chen, J., Li, W., Wang, X., Fei, J., Wu, L., Zhao, R., & Zheng, Y. (2022). MIAD: A Maintenance Inspection Dataset for Unsupervised Anomaly Detection. arXiv. https://arxiv.org/abs/2211.13968

Baradaran, A. H. (2025). Predictive Maintenance of Electric Motors Using Supervised Learning Models: A Comparative Analysis. *arXiv preprint arXiv:2504.03670*. https://arxiv.org/abs/2504.03670

Belim, M., Meireles, T., Gonçalves, G., & Pinto, R. (2024). Forecasting models analysis for predictive maintenance. *Frontiers in Manufacturing Technology*, *4*, 1475078. DOI: 10.3389/fmtec.2024.1475078

Carrasco, J., Markova, I., López, D., Aguilera, I., García, D., García-Barzana, M., Arias-Rodil, M., Luengo, J., & Herrera, F. (2021). Anomaly Detection in Predictive Maintenance: A New Evaluation Framework for Temporal Unsupervised Anomaly Detection Algorithms. arXiv. https://arxiv.org/abs/2105.12818

Damon, W. H., Bruce, D. D., & Clifford, H. G. (2008). *Reliability centered maintenance study on voltage regulators*. In 2008 IEEE Power and Energy Society General Meeting - Conversion and Delivery of Electrical Energy in the 21st Century, pp. 1–5. IEEE. https://doi.org/DOI: 10.1109/PES.2008.4596576

Devare, M. (2023). Predictive maintenance in Industry 4.0. In Gaurav, K. A., & Vishwakarma, D. K. (Eds.), *Computational intelligent security in wireless communications* (pp. 79–98). CRC Press., DOI: 10.1201/9781003323426-5

Juodelyte, D., Cheplygina, V., Graversen, T., & Bonnet, P. (2022). Predicting Bearings' Degradation Stages for Predictive Maintenance in the Pharmaceutical Industry. *arXiv preprint arXiv:2203.03259*. https://arxiv.org/abs/2203.03259 DOI: 10.1145/3534678.3539057

Kamariotis, A., Tatsis, K., Chatzi, E., Goebel, K., & Straub, D. (2023). A metric for assessing and optimizing data-driven prognostic algorithms for predictive maintenance. *arXiv preprint arXiv:2306.03759*. https://arxiv.org/abs/2306.03759

Kuzin, T., & Borovicka, T. (2016). Early Failure Detection for Predictive Maintenance of Sensor Parts, ITAT 2016 Proceedings, *CEUR Workshop Proceedings* Vol. 1649, 123–130.

Latrach, A. (2023). Application of deep learning for predictive maintenance of oilfield equipment. *arXiv preprint arXiv:2306.11040*. https://arxiv.org/abs/2306.11040

Levin, S. (2024). Enhancing predictive maintenance in the industrial sector: A comparative analysis of machine learning models. *AIP Conference Proceedings*, *3243*(1), 020035. DOI: 10.1063/5.0247617

Liang, F., Mukherjee, S., & West, M. (2007). The Use of Unlabeled Data in Predictive Modeling. *Statistical Science*, *22*(2), 189–205. DOI: 10.1214/088342307000000032

Malawade, A. V., Costa, N. D., Muthirayan, D., Khargonekar, P. P., & Al Faruque, M. A. (2021). Neuroscience-Inspired Algorithms for the Predictive Maintenance of Manufacturing Systems. arXiv. https://arxiv.org/abs/2102.11450

Nascimento, E. G. S., Liang, J. S., Figueiredo, I. S., & Guarieiro, L. L. N. (2022). T4PdM: A Deep Neural Network Based on the Transformer Architecture for Fault Diagnosis of Rotating Machinery. *arXiv preprint arXiv:2204.03725*. https://arxiv.org/abs/2204.03725

Ohalete, N. C., Aderibigbe, A. O., Ani, E. C., Ohenhen, P. E., & Akinoso, A. (2023). Advancements in predictive maintenance in the oil and gas industry: A review of AI and data science applications. *World Journal of Advanced Research and Reviews*, *20*(3), 167–181. DOI: 10.30574/wjarr.2023.20.3.2432

Sengupta, P., Mehta, A., & Rana, P. S. (2023). Predictive maintenance of armoured vehicles using machine learning approaches. *arXiv preprint arXiv:2307.14453*. https://arxiv.org/abs/2307.14453

Shah, S. S., Tan, D., & Kumar, S. C. (2024). RUL forecasting for wind turbine predictive maintenance based on deep learning. *arXiv preprint arXiv:2412.17823*. https://arxiv.org/abs/2412.17823

Siddique, A., Kundu, R. K., Mode, G. R., & Hoque, K. A. (2023). RobustPdM: Designing robust predictive maintenance against adversarial attacks. *arXiv preprint arXiv:2301.10822*. https://arxiv.org/abs/2301.10822

Thakkar, U., & Chaoui, H. (2022). Remaining Useful Life Prediction of an Aircraft Turbofan Engine Using Deep Layer Recurrent Neural Networks. *Actuators*, *11*(3), 67. DOI: 10.3390/act11030067

Vassiliadis, P., Kotsiantis, S., & Kanellopoulos, D. (2022). A Context-Aware Unsupervised Predictive Maintenance Solution for Fleet Management. *Journal of Intelligent Information Systems*, *60*, 521–547. https://link.springer.com/article/10.1007/s10844-022-00744-2

Wei, M., Chen, M., & Zhou, D. (2013). Multi-sensor Monitoring Information Based Decision Support Method for Optimal Predictive Maintenance Policy, *25th Chinese Control and Decision Conference (CCDC)*, 4864-4868. DOI: 10.1109/CCDC.2013.6561815

Zheng, H., Paiva, A. R., & Gurciullo, C. S. (2020). Advancing from Predictive Maintenance to Intelligent Maintenance with AI and IIoT. *arXiv preprint arXiv:2009.00351*. https://arxiv.org/abs/2009.00351

Chapter 9
Solution for Error and Attenuation Detection in Network Services Based on Time Series Decomposition Model:
Pattern Mining Based on Time Series Decomposition via REFII Model and Data Science Methods Usage

Goran Klepac
https://orcid.org/0009-0000-6298-6790
Algebra Bernays University, Croatia

ABSTRACT

This chapter introduces an innovative methodology for error and attenuation detection in network services by unifying the REFII time series model with Bayesian networks (BNs), and complementary data science techniques traditionally applied outside temporal analytics. The proposed framework addresses the critical challenge of identifying rare, high-impact events—such as signal degradation in mobile or fixed networks—that compromise service quality but are often obscured by complex temporal dependencies and imbalanced data distributions. This transformation facilitates non-temporal expansion, a novel interpolation process that enriches time series with contextual events and operational parameters lacking inherent

DOI: 10.4018/979-8-3693-7994-3.ch009

temporal markers. Proposed solution unites advanced analytical techniques with decision making process for pattern mining like Bayesian networks, decision trees, FTP, K-Means clustering.

1. INTRODUCTION

Error detection in modern fixed and mobile network infrastructure faces a dual challenge: the inherent volatility of temporal signal characteristics and their often-overlooked correlations with non-temporal system parameters. Conventional error detection mechanisms—ranging from cyclic redundancy checks (CRCs) to parity-based protocols—exhibit critical vulnerabilities when confronted with complex temporal dependencies, such as burst errors masked by periodic network traffic patterns or intermittent signal degradation influenced by environmental factors.

In the realm of network technology, the repetitive occurrence of technical errors can serve as a critical catalyst for initiating a deep investigation into the underlying causes.

These recurrent issues not only disrupt the seamless operation of network services but can also indicate deeper systemic problems that necessitate thorough examination.

Repetitive network technical errors refer to the recurrence of similar faults or malfunctions within a network infrastructure. These errors can manifest in various forms, including connectivity issues, latency spikes, packet loss, and service outages.

Their persistent nature suggests that there may be fundamental issues within the network that require attention. When technical errors occur repeatedly, it is imperative to conduct a comprehensive investigation to identify the root causes. Deep investigations involve the systematic analysis of network logs, performance metrics, and user reports to uncover patterns and correlations.

This process often requires the collaboration of network engineers, data analysts, and cybersecurity experts to ensure a holistic understanding of the problem. Several factors can contribute to the repetitive occurrence of network technical errors.

One prominent factor is the increasing usage of network services. As the user portfolio grows and diversifies, the demand on network infrastructure intensifies, leading to potential overloads and performance bottlenecks.

Specific user behaviors, such as high-frequency data transmissions or resource-intensive applications, can exacerbate these issues.

Another significant factor is the complexity of modern network architectures. The integration of various technologies, including cloud computing, IoT devices, and virtualized environments, adds layers of complexity that can introduce new points of failure. Inadequate network design, configuration errors, and insufficient maintenance can further compound these challenges.

The specifics of the user portfolio, including the number of users, their geographical distribution, and usage patterns, play a crucial role in the manifestation of network errors. For instance, a network servicing a global user base may experience different error patterns compared to a network catering to a localized community. Understanding these specifics is essential for tailoring solutions that address the unique needs of the user population.

The repetitive occurrence of network technical errors serves as a vital indicator for initiating deep investigations into the root causes of these issues. By examining factors such as increasing service usage and user portfolio specifics, network administrators can develop targeted strategies to mitigate these errors and enhance overall network resilience. Through continuous monitoring, analysis, and optimization, it is possible to maintain robust network performance and ensure a reliable user experience.

Advanced analytical techniques for time series analysis can significantly contribute to investigating patterns in fixed and mobile networks by considering temporal characteristics alongside interpolated non-temporal components.

Time series analysis enables the examination of data points collected over time intervals, revealing trends, seasonality, and cyclic behaviors inherent to network performance and usage.

In fixed and mobile networks, temporal characteristics such as diurnal patterns, weekly cycles, and long-term trends play a crucial role in understanding network behavior.

For instance, mobile networks may exhibit peak usage during commute hours and reduced activity during late-night periods. Similarly, fixed networks might show varying traffic patterns based on work-from-home trends or streaming surges during weekends. Advanced time series models capture these temporal dynamics, enabling more accurate predictions and informed decision-making.

Interpolated non-temporal components, such as geographical location, user demographics, and network topology, further enrich the analysis. By integrating these contextual variables, analysts can uncover deeper insights into usage patterns and performance bottlenecks. For example, combining time series data with spatial information can reveal how network load varies across different regions, helping to pinpoint areas requiring infrastructure upgrades. Additionally, demographic data can elucidate user behavior variations, guiding targeted service improvements.

The fusion of temporal and non-temporal data through techniques like spatio-temporal modeling and multivariate time series analysis enhances the granularity and reliability of network insights. This comprehensive approach facilitates the identification of correlated events, such as concurrent spikes in mobile network traffic and specific applications or services. Moreover, it enables the assessment of network resilience by evaluating how external factors, like weather or major events, impact network performance over time.

In conclusion, advanced analytical techniques for time series analysis offer a powerful toolkit for exploring patterns in fixed and mobile networks. By accounting for temporal characteristics and interpolated non-temporal components, these methods provide a holistic view of network dynamics, driving more effective management, optimization, and strategic planning.

2. BACKGROUND

The topic of error recognition patterns in networks has garnered significant attention in numerous research studies. Scholars and industry experts have explored various methodologies to enhance the detection, analysis, and correction of errors within both fixed and mobile networks. These studies often focus on leveraging advanced statistical models, machine learning algorithms, and neural network architectures to identify recurring error patterns and develop robust solutions.

For instance, Sun et al. (2025) introduced a neural network architecture that accounts for autocorrelated errors in time series forecasting, thus improving prediction accuracy. Similarly, Ahmed & Li (2025) proposed an enhanced CRC-Polar code hybrid for error detection and correction in 5G networks, which effectively reduces retransmission needs. These examples highlight the innovative approaches being applied to address the complexities of network errors.

Gupta et al. (2023): Develops a self-healing framework for conversational AI using error attribution and active retraining to resolve misclassifications.

Waheeb et al. (2016): Demonstrates that error feedback in polynomial neural networks improves time series forecasting accuracy by 23%.

Ding et al. (2023): Uses residual neural networks to detect structured light system errors, achieving 93% accuracy in industrial 3D measurements.

Chen et al. (2025): Implements a W3C-standardized client-side error logging framework to capture network failures invisible to server-side monitoring and Kumar & Singh (2024): Enhances polar codes with dynamic error pattern analysis for low-latency 5G applications which can be inspiring for new algorithm development.

Lee et al. (2024): Combines LSTM and autoencoders to detect network anomalies in real time, reducing false positives by 18%.

Müller et al. (2023): Addresses autocorrelated sensor errors in IIoT time series using ARIMA-integrated neural networks.

Zhang et al. (2024): Proposes SDN-based cross-layer error correction, improving throughput by 32% in data center networks.

Johnson et al. (2023): Analyzes social media sentiment during outages using BERT, identifying critical pain points for telecom operators.

Wang et al. (2025): Implements federated learning to predict network errors across distributed edge nodes while preserving data privacy.

Nguyen et al. (2024): Applies graph neural networks to localize root causes of errors in cellular tower networks.

Kim & Park (2025): Uses transformer attention layers to prioritize critical errors in multivariate network monitoring systems.

García et al. (2024): Quantifies revenue loss from cloud network errors, showing a 1.8% churn increase per hour of downtime. Sun et al. (2025) introduces a neural network architecture that accounts for autocorrelated errors in time series forecasting, improving prediction accuracy.

Ahmed & Li (2025) proposes an enhanced CRC-Polar code hybrid for error detection/correction in 5G networks, reducing retransmission needs and Gupta et al. (2023): Develops a self-healing framework for conversational AI using error attribution and active retraining to resolve misclassifications.

If we are talking about optimizations, Waheeb et al. (2016) Demonstrates that error feedback in polynomial neural networks improves time series forecasting accuracy by 23%, ding et al. (2023): Uses residual neural networks to detect structured light system errors, achieving 93% accuracy in industrial 3D measurements.

Errors have significant contributions to customer experience, and Horowitz (2022) links rapid error resolution to customer trust in digital banking, emphasizing real-time session replay for diagnostics, where Rossi et al. (2023) predicts customer churn during outages using survival analysis, with 89% precision in telecom datasets. Becker et al. (2022): Validates a cross-channel customer experience (CX) metric, enabling comparisons of CX across touchpoints and journey stages. Rossi et al. (2023): Predicts customer churn during outages using survival analysis, with 89% precision in telecom datasets.

Zhang et al. (2024): Proposes SDN-based cross-layer error correction, improving throughput by 32% in data center networks. Wang et al. (2025): Implements federated learning to predict network errors across distributed edge nodes while preserving data privacy, and Nguyen et al. (2024): Applies graph neural networks to localize root causes of errors in cellular tower networks.

As successful application of alternative approaches in time series modelling there are few successful implementations of REFII model.

Klepac (2002) Introduced the foundational REFII model as an open-system mathematical framework for pattern recognition in time series. The work established core principles for transforming temporal data into analytical structures using shape and area-under-curve metrics, enabling applications in seasonality detection, rule discovery, and clustering.

Additionally, Klepac (2007) showed expanded the REFII model into a comprehensive transformation framework for automated time series analysis. This work formalized methods for linking diverse data mining techniques (e.g., clustering, classification) through unified temporal representations, with demonstrated applications across finance, medicine, and signal processing.

Additionally Klepac (2023) showed advanced solution within REFII model by integrating non-temporal expansion, enabling error detection in technical systems through associative algorithms. The methodology mapped non-temporal events (e.g., weather conditions, operational states) onto time series data using interpolation, enhancing analytical accuracy in complex industrial diagnostics.

3. HOW TECHNICAL ASPECTS CHALLENGES CUSTOMER EXPERIENCE

In today's digital age, the reliability of both mobile and fixed networks is paramount to ensuring customer satisfaction.

Errors in these networks can significantly disrupt service quality, leading to user dissatisfaction, particularly during periods of high demand. Network errors encompass a range of issues, from minor glitches to major outages, affecting the performance of mobile and fixed network services.

These errors can stem from various sources, including hardware failures, software bugs, and external factors such as weather conditions or cyberattacks.

When these errors occur, they can lead to slow internet speeds, dropped calls, or complete service interruptions, all of which detract from the user experience.

The impact of network errors is particularly pronounced during periods of intensive need. These periods include events such as live streaming of popular sports matches, significant online shopping days like Black Friday, or even emergency situations where communication is critical.

During these times, the demand for network services surges, and any disruption can have severe consequences for customer satisfaction. Modern network infrastructures are highly complex, comprising numerous interconnected components and systems.

This complexity can make it challenging to pinpoint the exact cause of network errors. Additionally, the interdependence of different network elements means that a failure in one part of the system can cascade and lead to widespread issues.

Given the potential for network errors to undermine customer satisfaction, it is crucial to have robust mechanisms in place for their recognition and resolution. This involves implementing advanced monitoring tools that can detect anomalies and performance issues in real-time, leveraging big data analytics to identify patterns and predict potential network failures before they occur, developing automated response systems that can quickly address minor issues and prevent them from escalating, and ensuring that skilled technicians are available to handle complex problems that require human insight and expertise.

Proactive measures are essential to mitigating the impact of network errors on customer satisfaction. By anticipating potential issues and addressing them before they affect users, service providers can maintain a higher level of service reliability. This includes regular maintenance, updates to network infrastructure, and continuous improvement of monitoring and response systems.

In conclusion, network errors, whether in mobile or fixed networks, pose a significant threat to customer satisfaction, especially during periods of intensive need. Identifying these errors is essential, albeit difficult due to the intricacies of contemporary network infrastructures.

By employing a combination of real-time monitoring, data analytics, automated responses, and human intervention, service providers can effectively manage network errors and uphold the quality of their services. Proactive measures further ensure that networks remain reliable, minimizing disruptions and maintaining customer trust.

Pattern recognition in time series is a crucial aspect of identifying and addressing network errors, especially during periods of intensive service consumption. By analyzing temporal and non-temporal events, it is possible to discern regularities that are pivotal for effective problem-solving.

Time series data consists of observations collected sequentially over time. It is characterized by various temporal dependencies, such as trends, seasonality, and cyclic patterns. These inherent properties make time series analysis a powerful tool for detecting anomalies and predicting future occurrences. In the context of network performance, time series analysis can help identify periods when errors are more likely to occur due to increased demand.

During periods of high service consumption, network infrastructure may face heightened stress, leading to potential failures or degraded performance. By employing pattern recognition techniques, it is possible to monitor and analyze time series data to detect unusual patterns indicative of network errors. These techniques can be broadly classified into three categories: statistical methods, machine learning algorithms, and hybrid approaches.

Statistical methods, such as autoregressive integrated moving average (ARIMA) models, are commonly used for time series forecasting. These models rely on the assumption that future values are dependent on past values and can capture the underlying structure of the data. By fitting an ARIMA model to network performance metrics, it is possible to forecast future values and identify deviations from expected behavior, signaling potential errors.

Machine learning algorithms, such as recurrent neural networks (RNNs) and long short-term memory (LSTM) networks, have gained popularity for time series analysis due to their ability to learn complex temporal dependencies. These algorithms can process sequential data and uncover hidden patterns that may not be evident through traditional statistical methods. By training an RNN or LSTM model on historical network performance data, it is possible to predict future network behavior and detect anomalies indicative of network errors.

Hybrid approaches combine the strengths of statistical methods and machine learning algorithms to improve the accuracy and robustness of pattern recognition. For example, a hybrid model may use an ARIMA model to capture the linear dependencies in the data and an LSTM network to model the non-linearities. This combination can enhance the detection of network errors during periods of intensive service consumption.

In addition to temporal events, non-temporal events also play a significant role in network performance. Non-temporal events, such as hardware failures, software updates, or security breaches, can introduce abrupt changes in the network behavior. By incorporating non-temporal events into the analysis, it is possible to gain a comprehensive understanding of the factors influencing network performance and improve the accuracy of error detection.

To recognize both temporal and non-temporal events, a multi-dimensional approach is necessary. This involves integrating time series data with contextual information, such as system logs, network topology, and external events. By correlating patterns in the time series data with contextual information, it is possible to identify the root causes of network errors and develop effective mitigation strategies.

In conclusion, pattern recognition in time series is essential for identifying network errors during periods of intensive service consumption. By leveraging statistical methods, machine learning algorithms, and hybrid approaches, it is possible to detect both temporal and non-temporal events that exhibit regularities critical for problem-solving. Integrating time series data with contextual information provides a holistic view of the network performance, enabling proactive error detection and resolution.

4. SOLUTION PROPOSAL

4.1. REFII model

Target attribute as quality index, volume as key indicators over time is observed as time series $S(s_1, \ldots, s_n)$ first step is normalization .

The normalization procedure implies the transformation of a time series $S(s_1, \ldots, s_n)$ into a time series $T(t_1, \ldots, t_n)$ where each element of the array is subject to a min-max normalization procedure to the < 0, 1 > interval, with previous interpolation with 0 values form time dependent missing values.

Time series T is made up of elements $(t_1, \ldots, t_n)$, where t_i is calculated as $t_i = \frac{s_i - \min(S)}{\max(s) - \min(s)}$, where min($s$) and max($s$) are the minimum and maximum values of time series T.

Next step is transformation to REF notation according to the formula:

$T_r = t_{i+1} - t_i$, $T_r > 0 \rightarrow R$; $T_r < 0 \rightarrow F$; $T_r = 0 \rightarrow E$;, where the Y_i elements are members of the N_s series.

After that, slope calculation based on the angle should be performed for calculating angular deflection coefficients:

$$T_r > 0 \rightarrow R\ Coefficient = t_{i+1} - t_i$$

$$T_r < 0 \rightarrow F\ Coefficient = t_i - t_{i+1}$$

$$T_r = 0 \rightarrow E\ Coefficient = 0$$

Area below curve is next calculation which can be performed with usage of numerical integration by rectangle theory on following way (“a “is a time span in this particular case can be arbitrary defined):

$$p = \frac{(t_i * a) + (t_i * a)}{2}$$

Final stage is the creation of time indexes. A time index serves as a mathematical structure for arranging time-based information using evenly spaced, distinct sampling points on a continuous time axis.

Time indexes are structured as transformations of original time stamp values translated into string values where their position in string determines time span which will be used for analytical purposes.

Table 1. Shows example of time series decomposition based on presented methodology for variable volume for one base station.

Table 1. Transformation into REFII model

Time index	REF	Deflection category	Area category
2025.01.00.15	R	High rise	Low
2025.01.00.30	R	High rise	Low
2025.01.00.45	R	High rise	Medium
2025.02.00.00	F	High High	High
2025.02.00.15	F	High Fall	High
2025.02.00.30	E	Equal	Medium
…	…	…	…

The translation of time series data into the REFII model represents a significant advancement in the field of data analysis and pattern recognition.

Time series, which are sequences of data points typically measured at successive points in time, exhibit a wealth of information about the underlying processes generating the data.

However, raw time series data can be complex and challenging to analyze directly.

The REFII model provides a structured framework to transform these data into a more analyzable form. The process of translating time series into the REFII model involves several key steps. Initially, the raw time series data undergo pre-processing steps such as normalization, smoothing, and noise reduction to enhance data quality and consistency. Following pre-processing, the time series data are segmented into windows or intervals, which are then encoded into a set of numerical features or symbolic representations. These features encapsulate essential characteristics of the time series, such as trend, seasonality, and periodicity. The REFII model utilizes sophisticated algorithms to capture these characteristics and represent them in a compact, yet information-rich format.

This transformation allows for the identification and extraction of patterns, making the data more amenable to advanced analytical techniques. The REFII model serves as a robust platform for pattern recognition due to its ability to distill complex time series data into a structured and interpretable format. By representing the time series in a way that highlights key features and patterns, the REFII model facilitates the application of various pattern recognition algorithms. Machine learning algorithms, such as clustering, classification, and anomaly detection, can be effectively employed to identify recurring patterns, detect outliers, and classify different time series behaviors. Furthermore, the REFII model's standardized representation enables the use of transfer learning techniques, where patterns learned from one dataset can be applied to another, enhancing the model's generalizability and applicability across different domains.

One of the significant advantages of transforming time series data into the REFII model is that it prepares the data for the application of sophisticated analytical models, such as Bayesian networks. Bayesian networks are probabilistic graphical models that represent a set of variables and their conditional dependencies via a directed acyclic graph. These networks are particularly well-suited for handling uncertainty and incorporating prior knowledge into the analysis.

When time series data are translated into the REFII model, the resulting feature representations can be used as input nodes in a Bayesian network. This allows for the modeling of complex dependencies and interactions between different time series components.

Bayesian networks can infer the likelihood of various outcomes based on observed data and can be employed for tasks such as forecasting, causal inference, and scenario analysis. In conclusion, the translation of time series data into the REFII model provides a powerful platform for pattern recognition and advanced analytical modeling. By transforming raw time series into a structured format, the REFII model enhances the ability to detect patterns, apply machine learning algorithms, and utilize Bayesian networks for comprehensive analysis.

The concept of time series expansion within the REFII model presents a significant opportunity for interpolating non-temporal elements such as temperature, location, and weather conditions. By incorporating these diverse factors, the model enriches the dataset, thus allowing algorithms designed for pattern seeking to operate with a more holistic view of the information. This enhancement is particularly valuable for identifying reliable patterns that are critical to understanding the reasons behind errors and the specific conditions under which they occur.

Table 2. Transformation into REFII model and temporal expansion illustration

Time index	REF	Deflection category	Area category	Feature weather (Temporal expansion)
2025.01.00.15	R	High rise	Low	Cloudy
2025.01.00.30	R	High rise	Low	Cloudy
2025.01.00.45	R	High rise	Medium	Cloudy
2025.02.00.00	F	High High	High	Rainy
2025.02.00.15	F	High Fall	High	Rainy
2025.02.00.30	E	Equal	Medium	Cloudy
…	…	…	…	…

In traditional time series models, the primary focus is on temporal data points, often neglecting the potential insights that non-temporal elements might offer. However, the REFII model's innovative approach of integrating these elements provides a more comprehensive understanding of the dynamics at play. For instance, temperature variations can significantly influence the behavior of certain systems, while location and weather conditions may introduce environmental factors that are pivotal in error analysis.

By expanding the time series to include these non-temporal elements, pattern-seeking algorithms benefit from a richer dataset, which leads to more accurate and reliable pattern recognition. This multi-dimensional approach enables the identification of correlations and causations that might otherwise remain obscured in a purely temporal analysis. Consequently, this methodology facilitates a deeper understanding of why errors occur, helping to pinpoint the exact conditions that trigger such anomalies.

Moreover, the enriched dataset offers a robust foundation for predictive analytics. With a more diverse range of variables, predictive models can anticipate errors and adverse conditions with greater precision. This foresight is invaluable for preemptive measures, allowing for timely interventions that can mitigate or even prevent errors from occurring.

In summary, the time series expansion within the REFII model, by integrating non-temporal elements, significantly enhances the capability of pattern-seeking algorithms. This enrichment not only aids in uncovering reliable patterns and understanding the underlying causes of errors but also empowers predictive models to forecast potential issues more accurately. Thus, this approach represents a substantial advancement in the field of error analysis and predictive maintenance.

REFII is not a substitute for different analytical methods; rather, it serves as a crucial bridge that enables the application of these methods to time series data. Techniques such as Bayesian networks, K-mean clustering, and decision trees, among others, traditionally struggle to be directly applied to time series due to their inherent sequential and temporal complexities. REFII, however, transforms these time series into a format that is compatible with these methods, allowing analysts to utilize powerful tools that would otherwise be impractical without it. This bridging capability renders REFII indispensable for expanding the analytical toolkit available for time series, thereby enhancing the depth and breadth of insights that can be derived from such data.

4.2. Role of Bayesian Network

Bayesian networks utilize probabilistic dependencies between variables, enabling the detection of anomalies based on deviations from expected patterns.

In observed cases it can be used for finding patterns from REFII model.

Bayesian networks have proven to be a powerful tool for analyzing complex data structures, including transformed time series over REFII. By leveraging their probabilistic frameworks, these networks can effectively seek out temporal and non-temporal patterns, as well as patterns derived from time indexes. The transformed time series allows the Bayesian networks to capture the intricate dependencies and relationships inherent in the data, facilitating a comprehensive analysis that encompasses various dimensions of the dataset.

Moreover, the integration of patterns from time indexes enriches the network's ability to discern underlying trends and shifts over time. The final output of such an analysis is a set of patterns that cohesively unite temporal dynamics, non-temporal associations, and indexed time-based insights. This holistic approach enables a deeper understanding of the phenomena under study, providing a robust foundation for predictive modeling and decision-making.

Bayesian network model for finding patterns should be constructed in way that target variable contains probabilities for e.g. intensity of error occurrence. Parental nodes contain elements of REFII or non-temporal elements.

Formally, we can express it as:

Pr[Pj] = Pr[node parents]

Or more precisely as:

$$Pr\left[P_j\right] = \prod_{i=1}^{n} Pr\left[P_j \middle| Y_i, \dots, Y_n\right]$$

Where:

Pr[Pj] – represents risk probability for j event, error intensity .

Yi – represents feature characteristics within Bayesian network expressed as a state value of Bayesian network node.

By transforming these enumerated variables into the REFII model, which are in discrete form, they become an input table for learning Bayesian Networks. This transformation allows for the incorporation of time series data into a probabilistic framework, enhancing the ability to predict network performance and reliability. The discrete nature of the variables ensures that each state is clearly defined, facilitating the learning process and enabling more accurate analysis. Using Bayesian Networks, analysts can uncover relationships between different network parameters, identify potential issues, and make informed decisions to optimize communication systems.

4.3. Integration results from REFII with other methods

The REFII model provides a robust framework for examining the occurrence and interplay of events over time.

By integrating decision trees, frequent pattern trees, and k-mean clustering with the REFII model, we can extract significant insights into both temporal and non-temporal patterns.

Decision trees are a fundamental tool in machine learning and data analysis. They are particularly useful for classification and regression tasks. In the context of the REFII model, decision trees can be utilized to identify the key features that contribute to the occurrence of bad events.

Decision trees can be trained on a dataset containing both temporal and non-temporal features. Temporal features might include timestamps or intervals between events, while non-temporal features could encompass factors such as environmental conditions, user behavior, or system states.

By examining the decision paths within the tree, we can identify which features are most influential in predicting bad events and how these features interact with one another.

Frequent pattern trees (FP-Trees) are a data structure used for mining frequent patterns, particularly in large datasets.

They allow for the efficient discovery of recurring patterns within the data.

When applied to the REFII model, FP-Trees can help uncover temporal patterns that indicate frequent occurrences of bad events at specific times.

By constructing an FP-Tree from the event data, we can identify common sequences and time indices where bad events are prevalent.

This information can be invaluable for proactive measures and mitigation strategies.

K-mean clustering is a popular unsupervised learning algorithm used to partition data into distinct clusters based on similarity. This method is particularly useful for identifying natural groupings within the data.

Using k-mean clustering in conjunction with the REFII model allows for the segmentation of both temporal and non-temporal features into meaningful clusters.

By clustering events based on their features, we can identify groups of events that share similar characteristics. For temporal data, this might reveal periods of heightened risk, while for non-temporal data, it can highlight common factors contributing to bad events. By leveraging the time index within the REFII model, we can pinpoint the frequency of bad events at specific times.

This temporal analysis enables the identification of high-risk periods, allowing for targeted interventions. Suppose we have a dataset containing timestamps and various features associated with system failures.

By applying the REFII model, we can calculate the frequency of failures over time. Integrating this with FP-Trees, we can identify specific time intervals where failures are most frequent.

Decision trees can then be used to determine the significant non-temporal features contributing to these failures during high-risk periods. Non-temporal pattern analysis focuses on identifying the features that contribute to bad events, regardless of when they occur.

By combining the REFII model with decision trees and k-mean clustering, we can uncover these influential factors.

Consider the same dataset of system failures. By training a decision tree, we can identify the key features that predict failures across all time periods.

K-mean clustering can further segment the data into clusters based on these features, providing a clearer understanding of the common characteristics associated with failures. Integrating decision trees, frequent pattern trees, and k-mean clustering with the REFII model offers a powerful approach to analyzing temporal and non-temporal patterns.

By leveraging these methods, we can gain valuable insights into the frequency and contributing factors of bad events. This comprehensive analysis enables more effective prediction, prevention, and management strategies, ultimately enhancing system reliability and performance.

5. MODEL APPLYING

5.1. REFII transformation

In the transformation process using the REFII model, both temporal and non-temporal attributes play crucial roles. The Signal-to-Noise Ratio (SNR) and Bit Error Rate (BER) conditions are essential metrics for evaluating the quality of the signal transmission and identifying potential issues that might affect data integrity was firstly transformed inti REFII model. Traffic volume and error correction frequency are transformed into REFII model. Congestions, which may lead to delays and packet losses, are also transformed into REFII model.

Non-temporal attributes such as weather conditions and temperature are included to account for external factors that might influence signal propagation and equipment performance. Portfolio feature characteristics provide additional context, encompassing various attributes relevant. Time indexes are crucial for aligning data points with their corresponding temporal context, enabling accurate analysis and transformation.

These non-temporal features are utilized for the non-temporal expansion of time series data, forming the foundation for the pattern recognition process. By integrating both temporal and non-temporal attributes, the REFII model enhances the accuracy and reliability of the transformation, leading to more precise and meaningful insights.

For analytical purposes, various radio communication variables are crucial to understanding the performance and reliability of base stations. These variables can be transformed into time series data to monitor changes and trends over time, providing valuable insights into network behavior and optimization.

- Absolute Radio Frequency Channel Number (ARFCN): Specifies the frequency channel used by the base station for communication. In GSM networks, ARFCN typically ranges from 0 to 1023. Time series data for ARFCN can illustrate how channel allocation varies and impacts network performance over time.
- Received Signal Level (RxLevel): Measures the signal strength received by the base station from a mobile device, commonly scaled from 0 to 63. Transforming RxLevel into time series data allows analysis of signal strength fluctuations and their correlation with network load and environmental changes.
- Signal Quality: Assesses the quality of the radio link, often based on bit error rates or interference levels. By creating time series data, analysts can track signal quality variations and identify periods of degraded performance or high interference.
- Traffic Load: Represents the current number of active users or sessions being handled by the base station. Time series data helps in understanding traffic patterns, peak usage times, and the effects of load on network stability and user experience.
- Power Level: Indicates the transmission power used by the base station's radio equipment, impacting coverage and interference. Analyzing power level as time series data reveals trends in power usage and its relationship with coverage area and interference management.
- Sleep Mode Parameters: Variables controlling the sleep/wake cycles of mobile stations, such as wake-up intervals and activity levels. Time series data for sleep mode parameters can optimize power consumption and improve response times by identifying efficient sleep schedules.
- Paging Parameters: Includes variables related to the process of alerting mobile devices to incoming calls or messages, such as paging group numbers and timers. Time series analysis helps in evaluating the effectiveness and efficiency of paging mechanisms.

- Backhaul Connection: Refers to the type and capacity of the link connecting the base station to the core network, such as fiber or microwave. Monitoring backhaul connection as time series data reveals its impact on network latency and overall performance.
- Power Supply Status: Indicates the availability and type of power sources, including backup systems for reliability. Time series data can track power supply disruptions and their effect on network uptime and reliability.
- Controller and Processor Status: Represents the state and configuration of the control unit and baseband processor, which manage signal processing, resource allocation, and network coordination. Analyzing these variables over time helps in diagnosing hardware issues and optimizing resource management.
- Call Drop Rate: The percentage of calls that are unexpectedly terminated, indicating network reliability. Time series data for call drop rate allows identification of periods of poor performance and potential causes of call drops.
- Handover Success Rate: The rate at which handovers are successfully completed without dropping the connection. Time series analysis can highlight trends in handover efficiency and areas needing improvement.
- Paging Success Rate: Measures the effectiveness of the base station in alerting devices to incoming communications. Time series data helps in evaluating the reliability of the paging process.
- SDCCH Drop Rate: Measures the drop rate on the Standalone Dedicated Control Channel, impacting call setup reliability. By tracking this variable over time, analysts can diagnose issues in call setup and improve reliability.

By transforming these variables into time series, analysts can gain a comprehensive view of the network's performance, identify trends, and make data-driven decisions to enhance the reliability and efficiency of communication systems.

In the analysis of REFII elements, it has been determined that the common denominator across all variables is the use of time indexes. These time indexes serve as a unifying framework, linking together diverse features within a consistent temporal dimension. Specifically, the temporal resolution is set at 15-minute intervals, enabling a granular and synchronized examination of data points. This approach ensures that all features are consistently aligned and can be accurately compared and analyzed within these defined 15-minute snapshots, thereby facilitating a coherent and systematic understanding of the temporal dynamics at play.

Temporal indexes in REFII transformation can be leveraged as the primary dimension in a star schema, thus expanding the usability and analytical capacity of temporal data.

Temporal indexes play a pivotal role in the REFII model transformation process. They serve as a central point of reference, acting as unique keys that encapsulate all REFII elements of each decomposed time series as well as non-temporal features.

This approach not only optimizes data organization but also enhances the analytical capabilities of time series models. REFII transformation is a sophisticated data processing technique designed to handle complex time series data. By decomposing time series into multiple elements, it allows for more granular analysis and improved forecasting accuracy.

The temporal index is the link of this transformation, ensuring that each element is accurately referenced and integrated. The temporal index serves as a unique identifier for each data point in the time series, ensuring that all elements, whether they are part of the decomposed time series or non-temporal features, are correctly aligned.

This alignment is crucial for maintaining the integrity of the data and for performing accurate analyses. Each decomposed element of the time series, along with any associated non-temporal features, is linked to a specific temporal index. This linkage allows for a comprehensive representation of the data, capturing both temporal dynamics and static characteristics. The temporal index thus acts as a bridge, connecting different facets of the dataset in a coherent manner.

By using temporal indexes as unique keys, REFII transformation facilitates efficient data organization. This organized structure simplifies the process of data retrieval and manipulation, enabling analysts to perform complex queries with ease. Furthermore, it enhances the scalability of time series models, allowing them to accommodate large datasets without compromising on performance.

The integration of REFII elements via temporal indexes significantly enhances the analytical power of time series models.

By providing a unified framework for referencing all data points, it enables more sophisticated analyses and more accurate forecasts. They ensure that all elements of decomposed time series and non-temporal features are accurately referenced and integrated. This central role not only enhances data organization but also improves the analytical capabilities of time series models, making REFII transformation a powerful tool for handling complex datasets.

5.2. Bayesian network construction

The construction of the Bayesian network employed a comprehensive approach, incorporating various critical attributes to ensure robust sensitivity evaluation and thorough pattern investigation. Among these attributes, The Signal-to-Noise Ratio (SNR) and Bit Error Rate (BER), traffic volume and error correction frequency, weather conditions and temperature variations were also included, recognizing their significant influence on network stability and signal propagation. Time indexes were integrated to capture temporal variations and trends, essential for dynamic analysis and predictive modeling.

The final Bayesian network model, comprising over 40 interconnected components, serves as a sophisticated tool for sensitivity evaluation and pattern investigation. By leveraging these diverse attributes, the model offers a holistic view of the network's behavior, enabling the identification of key factors affecting performance and facilitating the development of optimized strategies for network management and improvement.

Bayesian networks by directed acyclic graphs (DAGs) represents variables and their conditional dependencies. By leveraging Bayesian networks, we got insights into how weather events influence network difficulties. For instance, adverse weather conditions like heavy rainfall, extreme temperatures can degrade network performance, along with insight on several non-temporal attributes.

Such solution predict the likelihood of network issues under specific weather conditions, enabling proactive measures to mitigate their impact. In a Bayesian network, weather variables such as temperature, precipitation, and wind speed are represented as nodes, while conditional probability tables (CPTs) capture the relationships between these weather variables and network performance indicators like response time and packet loss.

For example, historical data and expert knowledge might indicate that heavy rainfall and high wind speeds significantly increase the probability of packet loss. Additionally, Bayesian networks help identify risky areas within the network, especially during critical periods. By analyzing historical performance data and weather patterns, these networks can pinpoint regions more prone to difficulties under certain weather conditions which help to prioritize monitoring and maintenance efforts in those high-risk areas. Another benefit of Bayesian networks is their ability to account for critical time periods when network usage volumes are expected to surge, such as peak hours, seasonal events, or emergencies. By incorporating temporal variables and usage patterns into the model, Bayesian networks provide insights into how weather events might exacerbate network difficulties during these periods. Temporal nodes, representing different times of the day, days of the week, or specific seasons, allow us to examine the conditional dependencies between these temporal nodes, weather variables, and network performance indicators. For instance, the network might be more susceptible to disruptions during some specific time, when both temperature and usage volumes are high.

Building on these insights, the following expertise and ideas can further enhance the use of Bayesian networks in this context:

1. Integrate Bayesian networks into adaptive network management systems that dynamically adjust network configurations based on predicted weather impacts. For example, rerouting traffic, adjusting bandwidth allocations, or deploying redundant resources can help mitigate the adverse effects of weather events on network performance.
2. Utilize real-time monitoring of weather conditions and network performance to continuously update the Bayesian network model. This ensures accurate and relevant predictions, allowing for timely interventions and adjustments. Incorporating machine learning techniques can further enhance the model's adaptability and predictive power.
3. Foster collaboration between meteorologists, network engineers, and data scientists to develop more sophisticated and accurate Bayesian network models. Combining domain-specific knowledge with advanced analytical techniques can deepen our understanding of the complex interactions between weather events and network performance.
4. Perform scenario analysis using Bayesian networks to explore the potential impacts of different weather events and usage patterns on network performance. This helps develop robust contingency plans and optimize resource allocation to ensure network resilience under various conditions. In conclusion, Bayesian networks offer a robust framework for understanding the relationships between weather events and network difficulties.

By modeling these interactions, identifying risky areas, and accounting for critical time periods, network operators can enhance their preparedness and response strategies. The integration of real-time data, adaptive management, and cross-disciplinary collaboration further amplifies the benefits of this approach, ensuring reliable network performance in the face of unpredictable weather conditions.

5.3. Frequent pattern tree appalying

Frequent Pattern Trees (FP-trees) are a streamlined data structure utilized in data mining to capture frequent itemsets within a transactional database. Formed through the FP-growth algorithm, this method efficiently mines the complete set of frequent patterns without the need for candidate generation. The procedure starts with scanning the database to identify frequent items, which are then sorted in descending order of their frequency. These frequent items become the branches of the FP-tree, where each branch symbolizes a transaction. By sharing common prefixes, the tree effectively compresses the data by merging identical item paths.

An FP-tree is made up of a root node, labeled as "null," followed by subsequent nodes that represent items from transactions. Each node records the item it represents, its frequency count, and links to child nodes. Additionally, the tree includes a header table that holds pointers to the first occurrence of each item in the tree, facilitating efficient traversal and pattern mining.

The results from an FP-tree are generated by traversing the tree to uncover frequent itemsets. This involves recursively building conditional FP-trees for each item in the header table and generating all possible combinations of frequent items. These combinations are then united with their respective support counts to produce the final set of frequent patterns. The results are typically presented as a list of itemsets along with their corresponding frequencies or support values. This provides valuable insights into the relationships and associations within the data, making this method particularly useful in market basket analysis, where it helps identify common purchasing patterns among customers.

Combining FP-trees with the REFII model enhances the efficiency and effectiveness of time series analysis. FP-trees provide a compact representation of frequent itemsets, which can be directly utilized by the REFII model for indexing and enrichment purposes.

The integration process involves constructing an FP-tree from the time series data, identifying frequent item sets using the FP-tree structure, feeding these item sets into the REFII model for initial indexing, and recursively refining the index using additional patterns identified by the FP-tree.

This integration allows the REFII model to leverage the compactness and efficiency of the FP-tree, resulting in a more accurate and scalable time series analysis framework. The combined use of FP-trees and the REFII model has numerous applications across various domains, including financial forecasting, healthcare, environmental science, and manufacturing.

The key benefits of this approach include reduced computational complexity due to the compact representation of frequent patterns, the ability to handle large-scale time series data with high dimensionality, and enhanced prediction and anomaly detection capabilities through recursive enrichment. The integration of frequent pattern trees with the time series REFII model represents a significant advancement in the field of time series analysis. By leveraging the strengths of both techniques, this approach offers a powerful and efficient method for uncovering complex patterns and making accurate predictions in time-dependent data.

From the REFII elements, the FPT generated numerous valuable patterns that demonstrated the continuity of problems and a consistent decreasing trend in qualitative metrics over time. These patterns elucidated dependencies in long-term time series data, particularly highlighting suboptimal performance in specific areas during periods.

Furthermore, the FPT produced action point-oriented rules to address these issues effectively. These rules include monitoring key performance indicators (KPIs) regularly, implementing targeted interventions in identified areas of poor performance, conducting periodic reviews of qualitative metrics, and adjusting strategies in response to observed trends and dependencies in the data.

5.4. Decision tree applying

The application of a decision tree to the REFII model data provided significant insights into the patterns of error intensity occurrences.

By systematically analyzing the data, the decision tree revealed how various factors influence the frequency and magnitude of errors.

This method allowed for the identification of key variables that contribute to error generation, thus enabling a more nuanced understanding of error dynamics within the model. The insights gained from this analysis are crucial for optimizing the model's performance and mitigating potential risks associated with high error intensities.

The application of decision tree analysis and the REFII model has unveiled significant patterns that correlate time indexes with the demand for intensive traffic. These patterns predominantly highlight the oscillation in quality performance metrics. By examining the interrelationships within these variables, it becomes evident that fluctuations in traffic demand are closely linked to changes in service quality. Such insights are pivotal for further investigations into the primary hotspots that detrimentally affect performance and customer experience. Identifying these critical areas can help in devising strategic interventions aimed at enhancing overall service delivery and customer satisfaction.

5.5. K-means clustering applaying

K-means clustering is a widely used unsupervised machine learning algorithm designed to partition a dataset into K distinct, non-overlapping clusters. In this analysis, K-means clustering was applied to data transformed using the REFII method.

The primary objective was to identify clusters that exhibited mutual characteristics and investigate their implications in the context of performance and technological infrastructure.

The data underwent a transformation process using the REFII technique, which enhanced its features and made it more suitable for clustering analysis. Following this transformation, the K-means algorithm was employed to partition the data into clusters based on the similarity of their features.

The REFII transformation method was utilized to preprocess the data. This step involved normalizing and scaling the data to ensure that all features contributed equally to the clustering process. The transformed data presented a more structured format, making it amenable to effective clustering. K-means clustering was then performed on the REFII transformed data.

The algorithm iteratively assigned data points to one of the K clusters by minimizing the within-cluster variance. Each cluster contained data points that were more like each other than to those in other clusters.

The application of K-means clustering on the REFII transformed data resulted in the identification of several clusters, each with distinct characteristics. One notable cluster demonstrated a high concentration of snapshots depicting bad performance times. A deeper investigation into this cluster revealed shared attributes among the data points, particularly concerning the technological infrastructure responsible for these performance issues.

The cluster characterized by bad performance snapshots exhibited mutual characteristics that pointed to specific technological components. These components were consistently associated with performance degradation, suggesting a correlation between the technological infrastructure and the observed performance issues.

The clustering results provided a leading point for further investigation into the correlation between technological infrastructure and performance. By analyzing the mutual characteristics within this cluster, it became evident that certain technological elements were contributing to the performance problems.

This insight prompted a more focused examination of the technological infrastructure, guiding efforts to identify and address the root causes of performance degradation. The scientific application of K-means clustering on REFII transformed data successfully identified clusters with mutual characteristics. The discovery of a cluster with a high concentration of bad performance time snapshots highlighted a significant correlation between specific technological infrastructure and performance issues.

This analysis underscores the importance of clustering techniques in uncovering hidden patterns and relationships within complex datasets, ultimately aiding in the optimization and improvement of technological systems.

5.6. Lessons learned from applying different methods on REFII transformed data

The transformation of time series data using the REFII method plays a central role in the integration of various data science techniques. See figure 1.

Traditional models typically do not interface directly with time series data; however, the REFII transformation addresses this issue. By applying each model separately, specific problem types can be effectively solved. Additionally, results obtained from one model through the REFII transformation can serve as inputs for another model, enabling a synergistic effect in data analysis and problem-solving.

Figure 1 . REFII model as a central point for applying methods.

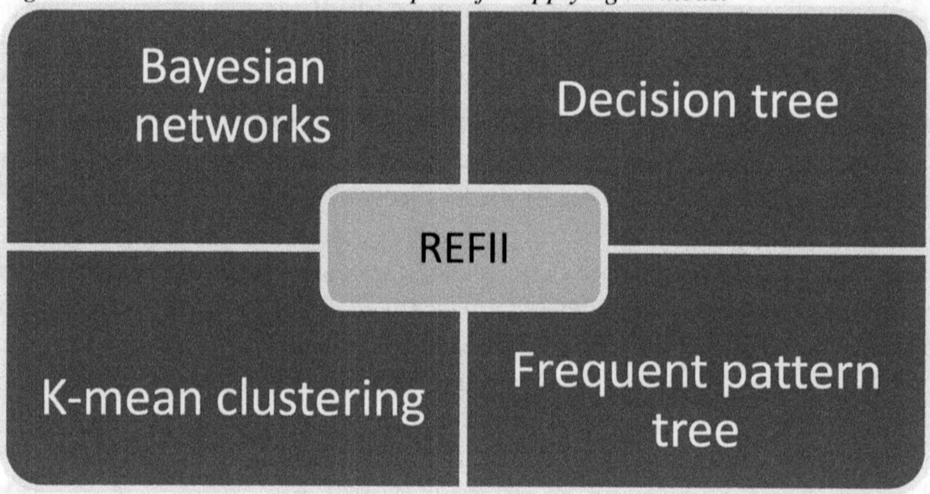

Results from Frequent Pattern Tree (FPT) mining can be effectively propagated to decision trees, providing a robust methodology for enhancing pattern mining. In this integrated approach, frequent patterns are identified and then transformed into new features.

These features can subsequently be incorporated into decision trees, augmenting the decision-making process with rich, pattern-based context. This fusion leverages the strengths of both techniques: the ability of FPT to uncover significant associations within the data, and the capacity of decision trees to utilize these associations for classification and prediction tasks.

By introducing frequent patterns as new features, the decision tree's ability to discern intricate patterns is enriched, leading to improved predictive performance and more insightful data analysis. This synergy fosters a more comprehensive and nuanced approach to pattern mining, employing both the depth of FPT and the structured decision-making framework of trees.

Cluster membership can be effectively utilized in decision trees and Bayesian networks, particularly when dealing with data that encompasses elements of time series in REFII format and includes time indexes. Decision trees leverage cluster membership to partition data into homogenous subsets, facilitating more accurate predictions by considering the temporal patterns and dependencies inherent in time series data.

Bayesian networks, on the other hand, utilize cluster membership to enhance probabilistic reasoning by incorporating the temporal structure and sequence patterns identified through frequent pattern tree (FPT) mining. This integration allows for a more nuanced understanding of the dependencies and conditional probabilities among variables over time.

By incorporating time indexes, both decision trees and Bayesian networks can account for the temporal dimension, thereby improving the robustness and reliability of predictive models in dynamic environments. The REFII format, which standardizes the representation of time series data, ensures consistency and interoperability, enabling seamless integration of cluster membership and temporal patterns in advanced analytical frameworks.

Time series analysis, a powerful tool in various scientific and engineering disciplines, focuses on analyzing sequences of data points collected over time. Its elements, when propagated to methods that traditionally lack access to temporal data, can contribute in novel and unexpected ways through synergy effects and the mining of patterns that encompass both temporal and associated non-temporal components.

Time series data is characterized by observations taken sequentially over a fixed interval, capturing the dynamic nature of the underlying processes. These temporal elements, such as trends, seasonality, and autocorrelation, provide a rich context that enhances the understanding of systems and phenomena. When these temporal components are integrated into methods that do not traditionally incorporate time series data, it creates a synergy effect—an emergent property where the combined effect of the two is greater than the sum of their individual effects. This synergy facilitates better model accuracy, improved forecasting, and deeper insights into complex systems.

Traditionally, machine learning models focus on static data, lacking the temporal dimension that time series data offers. However, by integrating time series elements, these models can harness the temporal dependencies and patterns inherent in the data as it is with usage of REFII model.

Moreover, the mining of patterns that contain both temporal and non-temporal components is a significant advancement facilitated by the integration of time series data. Temporal patterns, such as periodicity and seasonality (seasonality can be declared for shorter time periods like week), when combined with non-temporal attributes, such as categorical labels or numerical features, provide a comprehensive view of the data. This holistic approach enables the discovery of complex relationships and interactions that are not apparent when considering temporal or non-temporal components in isolation.

Furthermore, the integration of time series elements into data mining techniques, such as clustering and classification, enriches the analysis by capturing the temporal dynamics of the data. Temporal clustering, for example, groups time series data based on their temporal patterns, enabling the identification of similar temporal behaviors across different entities. This approach is particularly useful in applications such as anomaly detection, where the temporal context is critical for distinguishing normal from abnormal patterns like it is case in network error understanding.

In conclusion, the propagation of time series elements to methods that do not traditionally incorporate temporal data leads to significant advancements through synergy effects and the mining of patterns containing both temporal and non-temporal components. This integration enhances the analytical capabilities of various methods, providing deeper insights, improved accuracy, and a more comprehensive understanding of complex systems and phenomena. As the availability of time series data continues to grow, the opportunities for leveraging these elements in innovative ways will undoubtedly expand, driving further progress in numerous fields.

The REFII time series model can be effectively utilized for deep learning applications, particularly with Long Short-Term Memory (LSTM) networks. LSTM, a type of recurrent neural network (RNN), excels in learning from time series data due to its ability to retain long-term dependencies and manage sequential information. By transforming network performance variables into the REFII model, the discrete data can be fed into an LSTM network to predict future network states, identify patterns, and detect anomalies. This predictive capability is crucial for maintaining and optimizing communication systems.

Moreover, REFII serves as an essential bridge between traditional time series analysis and advanced deep learning techniques. By discretizing time series data, REFII enables LSTM networks to process and learn from the data more effectively, ensuring that each state is clearly defined and facilitating precise modeling. This integration of REFII with LSTM not only enhances predictive accuracy but also provides a robust framework for continuous improvement of network performance and reliability through deep learning insights.

5. FUTURE RESEARCH

The REFII model holds significant potential for advancing the fields of network fix and mobile error recognition. As networks and mobile systems become increasingly complex, the ability to effectively identify and rectify errors becomes paramount. Future research will undoubtedly focus on harnessing the strengths of the REFII model to enhance these processes. This document outlines key research directions that promise to propel the application of the REFII model to new heights.

The integration of Large Language Models (LLMs) with the REFII model presents an exciting avenue for future research. LLMs, such as GPT-4, have demonstrated remarkable prowess in understanding and generating human-like text.

By leveraging these capabilities, researchers can translate error identification results from the REFII model into comprehensible and actionable insights. This translation process involves several steps: interpreting error logs, system notifications, and user reports generated by the REFII model to provide context to the error, aiding in more precise diagnosis; generating fix recommendations based on the interpreted data, tailored to the specific language and understanding of the end-user, making them more accessible and actionable; and continuously learning and adapting over time from the feedback provided by users on the effectiveness of the recommended fixes.

This feedback loop enhances the model's accuracy and relevance, leading to more robust error resolution strategies.

Evolutionary methods, inspired by biological evolution, offer powerful techniques for optimizing complex systems. These methods can be applied to improve the performance of the REFII model in network fix and mobile error recognition.

Key research directions in this domain include employing genetic algorithms to optimize the parameters and structure of the REFII model by simulating the process of natural selection, evolving the model to achieve higher accuracy and efficiency in error recognition; using evolutionary strategies to help the REFII model adapt to changing network environments and error patterns, ensuring the model remains effective in dynamic and evolving systems; and integrating evolutionary methods with machine learning techniques to further enhance the REFII model, such as using evolutionary algorithms to pre-train machine learning models, providing a robust starting point for further refinement.

Future research on the REFII model will benefit greatly from interdisciplinary collaboration. Combining expertise from fields such as computer science, telecommunications, and artificial intelligence will foster innovative solutions. Key areas of collaboration include human-computer interaction (HCI) to inform the design of user-friendly interfaces that present error information and fix recommendations clearly and intuitively; integrating cybersecurity insights to enhance the REFII model's ability to identify and mitigate security-related errors, protecting networks and mobile systems from potential threats; and data science to develop advanced analytics tools to process and visualize the vast amounts of data generated by the REFII model, uncovering patterns and trends that inform error prevention strategies.

The application of the REFII model in network fix and mobile error recognition holds immense promise. By translating results on LLM models, leveraging evolutionary methods, and fostering interdisciplinary collaboration, researchers can unlock new levels of accuracy, adaptability, and user-friendliness. Future research in these directions will undoubtedly lead to more resilient and efficient network and mobile systems, enhancing the user experience and ensuring robust error resolution.

6. CONCLUSION

In conclusion, the REFII method introduces a significant advancement in the field of advanced analytics by enabling the application of traditional data science techniques directly to time series data.

This innovative approach not only broadens the scope of traditional methods but also enhances their effectiveness. The true value of the REFII solution is further elevated by its ability to create a synergistic effect, wherein the results produced by one analytical method seamlessly become the enriched input for another. This integration, augmented by temporal data, allows for more comprehensive and insightful analysis, ultimately leading to more robust and actionable outcomes.

The REFII method thus represents a pivotal development in advanced analytics, offering new opportunities for more precise and dynamic data interpretation.

Time series analytics play a pivotal role in the evaluation of network performance and the detection of errors. These analytics involve the study of data points collected or recorded at specific time intervals, allowing for the observation of trends, cycles, and other temporal structures in the data. Time series analysis can provide profound insights into the performance metrics of a network, such as latency, throughput, and packet loss, by identifying patterns and anomalies that might indicate underlying issues or potential areas for improvement.

However, the challenge lies in the limitations of traditional time series methods, such as ARIMA (AutoRegressive Integrated Moving Average), ARMA (AutoRegressive Moving Average), and trend extrapolation techniques. While these methods are well-established and have been widely used for forecasting and trend analysis, they often fall short when it comes to error detection in complex and dynamic network environments.

ARIMA and ARMA models rely on the assumption of linearity and stationarity in the data, which may not always hold true for network performance metrics that can exhibit non-linear and non-stationary behaviors due to fluctuations in network traffic, varying loads, and unexpected events. Furthermore, these models may not effectively capture sudden changes or anomalies, as they primarily focus on modeling the overall trend and seasonal components of the time series.

Trend extrapolation methods, on the other hand, extend the existing trend into the future but can lead to significant inaccuracies in the presence of abrupt changes or shifts in the data. These methods often lack the robustness needed to detect and adapt to new patterns or unexpected deviations that might indicate an error or a performance issue.

As a result, there is a need for more advanced and adaptive approaches to time series analysis for network performance evaluation and error detection. Techniques such as machine learning-based models, which can handle non-linear and non-stationary data, and more sophisticated anomaly detection algorithms, offer promising alternatives.

REFII is good solution, because it can learn from historical data, adapt to new patterns, and provide more accurate and timely detection of errors, ultimately leading to improved network performance and reliability with help of traditional methods applaying.

REFERENCES

Ahmed, Z., & Li, Q. (2025). Data transmission error detection and correction with cyclic redundancy check and polar code integration. *Applied Sciences (Basel, Switzerland), 15*(3), 1124. DOI: 10.3390/app15031124

Becker, L.. (2022). Customer experience measurement in omnichannel environments. *Journal of Service Research, 25*(4), 589–605. DOI: 10.1177/10946705221126590

Chen, Y.. (2025). Network error logging (NEL): A framework for client-side failure reporting. *IEEE Transactions on Network and Service Management.*

Ding, A.. (2023). Automatic error data recognition for structured light systems using residual neural networks. *Applied Sciences (Basel, Switzerland), 13*(5), 2920. DOI: 10.3390/app13052920

García, D.. (2024). Quantifying the business impact of network errors in cloud services. *IEEE Transactions on Cloud Computing, 12*(2), 1–14.

Gupta, R., et al. (2023). Self-healing through error detection, attribution, and re-training in conversational AI systems. Amazon Science Technical Report.

Horowitz, R. (2022). Error resolution and digital trust in financial services. *Journal of Digital Customer Experience, 8*(2), 45–59.

Johnson, M.. (2023). Customer sentiment analysis during network outages using NLP. *ACM Transactions on Social Computing, 6*(3), 1–22.

Kim, J., & Park, S. (2025). Attention mechanisms for multivariate network error detection. *Neural Networks, 165*, 832–843.

Klepac, G. (2002). REFII model-Model for recognition patterns in time series. *Conference Proceedings of the International Conference on Information Technology Interfaces.*

Klepac, G. (2007). Time series analysis using a unique model of transformation. *Journal of Information and Organizational Sciences, 31*(2), 1–13. DOI: 10.29133/jios

Klepac, G. (2023). Time series error detection solution based on REFII model and concept of nontemporal expansion. *CEUR Workshop Proceedings, 3900*, 1–9. https://ceur-ws.org/Vol-3900/

Kumar, V., & Singh, N. (2024). Polar code optimization for 5G networks using dynamic error pattern analysis. *Wireless Communications and Mobile Computing, 2024*, 1–15.

Lee, H.. (2024). Real-time network anomaly detection using LSTM-autoencoder hybrid models. *Computer Networks*, *242*, 110–125.

Müller, T.. (2023). Autocorrelation-aware deep learning for industrial IoT time-series forecasting. *IEEE Internet of Things Journal*, *10*(15), 13278–13289.

Nguyen, T.. (2024). Graph neural networks for root cause analysis in telecom networks. *Expert Systems with Applications*, *238*, 121–134.

Rossi, F.. (2023). Customer churn prediction during service disruptions using survival analysis. *Decision Support Systems*, *172*, 114–126.

Sun, F.-K., Lang, C. I., & Boning, D. S. (2025). Adjusting for autocorrelated errors in neural networks for time series. *Proceedings of the IEEE Conference on Neural Networks for Signal Processing*.

Waheeb, W., Ghazali, R., & Herawan, T. (2016). Ridge polynomial neural network with error feedback for time series forecasting. *PLoS One*, *11*(12), e0167248. DOI: 10.1371/journal.pone.0167248 PMID: 27959927

Wang, Y.. (2025). Federated learning for distributed network error prediction. *Future Generation Computer Systems*, *152*, 342–355.

Zhang, L.. (2024). Cross-layer error correction in software-defined networks. *IEEE/ACM Transactions on Networking*, *32*(1), 512–525.

Chapter 10
The Role of Digital Twins in AI-Driven Industry 4.0 and 5.0:
Enhancing Sustainability and Innovation

C.V. Suresh Babu
https://orcid.org/0000-0002-8474-2882
Hindustan Institute of Technology and Science, Chennai, India

M Noorul Afreed
https://orcid.org/0009-0009-4564-7194
Hindustan Institute of Technology and Science, India

V Sakthi Sree
https://orcid.org/0009-0007-5777-0750
Hindustan Institute of Technology and Science, India

J Mohamed Arshath
Hindustan Institute of Technology and Science, India

ABSTRACT

The convergence of Digital Twins and Industry 4.0/5.0, driven by advancements in artificial intelligence, offers transformative potential for smart industrial technologies. This chapter explores how Digital Twins, virtual representations of physical assets, enhance real-time monitoring, predictive maintenance, and operational optimization. The objective is to bridge interdisciplinary research gaps by fostering collaboration across AI, IoT, and sustainability domains. The methodological approach involves

DOI: 10.4018/979-8-3693-7994-3.ch010

a comprehensive literature review, analysis of existing implementations, and case studies demonstrating best practices. Key findings highlight the role of Digital Twins in improving efficiency, reducing environmental impact, and supporting decision-making. The chapter concludes with recommendations for adopting Digital Twins in next-generation industrial practices and outlines future research directions in AI-driven sustainable technologies.

1. INTRODUCTION

As industries worldwide transition towards more intelligent, efficient, and sustainable operations, Digital Twin (DT) technology emerges as a cornerstone of this digital transformation. This introductory section lays the groundwork by defining Digital Twins, exploring their historical evolution, and articulating the critical need for integrating Artificial Intelligence (AI) into DT systems. By highlighting current challenges and opportunities, the section sets the stage for the deeper explorations that follow.

1.1 Background of Digital Twin Technology

The fast-paced growth of Industry 4.0 and the still-to-be-thought-of concepts of Industry 5.0 are revolutionizing industrial industries with a focus on automation, data exchange, and human-centric strategies. Digital twins, or computerized copies of physical systems, are central to this revolution through real-time monitoring, simulation, and predictive analytics of systems. The term "Digital Twin" was first introduced by Grieves and later refined with Vickers as a model to mirror physical systems virtually (Grieves & Vickers, 2017). With AI, digital twins advance decision-making, operational excellence, and innovation. Such integration is critical in enabling sustainable industrial operations, enhancing resource optimization, and reducing environmental footprint. Digital twins would allow industries to enhance productivity, reduce waste.

1.2 Problem Statement

Although Industry 4.0 technologies are gaining greater adoption, there continue to be challenges for sustainable development to realize synergies to advance the pace of technology adoption. Digital Twins combined with AI represent innovative possibilities. but their deployment is fractional and lacks an overarching framework considering sustainability. Without understanding a systematic approach to incorporating AI-powered Digital Twins within industrial systems, they will have little to no influence on creating innovative practices and long-term environmental goals. This study addresses the important gap in a basic understanding of how to systematically employ Digital Twins to promote sustainability and innovation in AI-led industrial ecosystems.

1.3 Evolution of Industry 4.0 and 5.0

Recent manufacturing has been impacted by the Industrial revolutions. With the implementation of the IoT, cyber physical systems, cloud computing, big data, artificial intelligence, and robots in automation systems, real time decision making has now become possible in industry 4.0. Efficiency levels as well as precision has greatly improved. Sustainability, personalized manufacturing, human-machine collaboration, and innovation are the main objectives of industry 5.0 (Suresh Babu et. al., 2024).. Intelligent systems have further enabled improved performance of the environment. Facilitating the transformation for sustainable innovation is predictive maintenance and process optimization that the Digital Twins technology provides.

1.4 AI Integration and Model Lifecycle Management in Digital Twins

The use of Artificial Intelligence (AI) in digital twins (DT) is central to being predictive, adaptive, and autonomous in Industry 4.0 and 5.0, yet to ensure that AI-enabled systems and models are effective and reliable, a robust framework for the management of the AI model lifecycle is necessary.

This framework includes the following stages:

Model development and training

AI models used in DTs are trained using historical sensor data as well as simulation results. Development typically involves supervised or reinforcement learning methods of detecting patterns, predicting outcomes, or optimizing behavior. As part of the preprocessing pipeline, outliers, missing values, and normalization should be addressed for model robustness.

Deployment and inference

When models are trained, they are deployed to the DT (cloud-based, edge/sensor based, or on-site servers), depending on latency and compute constraints. Inference models should be built light (i.e., using TensorFlow Lite or ONNX Runtime) for edge-based inferencing to provide real-time responsiveness.

Ongoing Learning and Retraining

Static models deteriorate when the environment is dynamic and concept drift takes place. Hence, it is imperative to build a retraining pipeline that includes learning from scratch. For example, the retraining pipeline may work through scheduled time intervals or be triggered by performance thresholds. The process of either incremental learning or transfer learning will ensure that models can adapt piecemeal with minimal disruption.

Versioning and Governance

All AI models need to undergo versioning similar to the software code and use tools like MlFlow or DVC. Commercial verifiable versions are required for reproducibility and roll-back procedure in case of deployment failure. This may be operationally trivial but is necessary for traceability or in audited compliance, especially for regulated sectors.

Monitoring and Drift Detection

Monitoring model accuracy in an online workflow instantly is critical. Drift detection functions such as Kullback–Leibler divergence (KLD) or Hellinger distance, which can trigger internal alarms, should be developed to detect when the incoming data falls outside of trained distributions of training data and datasets' accuracy levels falls both outside of accuracy design limits. For example, at both the model and incoming data level (data not already covered in a closed dataset), retrain functionality should occur, combine data files together, or request human intervention if appropriate.

Explainability and Transparency

When trust and accountability are necessary (especially in decision-critical environments), the inclusion of explainable AI (XAI) such as SHAP, LIME and counterfactual explanations, they can enable stakeholders to understand the reasoning behind AI-generated outputs.

By integrating AI life-cycle management into digital twin ecosystems, organizations can create resilient, transparent, and adaptive systems that develop in accordance with their operational context, ensuring sustained reliability and innovation.

With an understanding of the background and motivations for Digital Twin adoption, the next section investigates how AI is practically integrated within these systems to enhance operational efficiency and sustainability across smart industrial ecosystems.

2. INTEGRATION OF AI IN DIGITAL TWIN SYSTEMS

Building on foundational concepts, this section delves into the practical integration of AI technologies within Digital Twin ecosystems. It discusses how AI-driven tools enhance automation, predictive maintenance, and sustainable outcomes, while outlining the structural alignment with Industry 4.0 and 5.0 paradigms. The reader is guided through key enablers, such as IoT, edge computing, and data governance frameworks, providing a comprehensive view of smart system integration.

2.1 Integration with AI-Driven Smart Industrial Technologies

Digital twins leverage AI-powered smart technologies to optimize industrial processes, including automation and predictive maintenance. Key components of this integration include IoT sensors for data collection, edge computing for real-time processing, and AI algorithms that drive system intelligence. These elements collectively improve production workflows and support sustainable manufacturing goals.

While this section outlines the high-level integration of AI, detailed guidance on AI model lifecycle management—including retraining, drift detection, version control, and explainability—is provided in Section 4.5.

2.2 Digital Twins and Sustainability Alignment

Merging the original content of Sections 2.2 and 3.3, this subsection consolidates how Digital Twins align with Sustainable Development Goals (SDGs) and drive energy and resource efficiency. AI-enhanced DTs facilitate predictive maintenance, real-time monitoring, and simulation-based insights, all of which contribute to:

- Reduced energy consumption and emissions
- Improved material and resource usage
- Quantifiable reductions in emissions and improved resource utilization metrics

By modeling and optimizing processes continuously, AI-augmented Digital Twins empower industries to meet global sustainability benchmarks.

2.3 Bridging Industry 4.0 and Industry 5.0

Digital Twins play a crucial role in the transition from Industry 4.0—focused on automation and data exchange—to Industry 5.0, which emphasizes human-machine collaboration and societal impact. Through adaptive control systems, customizable interfaces, and human-centric design, DTs facilitate:

- Collaboration between human operators and intelligent systems
- Personalized manufacturing workflows
- Innovation in sustainable production methods This bridging ensures a more inclusive, ethical, and resilient industrial future.

Having reviewed the core technologies and strategic role of AI-enhanced Digital Twins, we now shift focus to the primary research questions and problems addressed in this study. The following section outlines the thematic pillars that structure the inquiry.

3. RESEARCH QUESTIONS AND THEMATIC FOCUS AREAS

With a clearer understanding of the technological context, this section transitions into identifying the core problems, questions, and thematic contributions of the chapter. It articulates how Digital Twins contribute to various dimensions—operational efficiency, sustainability, and human-machine collaboration—while pinpointing the technological enablers and barriers to their adoption. The discussion sets the direction for methodological inquiry.

3.1 Contributions of Digital Twins to Industry 4.0 and 5.0

Digital Twins are changing the game for industrial operations by providing real time insights, supporting decision-making, and enabling automation.

- **Enhancing Manufacturing Processes:** Digital Twins foster real time simulation, process optimization, and fault detection, leading to less downtime and overall increase in production performance.
- How do Digital Twins help implement lean production workflows?
- What benefits do they bring to predictive maintenance and reducing machine failure?
- How might they facilitate quality control through virtual simulations?

- Enhancing Human-Machine Collaboration: Industry 5.0 focuses on human-centred automation, and Digital Twins can help close the gap between human intelligence and AI.
- How do Digital Twins assist human operators by providing intelligent aids and suggestions?
- What are the major contributions of Digital Twins in training and skills development?
- How do Digital Twins contribute to safety and risk management?

Fostering Innovation and Customization: Digital Twins increase innovation through speedy prototyping, mass customization, and flexible production approaches.

- How does rapid prototyping enable greater product innovation and R&D?
- What advantages does virtual prototyping provide in terms of shortening time-to-market?
- How might Digital Twins facilitate custom manufacturing due to consumer demand?

3.2 Technological Enablers for Digital Twins

AI and Machine Learning enhance Digital Twin systems through:

- Predictive analytics and anomaly detection
- Self-learning and adaptive responses to data
- Automated decision-making for system optimization These AI capabilities directly support the objectives of scalability, responsiveness, and autonomy in industrial ecosystems.

3.3 Sustainability and Energy Efficiency

The technology of the Digital Twin serves an important role to help facilitate sustainability in Industry 4.0 and 5.0 reflective of providing live, data-driven insights to optimize operations. Coupled with AI, industries are able to produce Digital Twin models to simulate real-world systems and processes to better understand their energy consumption, waste production, and all subsequent of the environment. Digital Twins present an accurate, live model of the physical assets, that allow for predictive maintenance, process optimization, and, ultimately accountability of energy efficiency (Suresh Babu et. al., 2025).

Energy Consumption Optimization: Digital Twins allow for industries to continuously refine their energy consumption. For example, a Digital Twin of a manufacturing plant could highlight overt or latent inefficiencies in consumption through monitoring and analysis, suggesting change in shift, machinery, etc. Dynamic energy pricing can be leveraged to optimize scheduling and reduce energy costs. Making change overall can lead to significant operational cost savings, and a decreased energy consumption, and in turn increased sustainability

Resource Management: Digital Twins will support the industries to better analyse and optimize resource usage - which can range from water, raw materials and energy. Digital Twins affords a business the opportunity to analyse resource lifecycle, that helps decrease waste and overuse of resources - an aspect of sustainability.

To translate sustainability goals into measurable outcomes, industries should track performance using metrics like Energy Efficiency Ratio (EER), Carbon Emission Reduction (CER), and Waste Reduction Rate, as detailed in Section 7.6. Establishing baselines and comparative benchmarks helps quantify and validate the environmental benefits attributed to Digital Twin adoption.

3.4 Challenges and Risks in Digital Twin Adoption

Although Digital Twin technology has the potential to change industries, challenges and risks must be addressed to successfully implement it.

Data Security and Privacy Issues: Digital Twins continuously acquire and transmit much data in real-time from physical assets. The use of IoT devices and sensors may create public concerns or data privacy, which may exposing sensitive data to potential cyberattacks and unauthorized access. Adopting adequate cybersecurity practices, policies and technology like encrypting data, secure communications and access control will improve risks. Additionally, adherence to data protection laws like GDPR will also help mitigate risks.

Integrity of Cost and Implementation: Digital Twin entails upfront costs in investing in hardware, software and infrastructure. Costs related to buying and installing equipment such as sensors and IoT devices while using more advanced analytical tools can be costly especially for smaller and medium size enterprises (SMEs). The ongoing costs around maintenance of the hardware and software, training of users will also incur costs. The accumulation of costs will affect budgets for industries whose budgets are already constrained, limiting the wide-reach adoption of Digital Twins

4. OBJECTIVES AND SCOPE OF THE CHAPTER

To ensure academic rigor and contextual clarity, this section outlines the objectives and boundaries of the study. It frames the research purpose, explains the scope in terms of industry and geography, and presents the technological focus. The section also introduces supporting architectural concepts and governance strategies critical for Digital Twin deployment.

4.1 Objectives of the Chapter

This chapter sets out to explore the convergence of Digital Twin (DT) technology and Artificial Intelligence (AI) within the context of Industry 4.0 and 5.0. It seeks to establish a structured framework for examining how these technologies can support sustainable, human-centric, and intelligent industrial ecosystems.

This chapter aims to answer the following research questions:

1. How can AI-augmented Digital Twins accelerate sustainability in Industry 4.0/5.0?
2. What are the enablers and inhibitors of real-time human-AI collaboration using Digital Twins?
3. How can explainable AI frameworks improve trust and accountability in critical industrial applications?

The novel contributions of this chapter are threefold:

- It introduces an original conceptual framework that maps AI-enhanced DT capabilities to sustainability and innovation outcomes.
- It integrates principles of Explainable AI (XAI) within the Digital Twin life-cycle, advocating for transparency and ethical deployment.
- It presents a structured roadmap for implementing sustainable smart manufacturing solutions through real-world case studies, governance models, and performance metrics.

By addressing these objectives, the chapter contributes both theoretical insights and practical guidance for researchers, practitioners, and policymakers engaged in next-generation industrial transformation.

4.2 Scope of Coverage

The scope of the study was intentionally set to maintain focus and clarity.

Industry Scope: The research was focused on industries that are using AI-based digital twins, avoiding sectors where digital twins were not widely used.

Technology Scope: The study was focused on AI-based integrated digital twins, avoiding classical simulation models.

Geographical Scope: The research was focused on technologically advanced geographic locations to evaluate established digital twin use.

4.3 Edge Computing Architecture for Digital Twin Deployment

To ensure effective deployment of Digital Twin workloads at the edge, it is critical to profile system resources relative to workload demands. Profiling helps determine whether edge hardware can support compute-heavy tasks like AI inference or high-frequency sensor processing while staying within memory and network bandwidth constraints.

Edge computing is a key technology that enables low latency decisions and real-time action in many industrial situations. Using DT at the edge is a high-value proposition. However, there are many requirements to successfully deploy DT workloads at the edge, including understanding resource limitations, compute location, and network factors.

An overall architecture of this process is illustrated in the below diagram, where the architecture is pruned back to illustrate the compute elements. Edge computing moves compute resources to the physical assets enabling near-real-time processing of sensor data and actuation. Edge computing enables the ability to run DT simulations and algorithms locally on an installed digital twin computing platform without reliance on cloud communication, while along the way will lower overall bandwidth, and enable timely action for important event scenarios such as predictive maintenance and remote or mobile robotic control.

Defining latency budgets is critical for responsive digital twin behavior. For example, industrial robotics and control systems often require latencies under 5 milliseconds, while predictive maintenance systems may operate within a 20–100 ms window. Sensor update frequencies typically range from 1 Hz to 10 Hz depending on the asset criticality. These thresholds should guide architectural decisions around edge deployment, inference timing, and communication protocols.

Resource Profiling and Constraints

Deployment of DT components at the edge involves resource profiling to understand the hardware limitations relative to DT workloads. The capabilities of a typical edge node (e.g.: NVIDIA Jetson, Intel NUC) explored in terms of:

- Compute Resources: Enough CPU/GPU resources to run AI inference and simulation models
- Memory: Enough RAM to handle high-frequency data streams and caching of intermediate data
- Storage: Enough persistent storage for local logging of data, especially in periodically disconnected environments
- Network: Uplink/downlink speeds to ensure enough throughput and bandwidth for synced data sources going to cloud backends or other DT components

Workload Partitioning

To mitigate performance- and efficiency-related aspects of DT workloads, we can partition to different layers to take advantage of the strengths of edge, fog, and cloud layers. It is important to execute immediate (real-time) analytics (e.g., anomaly detection) at the edge, while other computationally heavy tasks such as model retraining or simulations can be offloaded to the cloud.

Containerization and Orchestration

Lightweight containers (e.g., Docker) allows us to deploy DT services utilizing smaller modules on edge nodes. Middleware solutions like K3s (lightweight kubernetes) or Balena provide a means for multi-device orchestration to develop scalable and fault tolerant DT systems.

Security & Reliability

Edge devices often operate in an untrustworthy environment. Therefore, there must be secure boot mechanisms, hardware-based encryption (e.g., TPM), and over-the-air (OTA) update abilities to ensure a continued safe operation.

Use Case: Predictive Maintenance

With an industrial example, DTs at the edge can carry out real-time vibration and thermal analysis on machinery, provide advance alerts, and limit unplanned downtime. Edge devices can also reduce timing latencies from hundreds of milliseconds with cloud processing to less than 20 milliseconds with edge processing, allowing better responsiveness and reliability.

4.4 Data Governance Framework for Digital Twins: Provenance, Quality, and Compliance

In digital twin (DT) environments, data holds a significant value that governs the simulation realism, predictive reliability, and decision-making confidence. It is therefore essential to adopt a data governance framework with appropriate policies and practices to provide transparency, autonomy, and compliance throughout the digital twin lifecycle.

Data Provenance and Lineage

Provenance is about where the data came from, the history of transformations undertaken, and the context of use. DT platforms will need to preserve lineage metadata so that each data point can be tracked from sensor capture to AI driven insights, including:

- Metadata markers at the sensors
- Audit logs of each transformation as well as model inputs and outputs
- Version control of data provenance and configurations of simulations

Data Quality Control

Data quality is pivotal to the reliability of DTs, which includes, but is not limited to:

- Schema validations should ensure that records are not malformed, incomplete etc.
- Identifying outliers, and filtering data noise should be purposeful (i.e. statistical methods, machine learning)
- Timeliness and completeness of real-time data streams

Version Control of Models & Data

Digital Twin (DT) systems continuously evolve over time. Data snapshots, simulation parameters, and AI models must be versioned and stored for traceability and rollback. Tools like DVC (Data Version Control) or MLflow can be used as plugins to keep track of your model versions, training datasets, and surrounding metrics.

Compliance & Regulatory Readiness

In implementation of DT in regulated industries (e.g., healthcare, aerospace, manufacturing), must follow data legislation (e.g., GDPR, ISO 27001). There should be a governance layer to:

- Implement access control and anonymization policies;
- Store auditable records for compliance purposes; and
- Safeguard data at rest and in transient by applying role-based access control (RBACs) and encryption.

Data Lifecycle Management

Define clear policies for retention, archiving, and deletion of telemetry and simulations data including managing the life of this data. Long term storage should apply tiered cloud storage, and active datasets should remain in high-speed caches or edge nodes.

Tooling & integrations

Adding governance tools, like Apache Atlas, OpenLineage, or Great Expectations, will also further the automation of provenance tracking and validation workflows and will help bind the development of Data Engineering, AI and operational teams in large-scale DT deployments.

Verifying a comprehensive data governance framework not only guarantees technical correctness, but legality (defensibility), auditability, and end-users trust in analyses and reporting.

These governance practices not only ensure technical correctness and operational efficiency but are also essential for auditability, legal defensibility, and regulatory compliance—especially in industries like healthcare, aerospace, and finance. By embedding traceability and lineage tracking, organizations can confidently demonstrate data integrity and model reliability under strict compliance requirements.

4.5 AI Model Lifecycle and Explainability in Digital Twin Systems

Artificial Intelligence (AI) plays a pivotal role in the ability of Digital Twin (DT) systems to do predictive analytics, anomaly detection, and optimization of decisions. For effective implementation of AI, however, the full lifecycle management of AI models must be accounted for to guarantee effective performance, reliability, and explainability over a prolonged period of time.

This lifecycle includes:

- Continuous retraining and adaptation to dynamic industrial environments
- Real-time drift detection and model performance monitoring
- Integration of Explainable AI (XAI) frameworks like SHAP and LIME for transparency

Incorporating these AI-driven enhancements ensures that DT systems remain resilient, transparent, and aligned with human-centric decision-making.

To situate the study within the broader academic landscape, the next section surveys existing literature related to Digital Twins and their integration with AI. This review helps validate the research focus and identify critical knowledge gaps.

4.6 Human-Machine Interface and Visualization in Digital Twin Systems

Digital Twins must provide intuitive, real-time interfaces to bridge the digital and physical realms effectively. Visualization plays a crucial role in making insights actionable, especially for operators, engineers, and decision-makers.

Dashboard Design: Interactive dashboards (e.g., using Grafana, Power BI, or custom web apps) should support layered visualization of sensor data, KPIs, and simulation results. Features like drill-down analysis, historical replay, and real-time anomaly alerts enhance decision-making.

VR/AR Integration: Augmented Reality (AR) and Virtual Reality (VR) interfaces can project Digital Twin data directly onto physical assets through devices like Microsoft HoloLens or mobile AR apps. This enables technicians to perform guided maintenance, real-time diagnostics, and training simulations in context-rich environments.

3D & Time-Series Visualization: Integration with tools like Unity3D or Cesium enables immersive, spatial representation of Digital Twins in industrial facilities or smart cities. Time-series visualization ensures operators can trace asset behavior over periods, helping in root cause analysis.

Human-Centric Design: UI/UX design should prioritize usability, responsiveness, and accessibility. Interactive elements (e.g., sliders for simulation parameters, toggle controls for failure scenario testing) empower users to engage proactively with Digital Twins.

Together, these interfaces close the feedback loop between human and machine, essential for Industry 5.0's human-centric vision.

5. LITERATURE REVIEW

This section synthesizes current academic and industry research related to Digital Twins and AI. It categorizes findings across smart manufacturing, sustainability, and system interoperability, highlighting emerging trends and identifying gaps. The review not only validates the study's relevance but also informs the development of the proposed conceptual framework and methodology.

5.1 Systematic Literature Review

A Systematic Literature Review (SLR) was carried out to gather richer perspectives of digital twin technologies within both an academic and industrial context.

The objectives were to:

- Identify the current state of Digital Twin technologies in industrial contexts.
- Identify how AI has been integrated into Digital Twin frameworks. Identify the intersection of both Digital Twin technologies and AI toward sustainability and innovation.
- Search Strategy Databases: IEEE Xplore, ScienceDirect, SpringerLink, Scopus, and Google Scholar
- Keywords: "Digital Twins," "Artificial Intelligence," "Industry 4.0," "Industry 5.0," "Sustainability," "Smart Manufacturing," "Predictive Analytics"
- Inclusion Criteria: Peer-review articles published in the time of 2017 to 2024, conference papers, and white papers.
- Exclusion Criteria: Old articles (articles pre-2017), non-industrial application articles, or non-AI articles.

The review was thematically organized in terms of the integration of technologies, technology challenges, and applications in different industrial sectors (e.g. manufacturing, logistics, and energy).

5.2 Overview of Existing Research

In recent decades, smart manufacturing and automation have evolved rapidly due to the development of new technologies like Artificial Intelligence (AI), the Internet of Things (IoT), Cyber-Physical Systems (CPS), and Big Data Analytics. Current research emphasizes smart manufacturing topics and examines predictive maintenance, real-time monitoring, autonomous decision making, and digital twin technologies. Digital Twins in manufacturing have been categorized into various types based on functionality and complexity (Kritzinger et al., 2018).

Multiple papers examine the applications of digital twin to create a virtual representation of a physical asset, enabling simulation, real-time monitoring, and optimization in manufacturing processes (Suresh Babu et. al., 2025)..

Figure 1. **Conceptual Framework for AI-Driven Digital Twins in Industry 4.0/5.0*(Source:****Author-generated framework)***

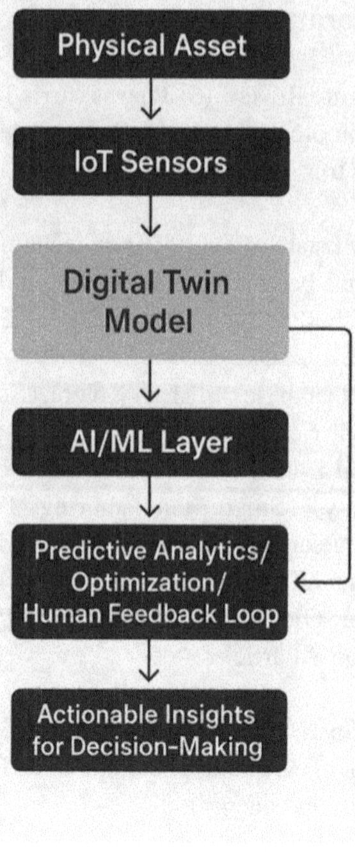

This figure illustrates the interaction between physical assets, IoT sensors, AI-enhanced Digital Twin models, and decision-making layers within smart manufacturing systems. It visually aligns with the discussion in Sections 2.1 and 4.5, representing the continuous data flow and feedback mechanisms integral to predictive analytics and sustainability.

Referenced in Text: Introduced in Section 2.1 and elaborated upon in the AI lifecycle discussion in Section 4.5.

5.3 Applications in Smart Manufacturing and Automation

Intelligent manufacturing merges new technologies to maximize production, increase product quality, and create efficiencies.

Here are some examples of applications:

1. Predictive Maintenance

Predictive Maintenance information involves the analysis of sensor parameters and historical data on how machines operate. This data provides an analysis of the predictive maintenance of your asset to predict the likelihood of failure to occur at a certain time in the future. Essentially, predictive maintenance is to support the intention to reduce, or eliminate, unplanned downtime and increase the run-time of the machine.

2. Real-time Monitoring and Control

IIOT sensors will enable the monitoring and tracking your temperature, pressure, humidity, and operating parameters all in real-time. This provides very important information to ensure optimal worker performance and helps if there is a fault immediately.

3. Robotic Process Automation (RPA)

AI robotic will change or assist assembly lines, material transport/storage and precision manufacturing. Robots will lead to decreased labour, be more accurate, and create better efficiency.

4. Digital Twins and Simulation

Digital Twins are virtual twins of physical machines that simulate a physical process and allows experimentation and optimization of processes without committing to a real environmental process. Digital Twins may lead to higher productivity, less wasted resource, or customization of products.

5.4 Identification of Gaps in the Literature

Despite considerable investigation, gaps exist in the current body of knowledge:

- **Interoperability and Standardization**

There is limited literature dealing with unified protocols and standards for continuous integration of smart manufacturing technologies. Current systems frequently face challenges with compatibility, due to vendor-specific architectures.

- **Scalability of AI and IoT Solutions**

While AI models provide improved efficiencies, scaling AI and IoT solutions across multiple manufacturing systems remains challenging. More investigative work is warranted on adaptive AI modelling, to adjust to the various industrial environments.

- **Cybersecurity and Data Privacy**

The increased risk of cyber threat is noted in research, however comprehensive frameworks for securing smart factories continue to be developed. There is a need for AI solutions for security of manufacturing data, that considers both product and data use.

- **Human-Machine Cooperation and Workforce Development**

Existing literature is focused on automation with much less emphasis on workforce reskilling and human-AI collaboration.

Research is needed to look at new modes of training for employees and models for human-machine sharing of labour.

- **Sustainability and Energy Efficiency**

There is limited literature on how smart manufacturing can promote energy efficient and sustainable production. More investigations are warranted on green AI, circular manufacturing and other eco-friendly automation modes.

These gaps in research could lead to future research into the evolution of smarter, safer and more sustainable manufacturing ecosystems.

5.5 Scalability and Fault-Tolerant Architectures for Digital Twin Systems

Digital Twin (DT) systems must be conceived in such a way to support enterprise-level deployments that exhibit scalability, redundancy, and availability. More specifically, scalability, redundancy, and availability are often important components of Industry 4.0 and 5.0 implementations because these implementations require the ability to manage multiple assets, across a number of locations, and across concurrently running streams of data with a desire to minimize downtime.

Scalability

DT systems must be capable of scaling horizontally (across DT assets) as well as vertically (high-load components). Achieving this requires making architectural decisions to support the following wishes: all components state-less microservices, data ingestion pipeline decoupled from data processing pipeline, data ingestion does not have dependencies with respect to other data ingestions, and telemetry (ex: sensors, tag readers, etc.) can be managed using a distributed data store (ex: Apache Kafka and Cassandra) for data ingestion when telemetry is to be acted upon at scale.

Performance Benchmarking and Load Testing

Before deployment, performance benchmarking should be conducted to identify throughput, latency, and resource utilization for multiple load situations, both at the overview level and detail level. This can be accomplished using commercially available solutions like: Apache JMeter, Locust, and Prometheus + Grafana, which are able to simulate multiple digital twin instances whilst tracking the real-time performance of the system with respect to latency, and transactions per second (TPS). The primary metrics / outcomes for benchmarking are:

- Transactions per second (TPS)
- End-to-end latency (ms)
- High-Availability Architecture

To enable uninterrupted operations, DT platforms must implement:

- Redundant Data Pipelines: Multiple ingestion pathways to prevent single points of failure.
- Load Balancers: Distribute traffic across replicas of processing components (transactional and analytical) in order to avoid performance bottlenecks.
- Failover Mechanisms: Automatically switch from active to standby nodes or clusters in case of node failure.
- Data Replication: Across zones or regions using cloud-native tools (AWS Multi-AZ, Azure Availability Zones).

Monitoring and Resilience

Real-time health monitoring can be accomplished with tools like the ELK Stack, Prometheus, and AlertManager. These help proactively identify performance issues or failures before users even notice them. Automated scaling policies (e.g., Kubernetes HPA, AWS Auto Scaling Service) dynamically match allotted resources to user loads - even as they fluctuate.

Example Use Case: Multi-Asset Deployment

In a smart manufacturing context of a hundred machines, Smart Manufacturing DT platforms should provide a unique DT for each machine. Each machine DT will be performing concurrent simulations of the machine, along with running parallel AI models and analytics. A container-based microservice architecture, specifically based on Kubernetes, will allow each DT operating independently of each machine. Each machine DT operates independently, but communicates with each other via any message broker (e.g., MQTT, AMQP) that can facilitate coordination among the different asset twins or sub-simulations.

These architectural measures—redundant pipelines, load balancing, failover mechanisms, and monitoring tools—are foundational for fault-tolerant Digital Twin deployments. They ensure continued operation despite hardware failures or network disruptions, supporting real-time resilience in critical industrial environments.

5.6. Quantitative Validation of Digital Twin Models

Validating digital twin (DT) models is an essential step to ensure these virtual mirrors of an asset's behavior are reflective of the real-world asset's behavior. Validation using quantitative approaches assures higher confidence that any decision or observation made within that environment will have a likelihood of achievable action in reality from that decision. Quantitative validation with applicable approaches and distilled from relatively conceptual modeling is a key distinction.

Model Calibration Approaches

- Calibration of the simulation model may include a range of options to validate the model's accuracy with empirical data as a reference or base data. For example:
- Calibration often takes the form of parameter tuning, either by using custom optimization algorithm sets (for example gradients, genetic algorithms, etc.)
- Bayesian calibration to reflect prior beliefs or knowledge and inherent uncertainty
- Using operational logs (also called traces) extracted from historical operations and system behavior to align time series

Tachymetry and Error Metrics

The accuracy of a DT simulation can be validated by many statistical and predictive error markets. The most common are measures of absolute error, predictive error, and fidelity error:

- Root Mean Square Error (RMSE), and Mean Absolute Error (MAE) as measures of comparative deviation from the observable empirical data.
- R^2 or Coefficient of Determination to assess the dispersion or variance the DT is explaining.
- Dynamic Time Warping (DTW) to compare time series operations sequences (vibration signals, flow rates, etc.)
- Benchmarked Using Published Datasets
- When possible, benchmarks of fidelity should be made using published, standard or public datasets. For example:
- NASA degradation datasets to measure fidelity of simulation with turbofan engines.
- UCI machine learning datasets for generalized industrial systems
- Real-time SCADA data logs from industrial sites for power grid or manufacturing systems

Scenario Based Validation & Stress Testing

DT simulation can be validated through a range of conditions of the system's standardized operating model.(For example):

- Normal vs. degraded system modes of operation
- Injection of failure events

Having established the theoretical underpinnings and research gaps, we now present the methodology employed in this study. This section connects directly to the research questions outlined earlier and forms the empirical backbone for the analysis and findings that follow.

Quantitative model validation is essential for ensuring that simulations reliably mirror real-world asset behavior. Without this fidelity, decisions based on DT outputs may be misaligned or even hazardous. Implementing robust calibration, error metrics, and scenario-based validation enhances the scientific credibility and industrial applicability of Digital Twin solutions.

6. METHODOLOGY

Transitioning from theoretical background to applied research, this section presents the mixed-method approach employed in the study. It details the qualitative and quantitative methods used to gather data and validate AI-enhanced DT systems, supported by real-world observations and software development techniques. The methodology ensures both empirical depth and practical relevance.

To address the research questions outlined in Section 4.1, a mixed-method research approach was adopted, focusing on the integration and validation of AI-driven Digital Twin (DT) systems in smart industrial environments. The methodology was designed to balance conceptual modeling with empirical validation to ensure both academic rigor and practical applicability.

Figure 2. Key Drivers Influencing the Adoption of Digital Twin Technology

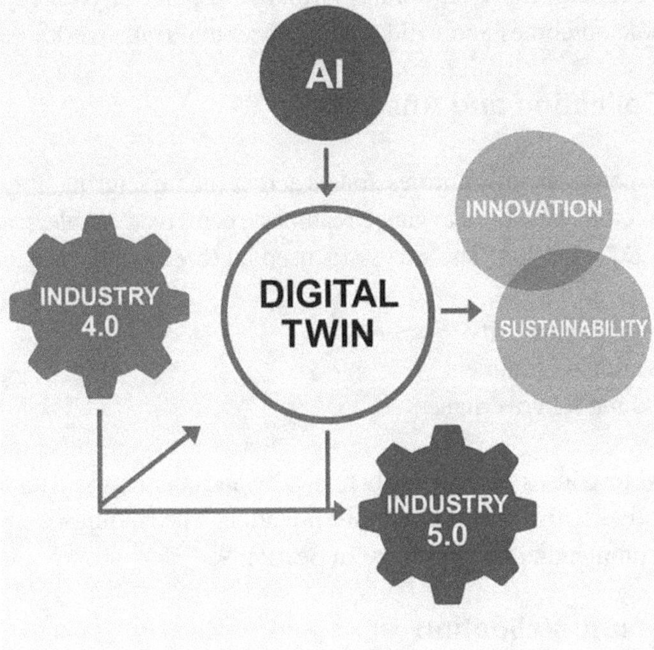

(Source: *Adapted from thematic synthesis of empirical and literature data; original visualization by authors).*

This figure summarizes the driving forces—technological, organizational, environmental, and economic—that impact the implementation of Digital Twins in industrial contexts. It complements the qualitative and quantitative research methods outlined in Section 6.

Referenced in Text: Discussed in Section 6 during the methodological exploration of influencing factors.

6.1 Research Design

The study employed a three-phase design:

1. **Literature Synthesis:** A comprehensive review of academic and industrial publications on AI-integrated DT systems, smart manufacturing, and sustainability was conducted to frame the theoretical foundation.
2. **Qualitative Exploration:** Semi-structured interviews with ten domain experts in AI, IoT, and industrial automation were conducted to identify current challenges, perceived benefits, and success criteria. Thematic analysis was used to derive cross-sector insights.

3. **Case Study Analysis:** Data from well-documented real-world implementations—Siemens Amberg, GE Aviation, and Virtual Singapore—were analyzed to extract measurable outcomes and validate the conceptual framework.

6.2 Data Collection and Analysis

Survey responses from 45 professionals across manufacturing, logistics, and infrastructure sectors were used to gauge readiness, perceived enablers, and inhibitors of AI-driven DT adoption. The survey focused on three key dimensions:

- Technological maturity
- Organizational alignment
- Data and model governance

Qualitative responses were coded thematically, and quantitative data were analyzed using descriptive statistics and trend interpretation. The findings were synthesized to inform recommendations presented in Section 9.

6.3 Toolset and Validation

Analytical tools (e.g., NVivo for qualitative coding and SPSS for survey analysis) were used to streamline data handling but are detailed in Appendix A to maintain focus on conceptual alignment.

This methodology directly supports the study's objectives by ensuring alignment between theoretical insights, field data, and the practical implications of deploying AI-augmented Digital Twins in industrial settings.

With the research methodology firmly established, the next section highlights how these methods were applied in the field. It presents real-world case studies, industry insights, and emerging best practices that demonstrate the tangible impact of AI-driven Digital Twin systems.

7. FIELD CONTRIBUTIONS AND CASE STUDIES

To demonstrate real-world applicability, this section compiles field contributions and documented case studies that exemplify the implementation of AI-driven Digital Twins. It includes insights from global leaders like Siemens, GE, and Virtual Singapore, providing empirical evidence of impact. The section focuses on technology integration and strategic deployment rather than reiterating general benefits already detailed earlier.

Table 1. Comparison of Industry 4.0 and Industry 5.0 with emphasis on Digital Twin capabilities.

Aspect	Industry 4.0	Industry 5.0
Primary Focus	Automation, digitalization, efficiency	Human-centric innovation, sustainability, resilience
Digital Twin Role	Operational control, predictive maintenance, process optimization	Personalized services, adaptive intelligence, collaborative environments
AI Integration	Machine learning for automation and data analytics	Explainable AI (XAI), ethical decision-making, human-AI interaction
Sustainability Emphasis	Energy efficiency, productivity	Circular economy, green AI, social sustainability
Technology Stack	IoT, Cloud, Big Data, AI, CPS	All from Industry 4.0 + Human-centric design, soft robotics, cognitive AI
Human Involvement	Minimal—focused on machine autonomy	High—humans as key decision-makers in harmony with machines
Societal Impact	Limited to economic efficiency	Includes ethical, social, and environmental impact
Typical Use of DTs	Factory automation, real-time monitoring, fault detection	Personalized medicine, citizen-centric smart cities, ethical robotics

7.1 Interoperability and Data Standards

7.1.1 Advancing Perspectives on Digital Twins

To ensure interoperability, Digital Twin architectures must adopt standardized protocols and schemas. Open standards such as OPC UA (Open Platform Communications Unified Architecture), MQTT (Message Queuing Telemetry Transport), and Digital Twin Definition Language (DTDL) provide semantic consistency and enable communication across heterogeneous systems. These standards avoid vendor lock-in, promote composability, and enable all-in-one system-of-systems integration across platforms.

7.2 Actionable Insights for Industry and Policy

The findings of this research have implications for industry leaders, policymakers, and regulatory agencies. Some important recommendations include:

For Manufacturing Industries: Utilize AI-supported automated processes to enhance efficiency and minimize operational costs. Adopt predictive maintenance practices using IoT sensors combined with digital twins to maximize machine uptimes.

Prioritize cybersecurity frameworks by integrating blockchain-based verification and AI-driven threat detection.

For Policymakers and Regulators: Establish international standards for smart manufacturing interoperability in order to support the continuous integration of technologies across the breadth of industries. Create policies that support workforce reskilling and retraining in AI, IoT, and robotics to support human-machine collaboration. Leverage the practice of incentivizing the use of sustainable and energy-efficient automation in the context of environmental goals and objectives. This information is important for bridging the gap between emerging technology innovations and reality, to help ensure an equitable smart manufacturing practice that is secure, scalable, and sustainable.

7.3 Recommended Approaches and New Trends

This subsection outlines forward-looking practices and technologies reshaping smart manufacturing. Emphasis is placed on:

- Predictive maintenance using DTs
- Real-time analytics and edge computing
- Ethical cybersecurity frameworks
- Workforce upskilling for AI collaboration

Emerging trends such as hyperautomation, green AI, and blockchain-backed transparency signal the future trajectory of industrial automation.

7.4 Real-World Case Studies

Figure 3. Lifecycle of a Digital Twin in Industrial Applications

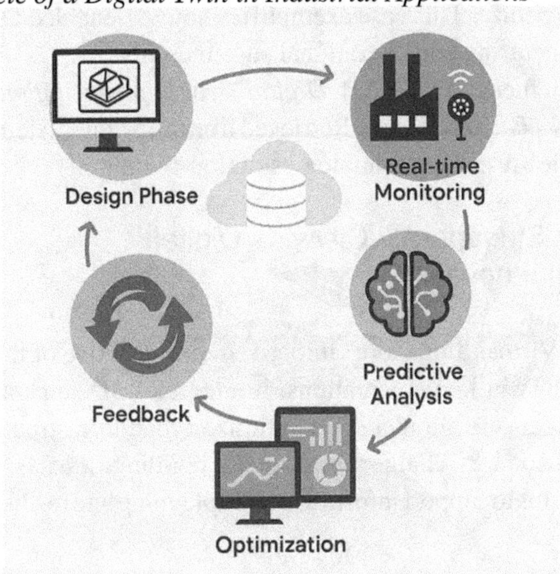

(**Source:** *Author-created schematic based on synthesized model lifecycle stages.*)

This figure presents the end-to-end lifecycle of a Digital Twin, from data acquisition and integration, through simulation and inference, to optimization and retraining. It supports the real-world case studies discussed in Section 7.4 and aligns with lifecycle management concepts in Section 4.5.

Referenced in Text: Introduced in Section 7.4 with detailed lifecycle references in Section 4.5.

Case Study 1: Siemens – Smart Manufacturing in Amberg, Germany

Siemens' electronics factory in Amberg is one of the most well-documented examples of a smart factory leveraging Digital Twin technology. The facility uses a full-scale Digital Twin of the production line for simulating operations, monitoring machine performance, and applying predictive analytics via AI. More than 75% of the production process is automated and digitally monitored, with over 1,000 embedded sensors providing data in real-time to optimize performance.

Results:

- 99.9% product quality rate

- 30% reduction in unplanned downtime
- Improved efficiency, throughput, and reduction in human error

Relevancy to Study: This case exemplifies how AI-enabled Digital Twins promote lean manufacturing and operational sustainability.

Reference: Siemens AG. (2022). *Digital Twin Implementation in Smart Manufacturing – Amberg Factory Report*. Retrieved from https://new.siemens.com/global/en/company/stories/research-technologies/digital-twin.html

Case Study 2: Singapore – Citywide Digital Twin for Urban Innovation

Singapore's "Virtual Singapore" initiative represents one of the most advanced city-scale Digital Twin implementations. It integrates 3D semantic modeling and real-time data streams to simulate and optimize urban infrastructure across transport, energy, and zoning. AI algorithms model building impacts, traffic flow, and emergency scenarios to support informed public planning and resilient infrastructure.

Outcomes:

- Improved traffic management and reduced congestion
- Enhanced zoning and construction policy
- Contributions toward SDG 11: Sustainable Cities and Communities

Relevance to Study: Demonstrates the scalability of Digital Twin technology for public-sector transformation and sustainability.

Reference: National Research Foundation Singapore. (2018). *Virtual Singapore: A 3D Smart City Model*. Retrieved from https://www.nrf.gov.sg/programmes/virtual-singapore

Case Study 3: GE Aviation – Predictive Maintenance for Jet Engines

GE uses Digital Twins to monitor the real-time performance of jet engines using thousands of sensor inputs. AI models analyze vibration, pressure, and thermal data to detect anomalies and predict degradation. Maintenance shifts from scheduled to condition-based, improving safety and reducing costs.

Results:

- 30% reduction in unplanned downtime
- Millions saved in maintenance and repair

- Enhanced engine efficiency and lifecycle management

Significance for the Study: This case highlights how Digital Twins extend asset longevity and support sustainable aviation operations.

Reference: General Electric. (2021). *GE Aviation's Digital Twin Applications.* Retrieved from https://www.ge.com/news/reports/why-digital-twins-are-saving-the-aviation-industry-billions

7.5 Cybersecurity Architecture and Threat Modeling in Digital Twin Systems

Digital twin (DT) environments are based on continuous data exchange of real-time data between physical assets, edge devices, cloud resources, and artificial intelligence (AI) models. This interconnectedness creates a large attack surface, which makes cybersecurity a fundamental obligation for trustworthy and resilient DT operations.

Threat Modeling and Risks

An explicit security strategy to protect DTs, starts with threat modeling, identifying the possible attack vectors, examples include:

- Privacy attacks, including sensor spoofing (e.g. GPS, sensor replay) and data injection
- Man-in-the-middle attacks on communications links
- Privilege boundary and unauthorized access to digital twin controllers (edge devices, AI enabling layer, upstream cloud)
- Adversarial ML attacks that compromise the integrity and reliability of predictive models based on the actual data
- Cybersecurity frameworks such as STRIDE (spoofing, tampering, repudiation, information disclosure, denial of service, elevation of privilege) can be used to catalogue risks.
- Secure Architecture Patterns

Defense-in-depth layers are essential security for DTs:

- Zero Trust Architecture (ZTA) to enforce identity verification and authentication prior to every access
- Network segmentation between operational technology (OT) and information technology (IT) layers

- Secure APIs and token-based authentication
- Data Protection Mechanisms
- End-to-end encrypted data streams via TLS/SSL encryption
- Leveraging blockchain or distributed ledgers as an immutable logging mechanism with tamper-proof audit trails
- Using digital signatures to verify provenance of data streams originating from sensors and actuators
- Real-time Intrusion Detection

In addition to threat modeling frameworks like STRIDE, cybersecurity assessments should incorporate quantitative risk metrics. Tools such as the Common Vulnerability Scoring System (CVSS) help prioritize security gaps by scoring severity and exploitability. Attack surface management, penetration testing, and risk scoring ensure that high-impact vulnerabilities in Digital Twin systems are identified and mitigated proactively.

7.6. Sustainability Metrics and SDG-Linked Impact of Digital Twins

Digital twin (DT) technologies can greatly increase the potential for sustainability initiatives in many fields, but when we hear "optimizing energy" or "reducing waste," the claims can only be made if they are backed up with real measurements and aligned to some global sustainability frameworks, particularly the United Nations Sustainable Development Goals (SDGs).

Key Performance Indicators (KPIs) for Sustainability

There should be measurable KPIs to determine the sustainability impact of the DTs such as:

- Energy Efficiency Ratio (EER): output per unit energy consumed, before and after optimization through the DT
- Carbon Emission Reduction (CER): tonnes of CO_2 equivalent saving accrued from using predictive maintenance or virtual testing
- Waste Reduction Rate (%): materials recovered through the process or defect prediction or simulation
- Water Usage Efficiency: measured as litres per unit process in smart agriculture or manufacturing
- Uptime Improvement (%):as a result of predicting failures and acting proactively

- Baseline/Measurement Comparison

For each of the KPIs above, you must establish the baseline condition and compare it to the outcome of using DT:

- Use historical SCADA data or operational logs to establish benchmark references
- Define your timeframes for consistent periods of impact evaluation (e.g. quarterly or annually)
- Establish measurement based on methodologies in Life Cycle Assessment (LCA) to assist with full sustainability profiling
- Alignment to the United Nations (UN) SDGs

Wherever possible, DT use cases should be mapped to specific SDGs and indicators. For example:

- SDG 7 (Affordable and Clean Energy): DTs can assist with load

8. ANALYSIS AND FINDINGS

This section analyzes the insights gleaned from case studies, literature, and empirical research. It explores how Digital Twins influence production efficiency, human-machine interaction, and sustainability. The findings help validate the proposed framework and provide a grounded assessment of DT capabilities in real-world industrial ecosystems.

Figure 4. Digital Twin as a Bridge between Industry 4.0, Industry 5.0, and AI for Sustainable Innovation.

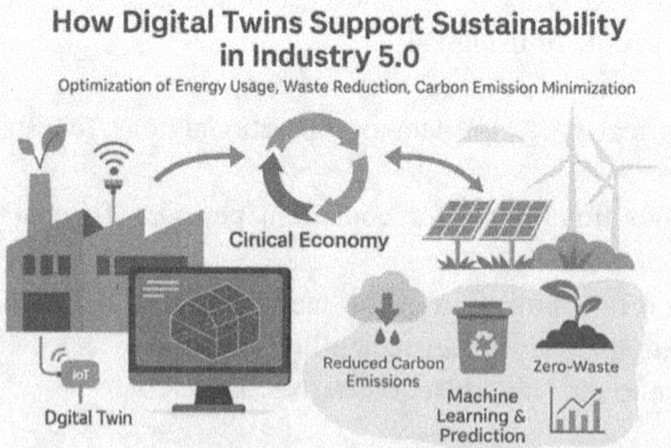

(Source: *Developed by authors; inspired by strategic frameworks from reviewed literature.)*

This figure conceptualizes how Digital Twin systems integrate core technologies from Industry 4.0 with human-centric and sustainable paradigms of Industry 5.0. It depicts AI's role in enabling this transformation, providing a visual roadmap of progression.

Referenced in Text: Highlighted in Section 8, Analysis and Findings.

8.1 Case Study Results

The case studies for the analysis focus on the real-world implementation of smart manufacturing technologies, including successes and challenges, as well as lessons learned from both. The case studies draw from multiple industry sectors: automotive, electronics and pharmaceuticals.

Key Case Study Observations

- **Predictive Maintenance with Digital Twins Observations from the case studies**

Predictive Maintenance with Digital Twins Case studies from automotive manufacturing plants indicated that the use of digital twins enabled a reduction of machine downtime by 30-40% via real-time monitoring and analysis (predictive). The use of AI-based diagnostics in automotive manufacturing plants helped to identify failing machine components early enough to avoid further costly repairs related to emergency repairs.

- **Impact of Industrial IoT (IIOT) on Production Efficiency**

The case studies indicated that smart factories that take advantage of IIoT sensors and cloud analytics has the potential to increase operational efficiency in the range of 20-35% due to improvements in supply chain workflows. It should be noted that the case studies also observed issues with data integration as non-standardized communication protocols of machinery between different IoT sensor vendors were utilized.

- **Cybersecurity Vulnerabilities in Smart Manufacturing**

Organizations that did not focus on strong cybersecurity measures had a higher risk of data breached which caused operational interruptions. blockchain authentication and AI-driven security platforms are effective tools to secure smart factory network These case study results demonstrate compelling empirical evidence of how smart automation technologies are changing industrial efficiency and suggest areas of required innovation.

8.2 Insights into human-machine collaboration

The evolution of smart manufacturing has brought renewed focus to the dynamics of human–machine collaboration (HMC). Rather than replacing human labor, Artificial Intelligence (AI) and automation technologies are increasingly being used to augment human capabilities. In particular, digital twin (DT) systems are enabling operators to shift from reactive, manual tasks to proactive, data-driven decision-making roles. For instance, where inspections once relied solely on human observation and post-failure analysis, AI-enabled DTs now facilitate predictive diagnostics and real-time monitoring, enhancing operational foresight and reducing cognitive load.

Empirical findings from industry case studies indicate that operators equipped with DT-supported interfaces report higher productivity and reduced error rates. AI systems assist in tasks such as failure forecasting, quality control, and anomaly detection, thereby allowing human workers to concentrate on high-level supervisory functions and strategic decision-making. Despite these advantages, gaps remain in workforce readiness. A recurring challenge identified across sectors is the lack of technical literacy among current employees, particularly in understanding the operational logic of AI-enhanced systems and interpreting automated insights.

Moreover, ethical considerations surrounding AI adoption—such as job displacement fears and trust in machine-generated outputs—necessitate thoughtful implementation strategies. Building trust in human–AI collaboration hinges on transparent system design, explainable AI outputs, and structured upskilling initiatives. Operators are more likely to embrace DT systems when provided with comprehensive training and tools that clarify how AI decisions are made and how they affect their tasks.

In essence, the integration of AI-enabled Digital Twins is transforming the human role in industrial ecosystems—from manual operators to informed supervisors and decision-makers. This transformation requires not only technological innovation but also a parallel commitment to human-centered design and workforce development.

8.3 Sustainability Framework

Sustainability plays an important role in the development of smart manufacturing. This research proposes a framework that incorporates principles of environmental, economic, and social sustainability into the elements of industry automation. Proposed Sustainability Framework:

1. Energy-Efficient Manufacturing AI-driven energy management systems can optimize energy use in factories by anticipating periods of peak demand and automating any processes that use large amounts of energy. Deployment of renewable sources of energy (i.e. solar, wind) to support factory operations.
2. Circular Economy and Waste Reduction Smart manufacturing allows for closed-loop production systems, which manage waste materials that are utilized and re-applied to production processes. Waste sorting and recycling systems based on AI can improve recovery rates of materials in industrial plants.
3. Green-AI and Low-Power Computing Encouraging the use of AI models that are lower power and are resource efficient, to minimize carbon footprints associated with digital twin simulations and automation systems. Leverage cloud and edge computing architectures to distribute workloads for AI models more efficiently.
4. Sustaining the Workforce and Inclusive Automation Upskill the workforce in anticipation of an industry-wide transition to AI. Provide fair labour policies and ethical automation standards that protect labour rights and job security.
5. Cybersecurity for Sustainable Digital Transformation Create cyber-resilient manufacturing ecosystems which protect business operations against cyberattacks while ensuring operational disruption. Using blockchain allows for transparent tracking within the supply chain.

9. STRATEGIC RECOMMENDATIONS FOR IMPLEMENTATION

Grounded in the findings and contributions discussed earlier, this section offers actionable recommendations for industries, academia, and policymakers. It presents structured strategies for implementing Digital Twins, fostering collaboration, and supporting responsible innovation. These recommendations aim to bridge gaps between research and application.

9.1 Phased Approach to Implementation of Digital Twins

Digital twins have begun to make significant contributions to smart manufacturing; however, in order to initiate a secure and scalable implementation of digital twins at the enterprise level it is necessary to adopt a systematic, phased approach. We propose a three phased approach to assist with the transition:

Phase 1:

The Pilot Phase and Proof of Concept Identify a set of target assets and target processes, that can yield the most value from digital twins Implement and develop digital twin prototypes on a smaller scale for monitoring and predictive maintenance; and, site-performance observation Focus on data collection, model verification, and proof of performance Ensure interoperability with existing IoT systems and factory automation systems

Phase 2:

System Integration and Scaling-Up Implement a larger digital twin solution across additional production lines and facilities Maximize the value of the digital twin solution by integrating AI based analytics and edge computing technologies to support real-time and live decision making Establish a standardized communication protocol for data-recorded performance stored across existing and new service systems to facilitate communication of data across systems Consider jurisdictional cybersecurity risks using blockchain authentication methods and/or secure cloud storage services

Phase 3:

Total Automation and AI Optimization Create self-learning digital twins, empowered by advanced AI algorithms that can improve manufacturing processes autonomously and continually. Establish autonomous feedback systems for machines to adjust themselves without human intervention. Utilize real-time simulation and modelling for demand and forecasting, optimizing supply chains, and sustainability. Adhere to global Industry 4.0 standards to enable future scalability and compliance. Following this structured phased approach will allow industries to minimize risk, have better ROI models, and enable ease of adoption of digital twins.

9.2 Industry-Academia Partnerships

Collaborative engagement between industry and academia is essential to drive innovation, bridge the skills gap, and accelerate the adoption of AI-enabled Digital Twin (DT) technologies. As smart manufacturing environments evolve rapidly, these partnerships play a pivotal role in aligning academic research with practical industrial needs.

One of the most effective mechanisms for fostering such collaboration is the establishment of **joint research laboratories**. These labs serve as innovation hubs where academia and industry co-develop solutions focused on digital twin architecture, AI integration, robotics, cybersecurity, and data governance. Co-funded and co-managed, these facilities enable translational research that moves beyond theoretical exploration into applied technological development.

Another key strategy involves the implementation of **industry-aligned certification programs**. Universities and technical institutions can partner with manufacturing firms to co-design curricula focused on AI model deployment, IoT integration, and data analytics in DT environments. Certification tracks—recognized by industry bodies—enhance employability and ensure graduates possess the skills needed to operate, maintain, and innovate within AI-driven ecosystems.

In addition, **open-access repositories** of industrial data, AI models, and validated digital twin case studies can facilitate cross-sector knowledge sharing. Curated by academic consortia with industry input, such platforms can accelerate R&D by enabling researchers to test and refine models using real-world datasets, while ensuring transparency and reproducibility.

By focusing on structured research collaboration, skills development, and data transparency, industry–academia partnerships can significantly strengthen the innovation pipeline and workforce readiness for Industry 4.0 and 5.0 transitions. These strategic linkages foster not only technological advancement but also sustainable, human-centric transformation across smart industrial systems.

9.3 Policy and Regulatory Recommendations

In order to safely, ethically, and sustainably address smart manufacturing technologies, policymakers need and will need to establish extensive regulatory frameworks.

The following policy recommendations refer to data security, standardization, workforce development, and sustainability:

1. Creating International Standards for Digital Twins and IIOT Standardize data models and interoperability standards to enable smooth adoption of digital twins as they are adopted in different industries. Adhere best practices in cybersecurity policies including zero-trust architectures and AI driven threat detection. Coordinate with various international organizations (ISO, IEEE, and NIST, etc.) to develop international standards for Industry 4.0.
2. Strengthening Cybersecurity Laws and Data Protections Publish strict data governance on how to store, control access to, and encrypt Industrial Data. Create compliance frameworks for AI-based decision making to prevent bias and be transparent accountability in a manufacturing automation system. Encourage industries to adopt blockchain-based verification in their supply chain and eliminate fraud while guaranteeing the product.
3. Workforce Development and Job Security Policies Governments should have incentives for training workers in AI, and automation as they migrate to support smart factories. Establish/promote public-private partnerships/groups to fund vocational programs in robotics, AI-driven quality control, and predictive analytics. Create ethics-based AI regulations to prevent job displacement and establish a human Centred approach within the technology.
4. Encouraging Green Manufacturing Policies Adopt tax credits and subsidies for businesses deploying energy-efficient AI models or principles of the circular economy. Establish carbon neutrality standards and protocols for AI-enabled automation in industry. Promote the use of renewable energy in smart factories by deploying energy monitoring systems based on the Internet of Things

By integrating policy constructs with technological advancements, government can promote innovation while also establishing security, sustainable workforce, and environmental stewardship.

10. CHALLENGES AND OUTLOOK

Recognizing that transformative technologies face practical and theoretical constraints, this section outlines the technical, organizational, and ethical challenges associated with AI-driven Digital Twins. It also highlights avenues for future research, such as explainable AI, cybersecurity, and sustainable DT systems, paving the way for continued innovation.

10.1 Technical Challenges

Data Integration and Interoperability

Despite their potential, Digital Twins face challenges in data interoperability, security, and real-time integration (Fuller et al., 2020). Industrial settings will deploy multiple platforms, from legacy machinery and SCADA systems, to IoT dashboards hosted in the cloud, often built on different vendor platforms. These systems will operate with different data formats, protocols, or metadata approach, making integration impossible. Digital Twin platforms cannot generate an integrated, real-time representation of the physical surroundings without shared data standards. We are left with "data silos" that decrease the analytical potential of AI, limit predictive possibilities, and slow process automation. Research on shared interoperability frameworks (i.e. OPC UA, Digital Twin Definition Language - DTDL) and middleware enables the facility of scaling DT implementations.

Model Accuracy and Real-Time Synchronization

Digital Twins rely on the fidelity of the real-time data to model, predict and optimize the operations of industries. Any delays, inconsistencies, or errors in the data to be used can result in false and dangerous system outputs especially since Digital Twin applications are typically used for higher risk applications such as autonomous vehicles, healthcare, or high manufacturing speed lines. Some of the common issues that occur include delays in communication from sensors, faulty equipment preventing the collection of complete data, and processing delays in the AI pipeline. Research into the use of edge computing, adaptive synchronization algorithms, and redundancy mechanisms that protect data integrity during varying network connection and hardware conditions is needed.

Cybersecurity and Data Privacy

Because they continually gather and process sensitive operational data (e.g. worker performance, energy use, proprietary processes), Digital Twins are prime targets for cyberattack as they require security exposures that could lead to breaches, data theft, disruption of operations, or manipulation of real time simulations all of which have serious operational, legal, and financial consequences. Data security risks multiply when AI is layered on top of the Digital Twin, particularly when AI systems are not adequately secured or if AI models are trained on sensitive data. Security protocols that protect systems from unauthorized access such as encryption, blockchain-based data integrity checks, role-based access control, and compliance with applicable privacy legislation (e.g. GDPR) are also critical from a data security standpoint in order to maintain the trust of involved stakeholders.

10.2 Organizational and Adoption Barriers

High Implementation Costs

Investing in Digital Twin ecosystems consists of a significant inflow of capital expenditures. There are costs incurred in different domains: IoT infrastructure (sensors, edge devices), cloud computing platform, AI development, and system integration. Coupled with operational costs, this represents a high rate of operational expenditure on the upkeep of digital models, the updating of simulations of the model (e.g., updating weather or rail schedules), and the training of human resources. SMEs often lack the budget and digital maturity to adopt large-scale Digital Twin solutions.. Government, investors and public-private partnership can help mitigate these costs via subsidy/grants for innovation initiatives, and shared infrastructure models.

Skill Gaps in the Workforce

Digital Twins involve a skillset that cuts across a range of disciplines: AI, data science, systems engineering, and industry operations. The present workforce is mostly untrained to deploy, interpret, and sustain Digital Twin platforms. This gap in knowledge can slow deployments, yield higher failure rates, and reduce returns on investment (ROI). Intentional investment into workforce development including technical training, university/industry partnerships, and the establishment of certification programs that blend AI and Digital Twin concepts will have long-term consequences. Human-centred research into tools and interfaces will also be beneficial in lowering the skills threshold for complex Digital Twin systems.

11. FUTURE RESEARCH DIRECTIONS AND CONCLUSION

The final section synthesizes the trajectory of Digital Twin development and proposes strategic areas for future research. It emphasizes interdisciplinary exploration, from scalable frameworks to smart cities, and urges further academic inquiry into AI-driven transformation in smart manufacturing and beyond.

In conclusion, this chapter advances the discourse on AI-driven Digital Twins by proposing an integrated framework that aligns with sustainability goals, outlines best practices, and identifies strategic gaps across smart industrial ecosystems. Through an exploration of enabling technologies, case-based evidence, and practical implementation pathways, the chapter contributes both conceptual and empirical insights that support sustainable, human-centric innovation.

Future research must extend this foundation by addressing several critical areas:

- **Scalable Explainable AI (XAI) Frameworks:** As AI systems become more autonomous, the need for scalable, domain-specific XAI tools integrated into DTs becomes vital for ensuring trust, accountability, and user comprehension.
- **Edge-AI Sustainability Metrics:** New models and frameworks are required to measure and benchmark the environmental performance of edge-deployed AI components within Digital Twin systems, especially as decentralization increases.
- **Low-Power Digital Twin Architectures for Emerging Economies:** Designing lightweight, resource-efficient DTs that are economically and infrastructurally viable for deployment in low-resource settings is essential for global adoption.

Additional areas for investigation include:

- Ethical governance of AI-DT interactions
- Cross-industry interoperability standards
- Integration with blockchain for enhanced traceability and transparency

By addressing these directions, researchers and practitioners can contribute to the evolution of AI-integrated Digital Twins as foundational tools for resilient, inclusive, and sustainable innovation in both developed and developing contexts

REFERENCES

Fuller, A., Fan, Z., Day, C., & Barlow, C. (2020). Digital twin: Enabling

General Electric. (2021). *GE Aviation's digital twin applications*. https://www.ge.com/news/reports/why-digital-twins-are-saving-the-aviation-industry-billions

Grieves, M., & Vickers, J. (2017). Digital twin: Mitigating unpredictable, undesirable emergent behavior in complex systems. In Kahlen, F.-J., Flumerfelt, S., & Alves, A. (Eds.), *Transdisciplinary Perspectives on Complex Systems* (pp. 85–113). Springer. DOI: 10.1007/978-3-319-38756-7_4

ISO. (2019). *Digital twin framework for manufacturing*. ISO/IEC JTC 1/SC 42 Artificial Intelligence. https://www.iso.org/standard/75066.html

Kritzinger, W., Karner, M., Traar, G., Henjes, J., & Sihn, W. (2018). Digital Twin in manufacturing: A categorical literature review and classification. *IFAC-PapersOnLine*, *51*(11), 1016–1022. DOI: 10.1016/j.ifacol.2018.08.474

National Research Foundation Singapore. (2018). *Virtual Singapore: A 3D smart city model*. https://www.nrf.gov.sg/programmes/virtual-singapore

Siemens, A. G. (2022). *Digital twin implementation in smart manufacturing – Amberg factory report*. https://new.siemens.com/global/en/company/stories/research-technologies/digital-twin.html

Suresh Babu, C. V., Anniyappa, C. S. A., & Raut, A. (2025). Toward seamless mobility: Integrating connected and autonomous vehicles in smart cities through digital twins. In Iyer, S., Nayyar, A., Paul, A., & Naved, M. (Eds.), *Digital twins for smart cities and villages* (pp. 169–187). Elsevier., DOI: 10.1016/B978-0-443-28884-5.00008-7

Suresh Babu, C. V., Auroshaa, A., Sowmi Saltonya, M., & Sathyanarayanan, A. S. (2024). Cloud-Enabled Fire Safety in Industry 5.0 Smart Factories: Leveraging IoT and Sensor Networks for Real-Time Monitoring and Proactive Prevention. In Hassan, A., Dutta, P., Gupta, S., Mattar, E., & Singh, S. (Eds.), *Emerging Technologies in Digital Manufacturing and Smart Factories* (pp. 150–166). IGI Global., DOI: 10.4018/979-8-3693-0920-9.ch009

Compilation of References

[] LaValle, S. M., & Kuffner, J. J. (2001). Rapidly-exploring random trees: Progress and prospects. *Algorithmic and computational robotics*, 303-307.

Acito, F. (2023). *Predictive Analytics with KNIME Analytics for Citizen Data Scientists*. Springer., DOI: 10.1007/978-3-031-45630-5

Ahmed, Z., & Li, Q. (2025). Data transmission error detection and correction with cyclic redundancy check and polar code integration. *Applied Sciences (Basel, Switzerland)*, *15*(3), 1124. DOI: 10.3390/app15031124

Al Mamlook, R. E., Obeidat, M. A., Elgeberi, N., & Nasayreh, A. (2024). Developing predictive maintenance for early warnings of machine failures using machine learning techniques in the fourth industrial era. In *Proceedings of the 2024 4th Interdisciplinary Conference on Electrics and Computer (INTCEC)*. DOI: 10.1109/INTCEC61833.2024.10602994

Alam, S. (2023). "Security Concerns in Smart Agriculture and Blockchain-based Solution," *2022 OPJU International Technology Conference on Emerging Technologies for Sustainable Development (OTCON)*, Raigarh, Chhattisgarh, India, 2023, pp. 1-6. DOI: 10.1109/OTCON56053.2023.10113953

Alexopoulos, K., Nikolakis, N., & Chryssolouris, G. (2020). Digital Twin-Driven Supervised Machine Learning for the Development of Artificial Intelligence Applications in Manufacturing. *International Journal of Computer Integrated Manufacturing*, *33*(5), 429–439. DOI: 10.1080/0951192X.2020.1747642

Alghananim, M. S., Feng, Y., & Ochieng, W. Y. (2025). *The first calibration model for bluetooth angle of arrival: Enhancing positioning accuracy in indoor environments* (arXiv:2501.08805). arXiv. https://doi.org//arXiv.2501.08805DOI: 10.48550

Ali, M. I., Lai, N. S., & Abdulla, R. (2021). Predictive Maintenance of Rotational Machinery Using Deep Learning. *International Journal of Electrical and Computer Engineering*, *14*(1), 1112–1121. . DOI: 10.11591/ijece.v14i1.pp1112-1121

Alizadeh, M., Moghaddam, M. M., & HosseinNia, S. H. (2021). A Novel Zero Delay Low Pass Filter: Application to Precision Positioning Systems. *ISA Transactions, 111*, 231–248. DOI: 10.1016/j.isatra.2020.11.013 PMID: 33250212

Alsouda, Y., Pllana, S., & Kurti, A. (2018) A Machine Learning Driven IoT Solution for Noise Classification in Smart Cities. *arXiv*, arXiv:1809.00238.

Alsouda, Y., Pllana, S., & Kurti, A. (2029) IoT-Based Urban Noise Identification Using Machine Learning: Performance of SVM, KNN, Bagging, and Random Forest. *InProceedings of the International Conference on Omni-Layer Intelligent Systems*, Crete, Greece, 5–7 May 2019; ACM: New York, NY, USA.

Aniba, Y., Bouhedda, M., Bachene, M., Seddiki, M., Tobbal, A., Hamrani, A., & Benyezza, H. (2024). Enhanced Digital Twin Development for a Conveyor Belt System: Integrating PLC Control, CNN Decision-Making, Gradio-based HMI, and Fuzzy Logic. *Brazilian Journal of Technology, 7*(4), 1–18. DOI: 10.38152/bjtv7n4-035

Antipov, K., Maslakov, M., & Yurenko, K. (2015). Improvement of the automated control systems for the development of the metallurgy. *Procedia Engineering, 129*, 1010–1014. DOI: 10.1016/j.proeng.2015.12.164

Aravindh, S., Nithisha, C., & Tanguturi, R. C. (2024). "Enhanced Agricultural Methods and Sustainable Farming Through IoT and AI Technology," *2024 Second International Conference on Intelligent Cyber Physical Systems and Internet of Things (ICoICI)*, Coimbatore, India, 2024, pp. 1206-1212.

Ardilla, F., Saputra, A. A., Besari, A. R. A., Naoki, D., Kohei, O., Obo, T., & Kubota, N. (2022, October). Topological Twin for Mobility Support Robots. In *2022 IEEE 8th World Forum on Internet of Things (WF-IoT)* (pp. 1-6). IEEE. DOI: 10.1109/WF-IoT54382.2022.10152223

Ardilla, F., Saputra, A. A., & Kubota, N. (2023). Multi-Scale Batch-Learning growing neural gas efficiently for dynamic data distributions. *International Journal of Automotive Technology, 17*(3), 206–216. DOI: 10.20965/ijat.2023.p0206

Arents, J., & Greitans, M. (2022). Smart Industrial Robot Control Trends, Challenges and Opportunities within Manufacturing. *Applied Sciences (Basel, Switzerland), 12*(2), 937. DOI: 10.3390/app12020937

Arima, K., Nagata, F., Shimizu, T., Otsuka, A., Kato, H., Watanabe, K., & Habib, M. K. (2023). Improvements of Detection Accuracy and Its Confidence of Defective Areas by YOLOv2 Using a Dataset Augmentation Method. *Artificial Life and Robotics, 28*(3), 625–631. DOI: 10.1007/s10015-023-00885-9

Arnob, S. S., AlQahtani, A. A. S., Alshayeb, T., & Alamleh, A. H. (2025). *Precision in Motion: An Approach to Indoor Localization and Navigation for Complex Spaces*. DOI: 10.20944/preprints202501.0356.v1

Arumugam, S. (2022). Investigations on Using White Noise as a Test Signal for Performing Frequency Response Measurements on Transformers. *Electric Power Systems Research, 202*, 107586. DOI: 10.1016/j.epsr.2021.107586

Asadi, F. (2023). *Essential Circuit Analysis using LTspice*. Springer Nature., DOI: 10.1007/978-3-031-09853-6

Attaran, S., Attaran, M., & Celik, B. G. (2024). Digital Twins and Industrial Internet of Things: Uncovering operational intelligence in industry 4.0. *Decision Analytics Journal, 10*(100398), 100398. DOI: 10.1016/j.dajour.2024.100398

Autran, J. V., Kuhn, V., Diguet, J.-P., Dubois, M., & Buche, C. (2024). AI4I-PMDI: Predictive maintenance datasets with complex industrial settings' irregularities. *Procedia Computer Science, 246*, 1201–1209. DOI: 10.1016/j.procs.2024.09.546

Babel, W. (2022). Fieldbuses, Communication Protocols, User Interfaces, Hardware ASICs. In *Industry 4.0, China 2025, IoT*. Springer., DOI: 10.1007/978-3-658-37852-3_5

Bai, H., Kuprat, J., Osório, C., Liu, C., Liserre, M., & Gao, F. (2024). Digital Twins for Modern Power Electronics: An investigation into the enabling technologies and applications. *IEEE Electrification Magazine, 12*(3), 50–67. DOI: 10.1109/MELE.2024.3423111

Bankins, S. (2021). The Ethical Use of Artificial Intelligence in Human Resource Management: A Decision-Making Framework. *Ethics and Information Technology, 23*(4), 841–854. DOI: 10.1007/s10676-021-09619-6

Bao, T., Chen, J., Li, W., Wang, X., Fei, J., Wu, L., Zhao, R., & Zheng, Y. (2022). MIAD: A Maintenance Inspection Dataset for Unsupervised Anomaly Detection. arXiv. https://arxiv.org/abs/2211.13968

Bapat, R. B. (2014). *Graphs and matrices*. Hindustan Book Agency

Baradaran, A. H. (2025). Predictive Maintenance of Electric Motors Using Supervised Learning Models: A Comparative Analysis. *arXiv preprint arXiv:2504.03670*. https://arxiv.org/abs/2504.03670

Baranwal, T., & Pateriya, P. K. (2016, January). Development of IoT based smart security and monitoring devices for agriculture. In 2016 6th international conference-cloud system and big data engineering (confluence) (pp. 597-602). IEEE.

Becker, L.. (2022). Customer experience measurement in omnichannel environments. *Journal of Service Research*, *25*(4), 589–605. DOI: 10.1177/10946705221126590

Beenakker, C. W. J., & Büttiker, M. (1992). Suppression of shot noise in metallic diffusive conductors. *Physical Review. B*, *46*(3), 1889–1892. DOI: 10.1103/PhysRevB.46.1889 PMID: 10003850

Belim, M., Meireles, T., Gonçalves, G., & Pinto, R. (2024). Forecasting models analysis for predictive maintenance. *Frontiers in Manufacturing Technology*, *4*, 1475078. DOI: 10.3389/fmtec.2024.1475078

Bello, S. A., Yu, S., Wang, C., Adam, J. M., & Li, J. (2020). Review: Deep Learning on 3D Point Clouds. *Remote Sensing (Basel)*, *12*(11), 1729. DOI: 10.3390/rs12111729

Bizon, I., Nimr, A., Fettweis, G., & Chafii, M. (2024). Indoor Positioning Using Correlation Based Signal Analysis and Convolutional Neural Networks. *2024 19th International Symposium on Wireless Communication Systems (ISWCS)*, 1–6. DOI: 10.1109/ISWCS61526.2024.10639139

Bongomin, O., Mwape, M. C., Mpofu, N. S., Bahunde, B. K., Kidega, R., Mpungu, I. L., Tumusiime, G., Awuor Owino, C., Goussongtogue, Y. M., Yemane, A., Kyokunzire, P., Malanda, C., Tigalana, D., Gumisiriza, O., & Ngulube, G. (2025). Digital Twin Technology Advancing Industry 4.0 and Industry 5.0 Across Sectors. In *Social Science Research Network*. https://doi.org/DOI: 10.2139/ssrn.5072457

Borenovic, M., & Neskovic, A. (2011, November). ANN based models for positioning in indoor WLAN environments. *2011 19thTelecommunications Forum (TELFOR) Proceedings of Papers*. 2011 19th Telecommunications Forum Telfor (TELFOR), Belgrade, Serbia. DOI: 10.1109/TELFOR.2011.6143551

Cai, W., Zhang, Q., & Cui, J. (2022). A Novel Fault Diagnosis Method for Denoising Autoencoder Assisted by Digital Twin. *Computational Intelligence and Neuroscience*, *2022*, 1–8. DOI: 10.1155/2022/5077134 PMID: 35909837

Can, O., & Turkmen, A. (2023). Digital Twin and Manufacturing. In Karaarslan, E., Aydin, Ö., Cali, Ü., & Challenger, M. (Eds.), *Digital Twin Driven Intelligent Systems and Emerging Metaverse*. Springer., DOI: 10.1007/978-981-99-0252-1_8

Cao, F., & Yang, X. (2023). UWB-Based Intelligent Positioning Method for Indoor Complex Environments. *2023 3rd International Conference on Mobile Networks and Wireless Communications (ICMNWC)*, 1–5. DOI: 10.1109/ICMNWC60182.2023.10435759

Carrasco, J., Markova, I., López, D., Aguilera, I., García, D., García-Barzana, M., Arias-Rodil, M., Luengo, J., & Herrera, F. (2021). Anomaly Detection in Predictive Maintenance: A New Evaluation Framework for Temporal Unsupervised Anomaly Detection Algorithms. arXiv. https://arxiv.org/abs/2105.12818

Carrilho, R., Yaghoubi, E., Lindo, J., Hambarde, K., Proença, H. (2024). Toward Automated Fabric Defect Detection: A Survey of Recent Computer Vision Approaches. Electronics, Vol. 13, No. 18, 3728, 20 pages. DOI: 10.3390/electronics13183728

Casanova-Marqués, R., Torres-Sospedra, J., Hajny, J., & Gould, M. (2023). Maximizing privacy and security of collaborative indoor positioning using zero-knowledge proofs. *Internet of Things : Engineering Cyber Physical Human Systems*, *22*, 100801. DOI: 10.1016/j.iot.2023.100801

Castro-Arvizu, J. M., Fernández Rubio, J. A., & Closas Gómez, P. (2017). *Robust indoor positioning in WLAN networks* [Universitat Politècnica de Catalunya]. DOI: 10.5821/dissertation-2117-107945

Chaiprabha, K., & Chancharoen, R. (2023). A Deep Trajectory Controller for a Mechanical Linear Stage Using Digital Twin Concept. *Actuators*, *12*(91), 1–14. DOI: 10.3390/act12020091

Chen, L. C., Zhu, Y., Papandreou, G., Schroff, F., & Adam, H. (2018). Encoder-Decoder with Atrous Separable Convolution for Semantic Image Segmentation. Procs. of European Conference on Computer Vision 2018 (ECCV2018), pp. 833–851. DOI: 10.1007/978-3-030-01234-2_49

Chen, M.-C., Cheng, Y.-T., & Chen, R.-W. (2022). A Novel Indoor Positioning Framework. *Computer Modeling in Engineering & Sciences*, *130*(3), 1459–1477. DOI: 10.32604/cmes.2022.015636

Chen, Y.. (2025). Network error logging (NEL): A framework for client-side failure reporting. *IEEE Transactions on Network and Service Management.*

Choudhary, K., DeCost, B., Chen, C., Jain, A., Tavazza, F., Cohn, R., Park, C. W., Choudhary, A., Agrawal, A., Billinge, S. J. L., Holm, E., Ong, S. P., & Wolverton, C. (2022). Recent Advances and Applications of Deep Learning Methods in Materials Science. *npj Computational Materials*, *8*(1), 59. DOI: 10.1038/s41524-022-00734-6

Coronado, E., Ueshiba, T., & Ramirez-Alpizar, I. G. (2024). A Path to Industry 5.0 Digital Twins for Human–Robot Collaboration by Bridging NEP+ and ROS. *Robotics (Basel, Switzerland)*, *13*(2), 28. DOI: 10.3390/robotics13020028

Csuros, M., & Csuros, C. (2010). Environmental sampling and analysis for metals. In *CRC Press eBooks*. DOI: 10.1201/9781420032345

Cui, G., Li, B., Tian, W., Liao, W., & Zhao, W. (2021). Dynamic modeling and vibration prediction of an industrial robot in manufacturing. *Applied Mathematical Modelling*, *105*, 114–136. DOI: 10.1016/j.apm.2021.12.031

Damioli, G., Van Roy, V., & Vertesy, D. (2021). The Impact of Artificial Intelligence on Labor Productivity. *Eurasian Bus Rev*, *11*(1), 1–25. DOI: 10.1007/s40821-020-00172-8

Damon, W. H., Bruce, D. D., & Clifford, H. G. (2008). *Reliability centered maintenance study on voltage regulators*. In 2008 IEEE Power and Energy Society General Meeting - Conversion and Delivery of Electrical Energy in the 21st Century, pp. 1–5. IEEE. https://doi.org/DOI: 10.1109/PES.2008.4596576

Dao, V.-L., & Salman, S. M. (2022). Deep Neural Network for Indoor Positioning Based on Channel Impulse Response. *2022 IEEE 27th International Conference on Emerging Technologies and Factory Automation (ETFA)*, 1–8. DOI: 10.1109/ETFA52439.2022.9921735

De Azambuja, A. J. G., Giese, T., Schützer, K., Anderl, R., Schleich, B., & Almeida, V. R. (2024). Digital Twins in Industry 4.0 – Opportunities and challenges related to Cyber Security. *Procedia CIRP*, *121*, 25–30. DOI: 10.1016/j.procir.2023.09.225

De Azambuja, A. J. G., Plesker, C., Schützer, K., Anderl, R., Schleich, B., & Almeida, V. R. (2023). Artificial Intelligence-Based Cyber Security in the Context of Industry 4.0—A Survey. *Electronics (Basel)*, *12*(8), 1920. DOI: 10.3390/electronics12081920

Deka, D., Kekatos, V., & Cavraro, G. (2023). Learning Distribution Grid Topologies: A tutorial. *IEEE Transactions on Smart Grid*, *15*(1), 999–1013. DOI: 10.1109/TSG.2023.3271902

Devagiri, J. S., Khanmohammadi Hazaveh, P., Rawashdeh, N., Dudipala, S., Desmuhk, P., & Karmarkar, A. (2024). PLC Multi-robot Integration via Ethernet for Human Operated Quality Sampling. Proceeding of 2024 ASEE Annual Conference & Exposition, Portland, Oregon. https://digitalcommons.mtu.edu/michigantech-p2/903

Devare, M. (2023). Predictive maintenance in Industry 4.0. In Gaurav, K. A., & Vishwakarma, D. K. (Eds.), *Computational intelligent security in wireless communications* (pp. 79–98). CRC Press., DOI: 10.1201/9781003323426-5

Ding, A.. (2023). Automatic error data recognition for structured light systems using residual neural networks. *Applied Sciences (Basel, Switzerland)*, *13*(5), 2920. DOI: 10.3390/app13052920

Doteguchi, N., & Kubota, N. (2022, May). Topological Tracking for Mobility Support Robots Based on Multi-scale Batch Learning Growing Neural Gas. In *Mobile Wireless Middleware, Operating Systems and Applications:10th International Conference on Mobile Wireless Middleware, Operating Systems and Applications (MOBILWARE 2021)* (pp. 17-31). Cham: Springer International Publishing. DOI: 10.1007/978-3-030-98671-1_2

Dubey, A. K., & Chahar, R. (2023). "Revolutionizing agriculture: harnessing IoT for productivity, sustainability, and innovation," 2023 6th International Conference on Recent Trends in Advance Computing (ICRTAC), Chennai, India, 2023, pp. 390-396. DOI: 10.1109/ICRTAC59277.2023.10480861

Dzedzickis, A., Subačiūtė-Žemaitienė, J., Šutinys, E., Samukaitė-Bubnienė, U., & Bučinskas, V. (2021). Advanced Applications of Industrial Robotics: New trends and possibilities. *Applied Sciences (Basel, Switzerland)*, *12*(1), 135. DOI: 10.3390/app12010135

Elamanov, S., Son, H., Flynn, B., Yoo, S. K., Dilshad, N., & Song, J. (2024). Interworking Between Modbus and Internet of Things Platform for Industrial Services. *Digital Communications and Networks*, *10*(2), 461–471. DOI: 10.1016/j.dcan.2022.09.013

El-Sheimy, N., & Li, Y. (2021). Indoor navigation: State of the art and future trends. *Satellite Navigation*, *2*(1), 7. Advance online publication. DOI: 10.1186/s43020-021-00041-3

Eppstein, D., Paterson, M. S., & Yao, F. F. (1997). On nearest-neighbor graphs. *Discrete & Computational Geometry*, *17*(3), 263–282. DOI: 10.1007/PL00009293

Fang, Q., Ibarra-Castanedo, C., Garrido, I., Duan, Y., Maldague, X. (2023). Automatic Detection and Identification of Defects by Deep Learning Algorithms from Pulsed Thermography Data. Sensors, Vol. 23, No. 9, 4444, 33 pages. doi:DOI: 10.3390/s23094444

Fang, S., Zhao, Q., Zhao, K., & Yang, L. (2025). Research on Bolt Loosening Detection Method Based on PointNet. Procs. of 2025 IEEE 8th Information Technology and Mechatronics Engineering Conference (ITOEC), Chongqing, China, 2025, pp. 1331-1335, DOI: 10.1109/ITOEC63606.2025.10967991

Farahsari, P. S., Farahzadi, A., Rezazadeh, J., & Bagheri, A. (2022). A Survey on Indoor Positioning Systems for IoT-Based Applications. *IEEE Internet of Things Journal*, *9*(10), 7680–7699. DOI: 10.1109/JIOT.2022.3149048

Felício, T., Postolache, O. A., Rodrigues, M. J., & Sebastião, P. (2023). "Vineyard Thermal Stress Assessment Through the Combination of In-Situ and Remote Sensing Technology," *2023 IEEE Sensors Applications Symposium (SAS)*, Ottawa, ON, Canada, 2023, pp. 1-6. DOI: 10.1109/SAS58821.2023.10254061

Fogel, D. B. (1995). *Evolutionary Computation: Towards a New Philosophy of Machine Intelligence*. IEEE Press.

Frick, J. (2023). Future of Industrial Asset Management: A Synergy of Digitalization, Digital Twins, Maintenance 5.0 / Quality 5.0, Industry 5.0 and ISO55000. [IJBMM]. *International Journal of Business Marketing and Management, 8*(4), 93–99.

Fritzke, B. (1994). A growing neural gas network learns topologies. *Advances in neural information processing systems, 7*. Vol. 7, 1995, pp. 625-632.

Fritzke, B. (1997). A self-organizing network that can follow non-stationary distributions. In *Artificial Neural Networks—ICANN'97:7th International Conference Lausanne,Switzerland,October 8–10, 1997Proceeedings 7* (pp. 613-618). Springer Berlin Heidelberg. DOI: 10.1007/BFb0020222

Fritzke, B. (1991, July). Unsupervised clustering with growing cell structures. In *IJCNN-91-seattle international joint conference on neural networks* (Vol. 2, pp. 531–536). IEEE. DOI: 10.1109/IJCNN.1991.155390

Fu, K. S., Gonzalez, R. C., Lee, C. G., & Freeman, H. (1987). *Robotics: control, sensing, vision, and intelligence*. McGraw-Hill.

Fukuda, T., & Kubota, N. (1999). An intelligent robotic system based on a fuzzy approach. *Proceedings of the IEEE, 87*(9), 1448–1470. DOI: 10.1109/5.784220

Fuller, A., Fan, Z., Day, C., & Barlow, C. (2020). Digital twin: Enabling

Gafurov, A. N., Kim, J., Kim, I., & Lee, T.-M. (2024, September 19). Web Tension AI Modeling and Reconstruction for Digital Twin of Roll-to-Roll System. *Journal of Intelligent Manufacturing*. Advance online publication. DOI: 10.1007/s10845-024-02488-y

Găitan, N. C., Zagan, I., & Găitan, V. G. (2024). Proposed Modbus Extension Protocol and Real-Time Communication Timing Requirements for Distributed Embedded Systems. *Technologies, 12*(10), 187. DOI: 10.3390/technologies12100187

García, D.. (2024). Quantifying the business impact of network errors in cloud services. *IEEE Transactions on Cloud Computing, 12*(2), 1–14.

General Electric. (2021). *GE Aviation's digital twin applications*. https://www.ge.com/news/reports/why-digital-twins-are-saving-the-aviation-industry-billions

Genuer, R., Poggi, J.-M., Tuleau-Malot, C., & Villa-Vialaneix, N. (2017). Random Forests for Big Data. *Big Data Research*, *9*, 28–46. DOI: 10.1016/j.bdr.2017.07.003

Gong, W., Akbar, M. F., Jawad, G. N., Shrifan, N. H. M. M., & Zhang, F. (2024). A Magnetic Field Concentration Method for Magnetic Flux Leakage Detection of Rail-Top Surface Cracks. *IEEE Access : Practical Innovations, Open Solutions*, *12*, 43245–43254. DOI: 10.1109/ACCESS.2024.3377692

Gonzalez, M. K., Theissen, N. A., Barrios, A., & Archenti, A. (2022). Online compliance error compensation system for industrial manipulators in contact applications. *Robotics and Computer-integrated Manufacturing*, *76*, 102305. DOI: 10.1016/j.rcim.2021.102305

Greer, C., Burns, M., Wollman, D., & Griffor, E. (2019). *Cyber-physical systems and internet of things*. DOI: 10.6028/NIST.SP.1900-202

Grieves, M., & Vickers, J. (2017). Digital twin: Mitigating unpredictable, undesirable emergent behavior in complex systems. In Kahlen, F.-J., Flumerfelt, S., & Alves, A. (Eds.), *Transdisciplinary Perspectives on Complex Systems* (pp. 85–113). Springer. DOI: 10.1007/978-3-319-38756-7_4

Gross, J. L., & Yellen, J. (2005). Graph Theory and its applications. In *Chapman and Hall/CRC eBooks*. DOI: 10.1201/9781420057140

Guidotti, D., Marchi, S., Antognelli, S., and Cruciani, A., (2019). "Water Management: Agricolus tools integration," 2019 Global IoT Summit (GIoTS), Aarhus, Denmark, 2019, pp. 1-5.

Gulay, C. F., & Fajardo, A. C. (2024). "Rethinking Soil Diagnostics: Why IoT-Based Soil Sensors are Essential for the Future of Precision Agriculture," *2024 IEEE Asia-Pacific Conference on Geoscience, Electronics and Remote Sensing Technology (AGERS)*, Manado, Indonesia, 2024, pp. 318-323. DOI: 10.1109/AGERS65212.2024.10932884

Gupta, R., et al. (2023). Self-healing through error detection, attribution, and re-training in conversational AI systems. Amazon Science Technical Report.

Haas, R., Lukachev, E., & Scheidl, R. (2014). An RC Filter for Hydraulic Switching Control with a Transmission Line Between Valves and Actuator. *Int. J. Fluid Power*, *15*, 139–151.

Hailu, T. G., Guo, X., Si, H., Li, L., & Zhang, Y. (2024). Theories and Methods for Indoor Positioning Systems: A Comparative Analysis, Challenges, and Prospective Measures. *Sensors (Basel)*, *24*(21), 6876. DOI: 10.3390/s24216876 PMID: 39517773

Han, T.-T., Nguyen Dinh, P., Nguyen Duc, T., Nguyen Long, V., & Tan, H. D. (2024). An Efficient Method for BLE Indoor Localization Using Signal Fingerprint. *EAI Endorsed Transactions on Industrial Networks and Intelligent Systems*, *12*(1). Advance online publication. DOI: 10.4108/eetinis.v12i1.6571

He, J., Deng, Z., Hu, E., Huang, Y., & Pan, T. (2024). Hybrid Positioning Framework with Innovation Based Judgment in Complex Indoor Environments. *2024 IEEE 14th International Conference on Electronics Information and Emergency Communication (ICEIEC)*, 1–4. DOI: 10.1109/ICEIEC61773.2024.10561735

Homma, S., Ida, Y., Ohira, Y., Kuroda, S., & Matsumoto, T. (2024). Study on Effective CSI Feature Extraction and Deep Learning Based Indoor Positioning Method. *2024 IEEE VTS Asia Pacific Wireless Communications Symposium (APWCS)*, 1–5. DOI: 10.1109/APWCS61586.2024.10679303

Horowitz, R. (2022). Error resolution and digital trust in financial services. *Journal of Digital Customer Experience*, *8*(2), 45–59.

Hsu, S. T., Whittier, R. J., & Mead, C. A. (1970). Physical model for burst noise in semiconductor devices. *Solid-State Electronics*, *13*(7), 1055–1056. DOI: 10.1016/0038-1101(70)90102-4

Hu, X., Chen, L., Lian, X., Wang, T., & Cai, J. (2023). Crowdsourcing-based high-precision Bluetooth indoor location method for adapting to environmental dynamics. *2023 IEEE 98th Vehicular Technology Conference (VTC2023-Fall)*, 1–7. DOI: 10.1109/VTC2023-Fall60731.2023.10333616

Huang, J., Junginger, S., Liu, H., & Thurow, K. (2023). Indoor Positioning Systems of Mobile Robots: A Review. *Robotics (Basel, Switzerland)*, *12*(2), 47. DOI: 10.3390/robotics12020047

Huang, Y., Ando, T., Sebastian, A., Chang, M.-F., Yang, J. J., & Xia, Q. (2024). Memristor-Based Hardware Accelerators for Artificial Intelligence. *Nat. Rev. Electr. Eng.*, *1*(5), 286–299. DOI: 10.1038/s44287-024-00037-6

Huang, Z., Fey, M., Liu, C., Beysel, E., Xu, X., & Brecher, C. (2023). Hybrid Learning-Based Digital Twin for Manufacturing Process: Modeling Framework and Implementation. *Robotics and Computer-integrated Manufacturing*, *82*(102545), 102545. DOI: 10.1016/j.rcim.2023.102545

Industry 5.0 - Towards a sustainable, human-centric and resilient European industry. (n.d.). Research and Innovation. https://research-and-innovation.ec.europa.eu/knowledge-publications-tools-and-data/publications/all-publications/industry-50-towards-sustainable-human-centric-and-resilient-european-industry_en

ISO. (2019). *Digital twin framework for manufacturing*. ISO/IEC JTC 1/SC 42 Artificial Intelligence. https://www.iso.org/standard/75066.html

Iwasa, M., Kubota, N., & Toda, Y. (2021, October). Multi-scale batch-learning growing neural gas for topological feature extraction in navigation of mobility support robots. In *The 7th International Workshop on Advanced Computational Intelligence and Intelligent Informatics (IWACIII 2021)*.

Jada, I., & Mayayise, T. O. (2024). The Impact of Artificial Intelligence on Organisational Cyber Security: An Outcome of a Systematic Literature Review. *Data and Information Management*, *8*(2), 100063. DOI: 10.1016/j.dim.2023.100063

Jamshidi, M., Lotfi, S., Siahkamari, H., Bleca, T., & Talla, J. (2024). An intelligent digital twinning approach for complex circuits. *Applied Soft Computing*, *154*, 111327. https://doi.org/. DOI: 10.1016/j.asoc.2024.111327

Jianyong, Z., Haiyong, L., Zili, C., & Zhaohui, L. (2014). RSSI based Bluetooth low energy indoor positioning. *2014 International Conference on Indoor Positioning and Indoor Navigation (IPIN)*, 526–533. DOI: 10.1109/IPIN.2014.7275525

Jin, G., Wu, H., Yin, Y., Zheng, L., & Zhuang, Y. (2024). A High-Accuracy RC Time Constant Auto-Tuning Scheme for Integrated Continuous-Time Filters. *Micromachines*, *15*(1), 166. DOI: 10.3390/mi15010166 PMID: 38276865

Ji, T., & Xu, X. (2025). Exploring the Integration of Cloud Manufacturing and Cyber-Physical Systems in the Era of Industry 4.0 – An OPC UA Approach. *Robotics and Computer-integrated Manufacturing*, *93*(102927), 102927. DOI: 10.1016/j.rcim.2024.102927

Johnson, J. B. (1971, February). Electronic noise: The first two decades. *IEEE Spectrum*, *8*(2), 42–46. DOI: 10.1109/MSPEC.1971.5217957

Johnson, M.. (2023). Customer sentiment analysis during network outages using NLP. *ACM Transactions on Social Computing*, *6*(3), 1–22.

Juodelyte, D., Cheplygina, V., Graversen, T., & Bonnet, P. (2022). Predicting Bearings' Degradation Stages for Predictive Maintenance in the Pharmaceutical Industry. *arXiv preprint arXiv:2203.03259*. https://arxiv.org/abs/2203.03259 DOI: 10.1145/3534678.3539057

Jurczenia, K., & Rak, J. (2022). A survey of Vehicular Network Systems for Road traffic management. *IEEE Access : Practical Innovations, Open Solutions*, *10*, 42365–42385. DOI: 10.1109/ACCESS.2022.3168354

Kaltsoukalas, K., Makris, S., & Chryssolouris, G. (2014). On generating the motion of industrial robot manipulators. *Robotics and Computer-integrated Manufacturing*, *32*, 65–71. DOI: 10.1016/j.rcim.2014.10.002

Kamariotis, A., Tatsis, K., Chatzi, E., Goebel, K., & Straub, D. (2023). A metric for assessing and optimizing data-driven prognostic algorithms for predictive maintenance. *arXiv preprint arXiv:2306.03759*. https://arxiv.org/abs/2306.03759

Karakusak, M. Z., Kivrak, H., Watson, S., & Ozdemir, M. K. (2023). Cyber-WISE: A Cyber-Physical Deep Wireless Indoor Positioning System and Digital Twin Approach. *Sensors (Basel)*, *23*(24), 9903. DOI: 10.3390/s23249903 PMID: 38139747

Karkaria, V., Tsai, Y. K., Chen, Y. P., & Chen, W. (2025). An Optimization-Centric Review on Integrating Artificial Intelligence and Digital Twin Technologies in Manufacturing. *Engineering Optimization*, *57*(1), 161–207. DOI: 10.1080/0305215X.2024.2434201

Kavitha, S., Visalini, D. S., Subasree, P., & Vedhika, A. L. (2025) Assessing Low Power Adder with Carbon Nanotube FET using LTSPICE simulations. *Proceeding of 2025 International Conference on Electronics and Renewable Systems (ICEARS)*, Tuticorin, India, 2025, pp. 406-409, DOI: 10.1109/ICEARS64219.2025.10940106

Kelemenová, T., Benedik, O., & Koláriková, I. (2020). Signal Noise Reduction and Filtering. *Acta Mechatronica*, *5*(2), 29–34. DOI: 10.22306/am.v5i2.65

Khafagy, A. A., & Abdellatif, M. M. (2024). VLC-IPS: A Visible Light Communications Based Indoor Positioning System. *2024 International Telecommunications Conference (ITC-Egypt)*, 172–177. DOI: 10.1109/ITC-Egypt61547.2024.10620517

Kharchenko, V., Illiashenko, O., Morozova, O., & Sokolov, S. (2020). Combination of digital twin and artificial intelligence in manufacturing using industrial IoT. *Proceeding of 2020 IEEE 11th International Conference on Dependable Systems, Services and Technologies (DESSERT)*, 196–201. DOI: 10.1109/DESSERT50317.2020.9125038

Khdoudi, A., Masrour, T., El Hassani, I., & El Mazgualdi, C. (2024). A Deep-Reinforcement-Learning-Based Digital Twin for Manufacturing Process Optimization. *Systems*, *12*(2), 38. DOI: 10.3390/systems12020038

Kim, D., & Choi, J.-Y. (2025). Development of a Common API for Multiple Ethernet Fieldbus Protocols in Embedded Slave Devices. *Electronics (Basel)*, *14*(3), 613. DOI: 10.3390/electronics14030613

Kim, J., & Park, S. (2025). Attention mechanisms for multivariate network error detection. *Neural Networks*, *165*, 832–843.

Kingma, D. P., & Ba, J. L. (2015). Adam: A method for stochastic optimization. Procs. of the 3rd International Conference on Learning Representations (ICLR2015), pp. 1– 15.

Kirdan, E., & Waedt, K. (2024). Modbus Covert Channels Within Industrial Automation and Control Systems. *Proceeding of IECON 2024 - 50th Annual Conference of the IEEE Industrial Electronics Society*, 1–6. DOI: 10.1109/IECON55916.2024.10905882

Klepac, G. (2002). REFII model-Model for recognition patterns in time series. *Conference Proceedings of the International Conference on Information Technology Interfaces.*

Klepac, G. (2007). Time series analysis using a unique model of transformation. *Journal of Information and Organizational Sciences*, *31*(2), 1–13. DOI: 10.29133/jios

Klepac, G. (2023). Time series error detection solution based on REFII model and concept of nontemporal expansion. *CEUR Workshop Proceedings*, *3900*, 1–9. https://ceur-ws.org/Vol-3900/

Klus, R., Talvitie, J., Torres-Sospedra, J., Quezada Gaibor, D. P., Casteleyn, S., Cabric, D., & Valkama, M. (2024). C2R: A Novel ANN Architecture for Boosting Indoor Positioning With Scarce Data. *IEEE Internet of Things Journal*, *11*(20), 32868–32882. DOI: 10.1109/JIOT.2024.3420122

Kohonen, T. (2001). Self-Organizing Maps. In *Springer series in information sciences*. DOI: 10.1007/978-3-642-56927-2

Kong, R., Li, R., Ma, Z., & Yu, Y. (2024). Indoor positioning systems based on visible light communication. *Applied and Computational Engineering*, *78*(1), 162–168. DOI: 10.54254/2755-2721/78/20240460

Kopytov, S. M., & Ulyanov, A. V. (2029) Development and Research of Marker Isolation Schemes from Low-Frequency PLC Signal. In *Proceedings of the 2019 International Multi-Conference on Industrial Engineering and Modern Technologies (FarEastCon)*, Vladivostok, Russia, 1–4 October 2019; IEEE: Piscataway, NJ, USA, 2019, 15, 1–4.

Koulamas, C., & Kalogeras, A. (2018). Cyber-Physical Systems and Digital Twins in the Industrial Internet of Things [cyber-physical systems]. *Computer*, *51*(11), 95–98. DOI: 10.1109/MC.2018.2876181

Krishnaveni, S., Chen, T. M., Sathiyanarayanan, M., & Amutha, B. (2024). CyberDefender: An Integrated Intelligent Defense Framework for Digital-Twin-Based Industrial Cyber-Physical Systems. *Cluster Computing*, *27*(6), 7273–7306. DOI: 10.1007/s10586-024-04320-x

Kritzinger, W., Karner, M., Traar, G., Henjes, J., & Sihn, W. (2018). Digital Twin in manufacturing: A categorical literature review and classification. *IFAC-PapersOnLine*, *51*(11), 1016–1022. DOI: 10.1016/j.ifacol.2018.08.474

Kubota, N. (2021, December). Multiscopic topological twin in robotics. In The 28th International Conference on Neural Information Processing (ICONIP2021), Bali, Indonesia, Online.

Kubota, N., Yuki, K., & Baba, N. (2009). Integration of intelligent technologies for simultaneous localization and mapping. *2009 ICCAS-SICE*, 4981–4986. https://ieeexplore.ieee.org/iel5/5305883/5332438/05334643.pdf

Kudari, J. M., & Samal, A. (2021). Smart Agriculture and Smart Farming using IoT Technology. *Journal of Physics: Conference Series*, *2089*(1), 012038. DOI: 10.1088/1742-6596/2089/1/012038

Kumar, V., & Singh, N. (2024). Polar code optimization for 5G networks using dynamic error pattern analysis. *Wireless Communications and Mobile Computing*, *2024*, 1–15.

Kuzin, T., & Borovicka, T. (2016). Early Failure Detection for Predictive Maintenance of Sensor Parts, ITAT 2016 Proceedings, *CEUR Workshop Proceedings* Vol. 1649, 123–130.

Ladegourdie, M., & Kua, J. (2022). Performance Analysis of OPC UA for Industrial Interoperability towards Industry 4.0. *IoT*, *3*(4), 507–525. DOI: 10.3390/iot3040027

Lai, P., Gonzalez, D., Madhusoodhanan, S., Sabbar, A., Ahmed, S., Dong, B., Wang, J., Mantooth, H. A., Yu, S.-Q., & Chen, Z. (2022). Development of LTCC-Packaged Optocouplers as Optical Galvanic Isolation for High-Temperature Applications. *Scientific Reports*, *12*(1), 1–11. DOI: 10.1038/s41598-022-15631-7 PMID: 35804033

Latrach, A. (2023). Application of deep learning for predictive maintenance of oilfield equipment. *arXiv preprint arXiv:2306.11040*. https://arxiv.org/abs/2306.11040

Lau, L. (2017). Wavelet Packets Based Denoising Method for Measurement Domain Repeat-Time Multipath Filtering in GPS Static High-Precision Positioning. *GPS Solutions*, *21*(2), 461–474. DOI: 10.1007/s10291-016-0533-1

Lee, H.. (2024). Real-time network anomaly detection using LSTM-autoencoder hybrid models. *Computer Networks*, *242*, 110–125.

Le, L. H., Ngo, T. Q., Toan, N. D., Nguyen, C. C., & Phong, B. H. (2024). Integration of Modbus-Ethernet Communication for Monitoring Electrical Power Consumption, Temperature, and Humidity. [JRC]. *Journal of Robotics and Control*, *5*(6), 1618–1625. DOI: 10.18196/jrc.v5i6.22456

Levin, S. (2024). Enhancing predictive maintenance in the industrial sector: A comparative analysis of machine learning models. *AIP Conference Proceedings*, *3243*(1), 020035. DOI: 10.1063/5.0247617

Li, Y., Zhang, R., Li, H., & Shao, X. (2020). Dynamic Attention Graph Convolution Neural Network of Point Cloud Segmentation for Defect Detection. Procs. of 2020 IEEE International Conference on Artificial Intelligence and Information Systems (ICAIIS), pp. 18–23. DOI: 10.1109/ICAIIS49377.2020.9194799

Liang, C. Y., Le, T. H., Ham, Y., Mantha, B. R. K., Cheng, M. H., & Lin, J. J. (2024). Ethics of Artificial Intelligence and Robotics in the Architecture, Engineering, and Construction Industry. *Automation in Construction*, *62*, 105369. Advance online publication. DOI: 10.1016/j.autcon.2024.105369

Liang, F., Mukherjee, S., & West, M. (2007). The Use of Unlabeled Data in Predictive Modeling. *Statistical Science*, *22*(2), 189–205. DOI: 10.1214/088342307000000032

Liu, J., Deng, Z., & Hu, E. (2024). An NLOS Ranging Error Mitigation Method for 5G Positioning in Indoor Environments. *Applied Sciences (Basel, Switzerland)*, *14*(9), 3830. DOI: 10.3390/app14093830

Liu, Y., Wang, Y., & Zhang, J. (2012). New Machine Learning Algorithm: Random Forest. In Liu, B., Ma, M., & Chang, J. (Eds.), Lecture Notes in Computer Science: Vol. 7473. *Information Computing and Applications. ICICA 2012*. Springer., DOI: 10.1007/978-3-642-34062-8_32

Li, X., Yan, Z., Huang, L., Chen, S., & Liu, M. (2020). High-Accuracy and Real-Time Indoor Positioning System Based on Visible Light Communication and Mobile Robot. *International Journal of Optics*, *2020*, 1–11. DOI: 10.1155/2020/3124970

Lo Brano, V., Ciulla, G., & Di Falco, M. (2014). Artificial Neural Networks to Predict the Power Output of a PV Panel. *International Journal of Photoenergy*, *2014*, 193083. DOI: 10.1155/2014/193083

Lyu, Z., Chan, T. T.-L., Hung, T. Y.-T., Ji, H., Leung, G., & Lun, D. P.-K. (2024). Deep Learning-Based Bluetooth Low-Energy 5.1 Multianchor Indoor Positioning with Attentional Data Filtering. *Advanced Intelligent Systems*, *6*(1), 2300292. DOI: 10.1002/aisy.202300292

Magsi, H., Sodhro, A. H., Chachar, F. A., & Abro, S. A. K. (2018) Analysis of Signal Noise Reduction by Using Filters. In Proceedings of the 2018 International Conference on Computing, Mathematics and Engineering Technologies (iCoMET), Sukkur, Pakistan, 3–4 March 2018; 1–6. DOI: 10.1109/ICOMET.2018.8346412

Malawade, A. V., Costa, N. D., Muthirayan, D., Khargonekar, P. P., & Al Faruque, M. A. (2021). Neuroscience-Inspired Algorithms for the Predictive Maintenance of Manufacturing Systems. arXiv. https://arxiv.org/abs/2102.11450

Malik, R. F., Gustifa, R., Farissi, A., Stiawan, D., Ubaya, H., Ahmad, M. R., & Khirbeet, A. S. (2019). The Indoor Positioning System Using Fingerprint Method Based Deep Neural Network. *IOP Conference Series. Earth and Environmental Science*, *248*, 012077. DOI: 10.1088/1755-1315/248/1/012077

Martinetz, T., & Schulten, K. (1991). A" neural-gas" network learns topologies. *Artificial Neural Networks*, *1*, 397–402.

Martínez, J. V. Y., Skarmeta, A. F., Zamora-Izquierdo, M. A., & Ramallo-González, A. P. (2020). "IoT-based data management for Smart Agriculture," *2020 Second International Conference on Embedded & Distributed Systems (EDiS)*, Oran, Algeria, 2020, pp. 41-46. DOI: 10.1109/EDiS49545.2020.9296443

Martins, A., Lucas, J., Costelha, H., & Neves, C. (2023). CNC Machines Integration in Smart Factories Using OPC UA. *Journal of Industrial Information Integration*, *34*(100482), 100482. DOI: 10.1016/j.jii.2023.100482

Massaro, A., Epicoco, N., Loseto, G., & Ardito, C. (2024 d), "IIOT Cyberattacks in Industrial Field Control Systems: PID Attack and Hardware Trojan Effects," *IEEE 20th International Conference on Automation Science and Engineering (CASE)*, Bari, Italy, 2024, pp. 1702-1707, DOI: 10.1109/CASE59546.2024.10711839

Massaro, A. (2021). *Electronics in Advanced Research Industries: Industry 4.0 to Industry 5.0 Advances*. Wiley., DOI: 10.1002/9781119716907

Massaro, A. (2022). Advanced Control Systems in Industry 5.0 Enabling Process Mining. *Sensors (Basel)*, *22*(22), 8677. DOI: 10.3390/s22228677 PMID: 36433272

Massaro, A. (2023 a). Process Mining in Production Management, Intelligent Control, and Advanced KPI for Dynamic Process Optimization: Industry 5.0 Production Processes. In *Perspectives and Considerations on the Evolution of Smart Systems* (pp. 1–17). IGI Global., DOI: 10.4018/978-1-6684-7684-0.ch001

Massaro, A. (2023d). Intelligent Materials and Nanomaterials Improving Physical Properties and Control Oriented on Electronic Implementations. *Electronics (Basel)*, *12*(18), 3772. DOI: 10.3390/electronics12183772

Massaro, A. (2024a). ANNs Predicting Noisy Signals in Electronic Circuits: A Model Predicting the Signal Trend in Amplification Systems. *AI*, *5*(2), 533–549. DOI: 10.3390/ai5020027

Massaro, A. (2024b). Artificial Intelligence Enabling Denoising in Passive Electronic Filtering Circuits for Industry 5.0 Machines. *Machines (Basel)*, *12*(8), 551. DOI: 10.3390/machines12080551

Massaro, A. (2024c). Artificial Intelligence Signal Control in Electronic Optocoupler Circuits Addressed on Industry 5.0 Digital Twin. *Electronics (Basel)*, *13*(22), 4543. DOI: 10.3390/electronics13224543

Massaro, A., Epicoco, N., Loseto, G., & Ardito, C. A. (2025a). AI-Based Detection and Simulation of Hardware Trojan Attacks in Industrial Applications. *IEEE Access : Practical Innovations, Open Solutions*, *13*, 50967–50979. DOI: 10.1109/ACCESS.2025.3552112

Massaro, A., Santarsiero, F., & Schiuma, G. (2025b). Advanced Electronic Controller Circuits Enabling Production Processes and AI-driven KM in Industry 5.0. *Journal of Industrial Information Integration*, *45*, 10841. DOI: 10.1016/j.jii.2025.100841

Matsuyama, H., Nagata, F., Miki, K., Kato, H., & Watanabe, K. (2021). Evaluation System Using 3D Convolutional Neural Network for Spark Test. Procs. of 22nd SICE SI Conference, pp. 328–331 (in Japanese).

Maturana, D., & Scherer, S. (2015). VoxNet:A 3D Convolutional Neural Network for Real-Time Object Recognition. Procs. of 2015 IEEE/RSJ International Conference on Intelligent Robots and Systems (IROS), Hamburg, Germany, pp. 922–928.

Mehmood. (2010). Indoor Positioning System Using Artificial Neural Network. *Journal of Computer Science*, *6*(10), 1219–1225. DOI: 10.3844/jcssp.2010.1219.1225

Miková, Ľ., Kelemen, M., Virgala, I., & Michna, M. (2014). Simulation model of manipulator for model based design. *Applied Mechanics and Materials*, *611*, 175–182. . DOI: 10.4028/www.scientific.net/AMM.611.175

Milosevic, Z., & van Schalkwyk, P. (2024). Towards Responsible Digital Twins. In Sales, T. P., de Kinderen, S., Proper, H. A., Pufahl, L., Karastoyanova, D., & van Sinderen, M. (Eds.), *Enterprise Design, Operations, and Computing. EDOC 2023 Workshops. EDOC 2023. Lecture Notes in Business Information Processing* (Vol. 498). Springer., DOI: 10.1007/978-3-031-54712-6_8

Modad, O. A. A., Ryska, J., Chehade, A., & Ayoub, G. (2024). Revolutionizing sheet metal stamping through industry 5.0 digital twins: A comprehensive review. *Journal of Intelligent Manufacturing*. Advance online publication. DOI: 10.1007/s10845-024-02453-9

Mokhtar, M., Robu, V., Flynn, D., Higgins, C., Whyte, J., Loughran, C., & Fulton, F. (2021). Prediction of Voltage Distribution Using Deep Learning and Identified Key Smart Meter Locations. *Energy AI*, *6*, 100103. DOI: 10.1016/j.egyai.2021.100103

Mondal, C., Azam, M. K., Ahmad, M., Hasan, S. M. K., & Islam, M. R. (2015) Design and Implementation of a Prototype Electrooculography Based Data Acquisition System. *InProceedings of the 2015 International Conference on Electrical Engineering and Information Communication Technology (ICEEICT)*, Savar, Bangladesh, 21–23 May 2015; IEEE: Piscataway, NJ, USA; 1–6. DOI: 10.1109/ICEEICT.2015.7307395

Monjur, M. M. R., Heacock, J., Calzadillas, J., Mahmud, M. D. S., Roth, J., Mankodiya, K., Sazonov, E., & Yu, Q. (2022). Hardware Security in Sensor and Its Networks. *Frontiers in Sensors*, *3*, 850056. DOI: 10.3389/fsens.2022.850056

Monjur, M. R., Sunkavilli, S., & Yu, Q. ADobf (2020) Obfuscated Detection Method against Analog Trojans on I2C Master-Slave Interface. *Proceedings of the 2020 IEEE 63rd International Midwest Symposium on Circuits and Systems (MWSCAS)*, Springfield, MA, USA, 9–12 August 2020; pp. 1064–1067. DOI: 10.1109/MWSCAS48704.2020.9184680

Monjur, M., Calzadillas, J., & Yu, Q. (2023). Hardware Security Risks and Threat Analyses in Advanced Manufacturing Industry. *ACM Transactions on Design Automation of Electronic Systems*, *28*(5), 1–22. DOI: 10.1145/3603502

Müller, T.. (2023). Autocorrelation-aware deep learning for industrial IoT time-series forecasting. *IEEE Internet of Things Journal*, *10*(15), 13278–13289.

Nagata, F., Hirono, M., Yonehara, H., Kato, H., & Watanabe, K. (2022). Steel Spark Test System Using Video Processing and 2D CNN. Procs. of The Institute of Industrial Applications Engineers (IIAE) Annual Conference 2022, pp. 41–42 (in Japanese).

Nagata, F., Abe, R., Sakata, S., Watanabe, K., & Habib, M. K. (2024). Hyper CLS Data-based Robotic Interface and Its Application to Intelligent Peg-in-Hole Task Robot Incorporating a CNN Model for Defect Detection. *Machines (Basel)*, *12*(11), 1–14. DOI: 10.3390/machines12110757

Nagata, F., Habib, M. K., & Watanabe, K. (2021). Design and implementation of convolutional neural network-based SVM technique for manufacturing defect detection. *International Journal of Mechatronics and Automation*, *8*(2), 53–61. DOI: 10.1504/IJMA.2021.115240

Nakata, T. (2012). Development of Automated Spark Testing Technique by Image Processing to Measure Carbon Content in Steel Materials. Procs. of IFAC Workshop on Automation in the Mining, Mineral and Metal Industries, Gifu, Japan, pp. 118–119. DOI: 10.3182/20120910-3-JP-4023.00029

Nascimento, E. G. S., Liang, J. S., Figueiredo, I. S., & Guarieiro, L. L. N. (2022). T4PdM: A Deep Neural Network Based on the Transformer Architecture for Fault Diagnosis of Rotating Machinery. *arXiv preprint arXiv:2204.03725*. https://arxiv.org/abs/2204.03725

National Research Foundation Singapore. (2018). *Virtual Singapore: A 3D smart city model*. https://www.nrf.gov.sg/programmes/virtual-singapore

Nguyen, S.; Abdelhakim, M. and Kerestes, R. (2021). Survey Paper of Digital Twins and their Integration into Electric Power Systems. *Proceeding of 2021 IEEE Power & Energy Society General Meeting (PESGM)*, 01–05. .DOI: 10.1109/PESGM46819.2021.9638011

Nguyen, T.. (2024). Graph neural networks for root cause analysis in telecom networks. *Expert Systems with Applications*, *238*, 121–134.

Obo, T., Kubota, N., Toda, Y., & Masuyama, N. (2024). Fast multi-scale Batch-Learning growing neural gas. In *Topics in intelligent engineering and informatics* (pp. 13–33). DOI: 10.1007/978-3-031-58257-8_2

Ogar, V. N., Hussain, S., & Gamage, K. A. A. (2023). The Use of Artificial Neural Network for Low Latency of Fault Detection and Localisation in Transmission Line. *Heliyon*, *9*(2), e13376. DOI: 10.1016/j.heliyon.2023.e13376 PMID: 36816249

Ohalete, N. C., Aderibigbe, A. O., Ani, E. C., Ohenhen, P. E., & Akinoso, A. (2023). Advancements in predictive maintenance in the oil and gas industry: A review of AI and data science applications. *World Journal of Advanced Research and Reviews*, *20*(3), 167–181. DOI: 10.30574/wjarr.2023.20.3.2432

Oshima, T., Maio, K., Hioe, W., Shibahara, Y., & Doi, T. (2003) Automatic Tuning of RC Filters and Fast Automatic Gain Control for CMOS Low-IF Transceiver. *InProceedings of the IEEE 2003 Custom Integrated Circuits Conference*, San Jose, CA, USA, 24 September 2003; 5–8. DOI: 10.1109/CICC.2003.1249348

Oshio, K., Kaneko, K., & Kubota, N. (2021, October). Multi-scopic simulation for human-robot interactions based on multi-objective behavior coordination. In *International Workshop on Advanced Computational Intelligence and Intelligent Informatics* (pp. 3-8).

Owolabi, O. S., Uche, P. C., Adeniken, N. T., Ihejirika, C., Islam, R. B., & Chhetri, B. J. T. (2024). Ethical implication of artificial intelligence (AI) adoption in financial decision making. *Computer and Information Science*, *17*(1), 49. DOI: 10.5539/cis.v17n1p49

Pandey, U., Vashisth, A., & Mishra, A. (2024). "AI and IoT Enabled Soil Health Monitoring System," 2024 Second International Conference Computational and Characterization Techniques in Engineering & Sciences (IC3TES), Lucknow, India, 2024, pp. 1-6. DOI: 10.1109/IC3TES62412.2024.10877563

Papacharalampopoulos, A., Sabatakakis, K., Karagianni, O. M., & Stavropoulos, P. (2025). Digital Twins of Manufacturing Processes Under Industry 5.0. In Alexopoulos, K., Makris, S., & Stavropoulos, P. (Eds.), *Advances in Artificial Intelligence in Manufacturing II. ESAIM 2024. Lecture Notes in Mechanical Engineering*. Springer., DOI: 10.1007/978-3-031-86489-6_1

Parmar, A., Katariya, R., & Patel, V. (2019). A Review on Random Forest: An Ensemble Classifier. In: Hemanth, J., Fernando, X., Lafata, P., Baig, Z. (eds) International Conference on Intelligent Data Communication Technologies and Internet of Things (ICICI) 2018. ICICI 2018. *Lecture Notes on Data Engineering and Communications Technologies*, vol. 26. Springer, Cham. https://doi.org/DOI: 10.1007/978-3-030-03146-6_86

Pascacio, P., Casteleyn, S., Torres-Sospedra, J., Lohan, E. S., & Nurmi, J. (2021). Collaborative Indoor Positioning Systems: A Systematic Review. *Sensors (Basel)*, *21*(3), 1002. DOI: 10.3390/s21031002 PMID: 33540703

Patel, S., Kovarovics, A., More. A., Chun, C., Fletcher, M., Tanveer, M.H. (2025). Towards Autonomous Robotic Grasping: 3D Point Cloud Analysis with Stereo Cameras in Unpredictable Environments. SoutheastCon 2025, Concord, NC, USA, pp. 520–525, .DOI: 10.1109/SoutheastCon56624.2025.10971505

Paul, R. P. (1981). *Robot manipulators: mathematics, programming, and control.* The MIT Press.

Pei, M., Liu, N., Zhao, B., & Sun, H. (2023). Self-Supervised Learning for Industrial Image Anomaly Detection by Simulating Anomalous Samples. International Journal of Computational Intelligence Systems, Vol. 16, No. 152, 15 pages. DOI: 10.1007/s44196-023-00328-0

Pero, C., Bakshi, S., Nappi, M., & Tortora, G. (2024). IoT-Driven Machine Learning for Precision Viticulture Optimization. *IEEE Journal of Selected Topics in Applied Earth Observations and Remote Sensing*, *17*, 2437–2447. DOI: 10.1109/JSTARS.2023.3345473

Pešić, S. (2024). *Modern Challenges in Indoor Positioning Systems: AI to the Rescue*. IntechOpen., DOI: 10.5772/intechopen.1005354

Pincheira, M., Donini, E., Giaffreda, R., & Vecchio, M. (2020). *"A Blockchain-Based Approach To Enable Remote Sensing Trusted Data," 2020 IEEE Latin American GRSS & ISPRS Remote Sensing Conference*. LAGIRS.

Prasanthi, T. S., Rayavarapu, S. M., Rao, G. S., Goswami, R. K., & Kumar, G. S. (2024). RADIO FREQUENCY BASED INPAINTING FOR INDOOR LOCALIZATION USING MEMORYLESS TECHNIQUES AND WIRELESS TECHNOLOGY. *Informatyka, Automatyka. Pomiary w Gospodarce i Ochronie Środowiska*, *14*(4), 10–15. DOI: 10.35784/iapgos.6236

Prunella, M., Scardigno, R. M., Buongiorno, D., Brunetti, A., Longo, N., Carli, R., Dotoli, M., & Bevilacqua, V. (2023). Deep Learning for Automatic Vision-Based Recognition of Industrial Surface Defects: A Survey. *IEEE Access : Practical Innovations, Open Solutions*, *11*, 43370–43423. DOI: 10.1109/ACCESS.2023.3271748

Puckdeevongs, A. (2021). Indoor Localization using RSSI and Artificial Neural Network. *2021 9th International Electrical Engineering Congress (iEECON)*, 479–482. DOI: 10.1109/iEECON51072.2021.9440373

Pu, S., Zhang, H., Mao, C., & Yang, G. (2021) A Classification Based on Random Forest for Partial Discharge Sources. *InProceedings of the 2021 33rd Chinese Control and Decision Conference (CCDC)*, Kunming, China, 22–24 May 2021; IEEE: Piscataway, NJ, USA; 2307–2311. DOI: 10.1109/CCDC52312.2021.9602056

Qi, C. R., Su, H., Mo, K., & Guibas, L. J. (2017). PointNet: Deep Learning on Point Sets for 3D Classification and Segmentation. Procs. of the IEEE Conference on Computer Vision and Pattern Recognition (CVPR), pp. 652–660.

Qiao, J., Yang, F., Liu, J., Huang, G., Zhang, W., & Li, M. (2024). Advancements in Indoor Precision Positioning: A Comprehensive Survey of UWB and Wi-Fi RTT Positioning Technologies. *Network (Bristol, England)*, *4*(4), 545–566. DOI: 10.3390/network4040027

Qi, L., Liu, Y., Yu, Y., Chen, L., & Chen, R. (2024). Current Status and Future Trends of Meter-Level Indoor Positioning Technology: A Review. *Remote Sensing (Basel)*, *16*(2), 398. DOI: 10.3390/rs16020398

Rahman, M. S., Ghosh, T., Aurna, N. F., Kaiser, M. S., Anannya, M., & Hosen, A. S. M. S. (2023). Machine Learning and Internet of Things in Industry 4.0: A Review. *Measurement. Sensors*, *28*, 100822. DOI: 10.1016/j.measen.2023.100822

Ramos, G. A., & Costa-Castelló, R. (2022). Energy Management Strategies for Hybrid Energy Storage Systems Based on Filter Control: Analysis and Comparison. *Electronics (Basel)*, *11*(10), 1631. DOI: 10.3390/electronics11101631

Redmon, J., & Farhadi, A. (2017). YOLO9000: Better, Faster, Stronger. Procs. *2017 IEEE Conference on Computer Vision and Pattern Recognition*, pp. 6517–6525.

Riedmiller, M., & Braun, H. (1993) A direct adaptive method for faster backpropagation learning: theRPROP algorithm, *Proceedings of the IEEE International Conference on Neural Networks (ICNN),* vol. 16, pp. 586-591, Piscataway, 1993, NJ: IEEE. DOI: 10.1109/ICNN.1993.298623

Rodrigues, C. H. A., Maia, O. B., & Bessa, A. R. R.; W. Júnior S. S. and Carvalho, C. B. (2025) Communication Protocols for Digital Twins. Proceeding of 2025 IEEE International Conference on Consumer Electronics (ICCE), Las Vegas, NV, USA, pp. 1-5, DOI: 10.1109/ICCE63647.2025.10929808

Rodriguez-Gallo, Y., Cañas, H., & Cardona, M. (2024). "Agriculture 4.0 in Central America: Current Status, Challenges, and Future Trends," 2024 IEEE 42nd Central America and Panama Convention (CONCAPAN XLII), San Jose, Costa Rica, 2024, pp. 1-6. DOI: 10.1109/CONCAPAN63470.2024.10933593

Ronneberger, O., Fischer, P., & Brox, T. (2015). U-Net: Convolutional Networks for Biomedical Image Segmentation, Medical Image Computing and Computer-Assisted Intervention (MICCAI), LNCS, Vol. 9351, pp. 234–241.

Rossi, F.. (2023). Customer churn prediction during service disruptions using survival analysis. *Decision Support Systems*, *172*, 114–126.

Russell, S., Norvig, P., & Intelligence, A. (1995). A modern approach. *Artificial Intelligence. Prentice-Hall. Egnlewood Cliffs*, *25*(27), 79–80.

Sahnoun, S., Souissi, R., Chiboub, S., Chabchoub, A., Baazaoui, M. K., Fakhfakh, A., & Derbel, F. (2025). Enhancing Localization Accuracy and Reducing Processing Time in Indoor Positioning Systems: A Comparative Analysis of AI Models. *Sensors (Basel), 25*(2), 475. DOI: 10.3390/s25020475 PMID: 39860845

Salari, O., Zaad, K. H., Bakhshai, A., & Jain, P. (2018) Filter Design for Energy Management Control of Hybrid Energy Storage Systems in Electric Vehicles. In *Proceedings of the 2018 9th IEEE International Symposium on Power Electronics for Distributed Generation Systems (PEDG)*, Charlotte, NC, USA, 25–28 June 2018; IEEE: Piscataway, NJ, USA; 1–7. DOI: 10.1109/PEDG.2018.8447608

Saputra, A. A., Botzheim, J., Sulistijono, I. A., & Kubota, N. (2015). Biologically inspired control system for 3-D locomotion of a humanoid biped robot. *IEEE Transactions on Systems, Man, and Cybernetics. Systems, 46*(7), 898–911. DOI: 10.1109/TSMC.2015.2497250

Saputra, A. A., Chin, W. H., Toda, Y., Takesue, N., & Kubota, N. (2019). Dynamic density Topological structure generation for Real-Time ladder affordance detection. *2021 IEEE/RSJ International Conference on Intelligent Robots and Systems (IROS)*, 3439–3444. DOI: 10.1109/IROS40897.2019.8968003

Saputra, A. A., Toda, Y., Botzheim, J., & Kubota, N. (2017). Neuro-Activity-Based Dynamic Path Planner for 3-D rough terrain. *IEEE Transactions on Cognitive and Developmental Systems, 10*(2), 138–150. DOI: 10.1109/TCDS.2017.2711013

Sarjerao, J. S., & Sudhagar, G. (2024). "Integration of Remote Sensing and IoT for Real-Time Monitoring of Irrigation in Smart Farming," 2024 1st International Conference on Innovative Sustainable Technologies for Energy, Mechatronics, and Smart Systems (ISTEMS), Dehradun, India, 2024, pp. 1-6. DOI: 10.1109/ISTEMS60181.2024.10560275

Sasaki, H., Kubota, N., & Taniguchi, K. (2008). Topological map and cell space map for SLAM of a mobile robot. *GESTS International Transactions on Computer and Engineering, 45*(3).

Sasikala, M., Mahaboob John, Y. M., & Jothi, B. (2024) Integrating Digital Twins with AI for Real-Time Intrusion Detection in Smart Infrastructure Networks. *Proceeding of 2024 International Conference on Intelligent Algorithms for Computational Intelligence Systems (IACIS)*, Hassan, India, 2024, pp. 1-6, DOI: 10.1109/IACIS61494.2024.10721892

Satomi, M., Masuta, H., & Kubota, N. (2009, March). Hierarchical growing neural gas for information structured space. In *2009* IEEE Symposium Series on Computational Intelligence 2009 (pp. 54-59). IEEE. DOI: 10.1109/RIISS.2009.4937906

Sawant, S. A., Adinarayana, J., & Durbha, S. S. (2014). "KrishiSense: A semantically aware web enabled wireless sensor network system for precision agriculture applications," *2014 IEEE Geoscience and Remote Sensing Symposium*, Quebec City, QC, Canada, 2014, pp. 4090-4093. DOI: 10.1109/IGARSS.2014.6947385

Scime, L., Singh, A., & Paquit, V. (2022). A Scalable Digital Platform for the Use of Digital Twins in Additive Manufacturing. *Manufacturing Letters*, *31*, 28–32. DOI: 10.1016/j.mfglet.2021.05.007

Sengupta, P., Mehta, A., & Rana, P. S. (2023). Predictive maintenance of armoured vehicles using machine learning approaches. *arXiv preprint arXiv:2307.14453*. https://arxiv.org/abs/2307.14453

Sethi, H. S., & Prasanth, P. V. (2023). "Optimizing Resource Allocation in Agriculture through IoT and AI," 2023 7th International Conference on Electronics, Communication and Aerospace Technology (ICECA), Coimbatore, India, 2023, pp. 1573-1578.

Shafi, U., Mumtaz, R., Iqbal, N., Zaidi, S. M. H., Zaidi, S. A. R., Hussain, I., & Mahmood, Z. (2020). A Multi-Modal Approach for Crop Health Mapping Using Low Altitude Remote Sensing, Internet of Things (IoT) and Machine Learning. *IEEE Access : Practical Innovations, Open Solutions*, *8*, 112708–112724. DOI: 10.1109/ACCESS.2020.3002948

Shah, S. S., Tan, D., & Kumar, S. C. (2024). RUL forecasting for wind turbine predictive maintenance based on deep learning. *arXiv preprint arXiv:2412.17823*. https://arxiv.org/abs/2412.17823

Sharma, S., Rani, S., Sharma, A., & Dogra, A. (2024). "Advancements in Image Fusion Techniques for IoT Systems: Current Trends and Applications," *2024 Second International Conference on Intelligent Cyber Physical Systems and Internet of Things (ICoICI)*, Coimbatore, India, 2024, pp. 368-373. DOI: 10.1109/ICoICI62503.2024.10696732

Shaterian, Z., & Mrozowski, M. (2022). Crack Detection in Metallic Surfaces Based on Dumbbell-Shaped Defected Ground Structures in Microstrip Technology. Procs. of 2022 24th International Microwave and Radar Conference (MIKON), Gdansk, Poland, pp. 1–4, DOI: 10.23919/MIKON54314.2022.9924958

Shen, Z., Arraño-Vargas, F., & Konstantinou, G. (2023). Artificial Intelligence and Digital Twins in Power Systems: Trends, Synergies and Opportunities. *Digital Twin*, *2*, 11. DOI: 10.12688/digitaltwin.17632.2

Shingare, S. S., Khampariya, P., & Bakre, S. M. (2023). Efficient Fault Detection and Location in Extra High Voltage Networks: An Artificial Neural Network (ANN)-Based Approach. *Int. J. Intell. Syst. Appl. Eng.*, *11*, 1051–1060.

Siddique, A., Kundu, R. K., Mode, G. R., & Hoque, K. A. (2023). RobustPdM: Designing robust predictive maintenance against adversarial attacks. *arXiv preprint arXiv:2301.10822*. https://arxiv.org/abs/2301.10822

Siemens, A. G. (2022). *Digital twin implementation in smart manufacturing – Amberg factory report.* https://new.siemens.com/global/en/company/stories/research-technologies/digital-twin.html

Singh, P. N., Saxena, U., & Chaudhary, M., and Pravartan, (2024). "Smart Agriculture: IoT-Enabled Precision Environmental Monitoring and Management," *2024 International Conference on Electrical Electronics and Computing Technologies (ICEECT)*, Greater Noida, India, 2024, pp. 1-6.

Singh, R. K., Berkvens, R., & Weyn, M. (2021). AgriFusion: An Architecture for IoT and Emerging Technologies Based on a Precision Agriculture Survey. *IEEE Access : Practical Innovations, Open Solutions*, *9*, 136253–136283. DOI: 10.1109/ACCESS.2021.3116814

Soori, M., Arezoo, B., & Dastres, R. (2023). Digital Twin for Smart Manufacturing, A Review. *Sustainable Manufacturing and Service Economics*, *2*(100017), 100017. DOI: 10.1016/j.smse.2023.100017

Soudan, B. (2024) Cybersecurity of Digital Twins in Industrial IoT Environments. *Proceeding of Advances in Science and Engineering Technology International Conferences (ASET)*, Abu Dhabi, United Arab Emirates, 2024, 1-6, DOI: 10.1109/ASET60340.2024.10708640

Stergiou, C. L., & Psannis, K. E. (2022). Digital Twin Intelligent System for Industrial Internet of Things-based Big Data Management and Analysis in Cloud Environments. *Virtual Reality & Intelligent Hardware*, *4*(4), 279–291. DOI: 10.1016/j.vrih.2022.05.003

Sun, F.-K., Lang, C. I., & Boning, D. S. (2025). Adjusting for autocorrelated errors in neural networks for time series. *Proceedings of the IEEE Conference on Neural Networks for Signal Processing*.

Sung, G.-M., Li, Z.-Y., & Yu, C.-P. (2024). Ethernet–USB Bridge Application-Specific Integrated Circuit Incorporating the User Datagram Protocol and Address Resolution Protocol. *IEEE Access : Practical Innovations, Open Solutions*, *12*, 106874–106883. DOI: 10.1109/ACCESS.2024.3436564

Sun, M., Cai, Z., & Zhao, N. (2022). Design of Intelligent Manufacturing System Based on Digital Twin for Smart Shop Floors. *International Journal of Computer Integrated Manufacturing*, *36*(4), 542–566. DOI: 10.1080/0951192X.2022.2128212

Sun, Y., Tang, X., Sun, X., Jia, D., Cao, Z., Pan, J., & Xu, B. (2019). Model Predictive Control and Improved Low-Pass Filtering Strategies Based on Wind Power Fluctuation Mitigation. *Journal of Modern Power Systems and Clean Energy*, *7*(3), 512–524. DOI: 10.1007/s40565-018-0474-5

Suresh Babu, C. V., Anniyappa, C. S. A., & Raut, A. (2025). Toward seamless mobility: Integrating connected and autonomous vehicles in smart cities through digital twins. In Iyer, S., Nayyar, A., Paul, A., & Naved, M. (Eds.), *Digital twins for smart cities and villages* (pp. 169–187). Elsevier., DOI: 10.1016/B978-0-443-28884-5.00008-7

Suresh Babu, C. V., Auroshaa, A., Sowmi Saltonya, M., & Sathyanarayanan, A. S. (2024). Cloud-Enabled Fire Safety in Industry 5.0 Smart Factories: Leveraging IoT and Sensor Networks for Real-Time Monitoring and Proactive Prevention. In Hassan, A., Dutta, P., Gupta, S., Mattar, E., & Singh, S. (Eds.), *Emerging Technologies in Digital Manufacturing and Smart Factories* (pp. 150–166). IGI Global., DOI: 10.4018/979-8-3693-0920-9.ch009

Thakkar, U., & Chaoui, H. (2022). Remaining Useful Life Prediction of an Aircraft Turbofan Engine Using Deep Layer Recurrent Neural Networks. *Actuators*, *11*(3), 67. DOI: 10.3390/act11030067

Thanasit Rithanasophon & Chatchai Wannaboon. (2023). *Integration_of_Machine_Learning_and_Kalman_Filter_Approach_for_Fingerprint_Indoor_Positioning.pdf*. DOI: 10.1109/ECTI-CON58255.2023.10153162

The MathWorks, Inc. (2025-1). https://www.mathworks.com/help/vision/ug/getting - started-with-yolo-v3.html Accessed on 4 February 2025

The MathWorks, Inc. (2025-2). https://www.mathworks.com/help/vision/ug/object -detection-using-yolo-v3-deep-learning.html Accessed on 4 February 2025

The MathWorks, Inc. (2025-3). https://jp.mathworks.com/help/vision/ref/unetlayers .html Accessed on 4 February 2025

The MathWorks, Inc. (2025-4). https://cvml-expertguide.net/terms/dl/inter-layer - connection/skip-connection/ Accessed on 4 February 2025

Thrun, S., Burgard, W., & Fox, D. (2005). *Probabilistic Robotics*. The MIT Press.

Thürer, M., Li, S. S., & Qu, T. (2022). Digital Twin Architecture for Production Logistics: The Critical Role of Programmable Logic Controllers (PLCs). *Procedia Computer Science, 200*, 710–717. DOI: 10.1016/j.procs.2022.01.269

Tian, D., Ke, C., Sun, B., Liang, H., Qian, Z., Wen, Q., Chen, X., Yang, C., Xu, M., & Zhao, Y. (2025). Quantum dot-based memristors for information processing and artificial intelligence applications. *Nanoscale, 17*(17), 10485–10505. Advance online publication. DOI: 10.1039/D5NR00136F PMID: 40211906

Toda, Y., & Kubota, N. (2013, August). Feature extraction based on hierarchical growing neural gas for informationally structured space. In *The 2013 International Joint Conference on Neural Networks (IJCNN)* (pp. 849-855). IEEE. DOI: 10.1109/IJCNN.2013.6706825

Toda, Y., Chin, W., & Kubota, N. (2017, December). Unsupervised neural network based topological learning from point clouds for map building. In *2017 International Symposium on Micro-NanoMechatronics and Human Science (MHS)* (pp. 1-6). IEEE. DOI: 10.1109/MHS.2017.8305188

Toda, Y., & Kubota, N. (2013). Self-Localization based on multiresolution map for remote control of multiple mobile robots. *IEEE Transactions on Industrial Informatics, 9*(3), 1772–1781. DOI: 10.1109/TII.2013.2261306

Toda, Y., & Kubota, N. (2018). Topological structure learning based enclosing formation behavior for monitoring system. *2022 IEEE International Conference on Systems, Man, and Cybernetics (SMC)*, 831–836. DOI: 10.1109/SMC.2018.00149

Toda, Y., Matsuno, T., & Minami, M. (2021). Multilayer batch learning growing neural gas for learning multiscale topologies. *Journal of Advanced Computational Intelligence and Intelligent Informatics, 25*(6), 1011–1023. DOI: 10.20965/jaciii.2021.p1011

Tran, D., Bourdev, L., Fergus, R., Torresani, L., & Paluri, M. (2015). Learning Spatiotemporal Features with 3D Convolutional Networks. Procs. of the IEEE International Conference on Computer Vision (ICCV2015), pp. 4489–4497.

Tuncer, S., & Tuncer, T. (2015). Indoor localization with Bluetooth technology using Artificial Neural Networks. *2015 IEEE 19th International Conference on Intelligent Engineering Systems (INES)*, 213–217. DOI: 10.1109/INES.2015.7329709

Usamentiaga, R., Lema, D. G., Pedrayes, O. D., & Garcia, D. F. (2022). Automated Surface Defect Detection in Metals: A Comparative Review of Object Detection and Semantic Segmentation Using Deep Learning. *IEEE Transactions on Industry Applications, 58*(3), 4203–4213. DOI: 10.1109/TIA.2022.3151560

Vassiliadis, P., Kotsiantis, S., & Kanellopoulos, D. (2022). A Context-Aware Unsupervised Predictive Maintenance Solution for Fleet Management. *Journal of Intelligent Information Systems*, *60*, 521–547. https://link.springer.com/article/10.1007/s10844-022-00744-2

Veeramanikandan, P., & Kumar, N. A. (2024). "Design and Implementation of an IoT-Based Control System for Precision Food Manufacturing," in Rathore, V.S., Piuri, V., Babo, R., S, K. (eds) Universal Threats in Expert Applications and Solutions. UNI-TEAS 2024. *Lecture Notes in Networks and Systems*, vol 1007, Springer, Singapore.

Ventuneac, C., & Gaitan, V. G. (2024). Industrial Internet of Things Gateway with OPC UA Based on Sitara AM335X with ModbusE Acquisition Cycle Performance Analysis. *Sensors (Basel)*, *24*(7), 2072. DOI: 10.3390/s24072072 PMID: 38610283

Verdugo-Cedeño, M., Jaiswal, S., Ojanen, V., Hannola, L., & Mikkola, A. (2023). Simulation-Based Digital Twins Enabling Smart Services for Machine Operations: An Industry 5.0 Approach. *International Journal of Human-Computer Interaction*, *40*(20), 6327–6343. DOI: 10.1080/10447318.2023.2254607

Victor, N., Maddikunta, P. K. R., Mary, D. R. K., Murugan, R., Chengoden, R., Gadekallu, T. R., Rakesh, N., Zhu, Y., & Paek, J. (2024). Remote Sensing for Agriculture in the Era of Industry 5.0—A Survey. *IEEE Journal of Selected Topics in Applied Earth Observations and Remote Sensing*, *17*, 5920–5945. DOI: 10.1109/JSTARS.2024.3370508

Voss, R. F., & Clarke, J. (1976). Flicker (1f) Noise: Equilibrium Temperature and Resistance Fluctuations. *Physical Review. B, Solid State*, *13*(2), 556–573. DOI: 10.1103/PhysRevB.13.556

Waheeb, W., Ghazali, R., & Herawan, T. (2016). Ridge polynomial neural network with error feedback for time series forecasting. *PLoS One*, *11*(12), e0167248. DOI: 10.1371/journal.pone.0167248 PMID: 27959927

Wang, Z., Zhu, H., Jia, X., Bao, Y., Wang, C. (2022). Surface Defect Detection with Modified Real-Time Detector YOLOv3. Journal of Sensors, Vol. 2022, No. 1, Article ID 8668149, 10 pages, doi:DOI: 10.1155/2022/8668149

Wang, F. (2010). The emergence of intelligent Enterprises: From CPS to CPSS. *IEEE Intelligent Systems*, *25*(4), 85–88. DOI: 10.1109/MIS.2010.104

Wang, P., Yang, L. T., Li, J., Chen, J., & Hu, S. (2018). Data fusion in cyber-physical-social systems: State-of-the-art and perspectives. *Information Fusion*, *51*, 42–57. DOI: 10.1016/j.inffus.2018.11.002

Wang, Y.. (2025). Federated learning for distributed network error prediction. *Future Generation Computer Systems*, *152*, 342–355.

Wang, Y., Zou, M., Peng, L., Chen, Z., & Zhang, L. (2023). Adaptive High-Precision Positioning Method Based On BLE RSSI/INS/ODO Tightly-Coupled Integrated Navigation System. *2023 China Automation Congress (CAC)*, 1881–1886. DOI: 10.1109/CAC59555.2023.10452079

Warnakulasuriya, H. N., Dorji, T., & Horanont, T. (2024). Indoor Positioning Based on Time of Flight and Kalman Filtering Using Ultra-Wideband Sensors. *2024 IEEE International Conference on Big Data and Smart Computing (BigComp)*, 117–123. DOI: 10.1109/BigComp60711.2024.00027

Warr, W. A. (2012). Scientific workflow systems: Pipeline Pilot and KNIME. *Journal of Computer-Aided Molecular Design*, *26*(7), 801–804. DOI: 10.1007/s10822-012-9577-7 PMID: 22644661

Watanabe, M., Obo, T., & Kubota, N. (2025, July). Topological mapping based on multi-layer growing neural gas for topological twin. In *2025 international joint conference on neural networks (IJCNN)* (pp. 1-8). IEEE.

Wei, M., Chen, M., & Zhou, D. (2013). Multi-sensor Monitoring Information Based Decision Support Method for Optimal Predictive Maintenance Policy, *25th Chinese Control and Decision Conference (CCDC)*, 4864-4868. DOI: 10.1109/CCDC.2013.6561815

Werth, A. W., & Morris, T. H. (2023). A Digital Twin Internal to a PLC to Detect Malicious Commands and Ladder Logic that Potentially Cause Safety Violations. *Journal of Cyber Security Technology*, *7*(2), 53–82. DOI: 10.1080/23742917.2023.2171538

Woo, J., Kubota, N., & Lee, B. (2010). Steady-State genetic algorithms for growing topological mapping and localization. In *Pacific Rim International Conference on Artificial Intelligence (PRICAI 2010),* (pp. 558–569). DOI: 10.1007/978-3-642-15246-7_51

Wu, C., Wang, Y., Ke, W., & Yang, X. (2024). A Dual-Branch Convolutional Neural Network-Based Bluetooth Low Energy Indoor Positioning Algorithm by Fusing Received Signal Strength with Angle of Arrival. *Mathematics*, *12*(17), 2658. DOI: 10.3390/math12172658

Wu, C., Wong, I.-C., Wang, Y., Ke, W., & Yang, X. (2024). Experimental Study of Bluetooth Indoor Positioning Using RSS and Deep Learning Algorithms. *Mathematics*, *12*(9), 1386. DOI: 10.3390/math12091386

Wu, R., Zhang, K., Xu, Y., & Zhou, M. (2023). *Application of Optocoupler Isolation in Rotation Speed Measuring Circuit of Maglev Blower. Lecture Notes in Electrical Engineering*. Springer Nature.

Xiao, K., Hao, F., Zhang, W., Li, N., & Wang, Y. (2024). Research and Implementation of Indoor Positioning Algorithm Based on Bluetooth 5.1 AOA and AOD. *Sensors (Basel), 24*(14), 4579. DOI: 10.3390/s24144579 PMID: 39065976

Xu, H., Wu, J., Pan, Q., Guan, X., & Guizani, M. (2023). A Survey on Digital Twin for Industrial Internet of Things: Applications, Technologies and Tools. *IEEE Communications Surveys and Tutorials*, *25*(4), 2569–2598. DOI: 10.1109/COMST.2023.3297395

Xu, M., Luo, Z., Liu, B., Wang, Z., Tang, X., & Song, T. (2025). Automatic Landmark Detection in Severe Craniomaxillofacial Deformities via Three-Dimensional (3D) Point Cloud Deformation Model and Deep Learning Networks. *IEEE Access : Practical Innovations, Open Solutions*, *13*, 32724–32740. DOI: 10.1109/ACCESS.2025.3541601

Yamashita, H., & Kobayashi, H. (2014). Automatic Identification of Carbon Steel on Spark Analysis -Automation of Spark Examination-. Procs. of the 2014 JSME Conference on Robotics and Mechatronics, 2A1-V03, 3 pages (in Japanese).

Yanan, S., Hui, Z., Li, L., & Hang, Z. (2018). Rail Surface Defect Detection Method Based on YOLOv3 Deep Learning Networks. Procs. of 2018 Chinese Automation Congress (CAC), pp. 1563–1568. DOI: 10.1109/CAC.2018.8623082

Yashika, A., Patra, N. S., & Das, (2024) Human-Machine Interaction in Industrial Automation: Gesture-Based PLC Control. *Proceeding of 15th International Conference on Computing Communication and Networking Technologies (ICCCNT)*, Kamand, India, 2024, pp. 1-7, DOI: 10.1109/ICCCNT61001.2024.10726180

Ye, J., Chen, B., Liu, Q., & Fang, Y. (2013). "A precision agriculture management system based on Internet of Things and WebGIS," 2013 21st International Conference on Geoinformatics, Kaifeng, China, 2013, pp. 1-5. DOI: 10.1109/Geoinformatics.2013.6626173

Ye, J., Liu, X., Madhusudanan, H., Wang, Y., Zhu, J., Wang, Y., Ru, C., Liu, X., & Sun, Y. (2025). Automatic Point Cloud Clustering for Surface Defect Diagnosis. *IEEE Transactions on Automation Science and Engineering*, *22*, 12538–12547. DOI: 10.1109/TASE.2025.3542076

Yin, R. (2021). Review on the study of metallurgical process engineering. *International Journal of Minerals Metallurgy and Materials*, *28*(8), 1253–1263. DOI: 10.1007/s12613-020-2220-z

Yoshioka, T., Ohba, Y., & Kobayashi, H. (2019). Development of a Steel Identification System by Spark Image Analysis Method [in Japanese]. *Sanyo Technical Report*, *26*(1), 23–27.

Yuan, L., Zhu, X., & Hong, K. (2020). Detection of Material Surface Cracks by Infrared Non-destructive Testing. Procs. of 2020 11th International Conference on Prognostics and System Health Management (PHM-2020 Jinan), Jinan, China, pp. 587–591, DOI: 10.1109/PHM-Jinan48558.2020.00114

Yu, Y., Xiao, M., Fieser, D., Zhou, W., & Hu, A. (2024). Nanoscale Memristor Devices: Materials, Fabrication, and Artificial Intelligence. *Journal of Materials Chemistry. C, Materials for Optical and Electronic Devices*, *12*(11), 3770–3810. DOI: 10.1039/D3TC04510B

Yu, Y., & Zhang, Y. (2021). Collision avoidance and path planning for industrial manipulator using slice-based heuristic fast marching tree. *Robotics and Computer-integrated Manufacturing*, *75*, 102289. DOI: 10.1016/j.rcim.2021.102289

Zafar, M. H., Langås, E. F., & Sanfilippo, F. (2024). Exploring the Synergies Between Collaborative Robotics, Digital Twins, Augmentation, and Industry 5.0 for Smart Manufacturing: A State-of-the-Art Review. *Robotics and Computer-integrated Manufacturing*, *89*(102769), 102769. DOI: 10.1016/j.rcim.2024.102769

Zhang, G., Cui, K., Hung, T. Y., & Lu, S. (2021). Defect-GAN: High-Fidelity Defect Synthesis for Automated Defect Inspection. Procs. of 2021 IEEE Winter Conference on Applications of Computer Vision (WACV), Waikoloa, HI, USA, pp. 2523–2533. DOI: 10.1109/WACV48630.2021.00257

Zhang, H., Wang, Q., Li, Z., Mi, J., & Zhang, K. (2023). Research on High Precision Positioning Method for Pedestrians in Indoor Complex Environments Based on UWB/IMU. *Remote Sensing (Basel)*, *15*(14), 3555. DOI: 10.3390/rs15143555

Zhang, L.. (2024). Cross-layer error correction in software-defined networks. *IEEE/ACM Transactions on Networking*, *32*(1), 512–525.

Zhao, H., Chen, Z., Shu, X., Shen, J., Liu, Y., & Zhang, Y. (2023). Multi-Step Ahead Voltage Prediction and Voltage Fault Diagnosis Based on Gated Recurrent Unit Neural Network and Incremental Training. *Energy*, *266*, 126496. DOI: 10.1016/j.energy.2022.126496

Zhao, S., Blaabjerg, F., & Wang, H. (2021, April). An Overview of Artificial Intelligence Applications for Power Electronics. *IEEE Transactions on Power Electronics*, *36*(4), 4633–4658. DOI: 10.1109/TPEL.2020.3024914

Zheng, H., Paiva, A. R., & Gurciullo, C. S. (2020). Advancing from Predictive Maintenance to Intelligent Maintenance with AI and IIoT. *arXiv preprint arXiv:2009.00351*. https://arxiv.org/abs/2009.00351

Zheng, Y., Li, S., Xiang, Y., & Zhu, Z. (2023). Crack Defect Detection Processing Algorithm and Method of MEMS Devices Based on Image Processing Technology. *IEEE Access : Practical Innovations, Open Solutions*, *11*, 126323–126334. DOI: 10.1109/ACCESS.2023.3331152

Zou, H., Chen, Z., Jiang, H., Xie, L., & Spanos, C. (2017). Accurate indoor localization and tracking using mobile phone inertial sensors, WiFi and iBeacon. *2017 IEEE International Symposium on Inertial Sensors and Systems (INERTIAL)*, 1–4. DOI: 10.1109/ISISS.2017.7935650

About the Contributors

Maki Habib, From his early beginnings at the University of Technology – Baghdad, where he fostered a deep interest in Control, Systems, and Computer EFrom his early beginnings at the University of Technology – Baghdad, where he fostered a deep interest in Control, Systems, and Computer Engineering, Professor Maki Habib's journey has been marked by a relentless pursuit of innovation in Control, Computer Systems, Mechatronics, and Robotics. Professor Maki Habib earned his B.Sc. in Control and Systems Engineering and Master of Science in Computer Engineering from the University of Technology - Baghdad, Iraq. He has earned his Doctor of Engineering Sciences in Intelligent and Autonomous Robotics from the University of Tsukuba, Japan. He has established himself as a specialist in Robotics and Mechatronics. His career began with roles as a research scientist at RIKEN, Japan, and as a senior researcher at RISO-Laboratories, Japan. His expertise was further recognized during his academic and professional career when he joined the team as a visiting researcher at EPFL, Switzerland. Professor Habib's influence extended to Malaysian industries as a visiting expert under the Asian Development Bank (ADB) and later as an Associate Professor at UTM, Malaysia, and Senior Manager at MCRIA, Malaysia. His leadership qualities were evident when he led the Telecooperation research group as a Senior Researcher and group leader at GMD-Japan in Japan. At Monash University (Australia/Malaysia), he joined as an Associate Professor. He played a central role in developing and leading the Monash Mechatronics Engineering program. Elevating his career further, Professor Habib was appointed Full Professor of Robotics and Mechatronics at Swinburne University (Australia/Malaysia) and subsequently at AUC as a Full Professor of Robotics and Mechatronics since September 2007. He has also enriched academic circles globally, serving as an invited Professor at the Korean Advanced Institute of Science and Technology KAIST, South Korea, and Resident Professor at Saga University, Japan. Besides, he was a visiting professor at another number of global Universities, such as Nagasaki University - in Japan; Orebro University – in Sweden; University of

Science and Technology – in Norway; Hokkaido University – in Japan; University of Duisburg-Essen – in Germany; International Islamic University, Malaysia (UIA). Malaysia, Also as a Guest Professor at the University of Science Malaysia (USM) – Malaysia, Beyond academia, Professor Habib has been a vital consultant and technical advisor to numerous industrial giants, including Toyota, Denso, and ABB. His work with more than one European research consortium has significantly contributed to advancements in Robotics, IoTs, CPS, Sensors for humanitarian demining, and sustainable agriculture. His notable projects include leading the development of advanced robots—control, sensor, and navigation technology for humanitarian demining. Also, the development of portable IoT kits facilitates the collection of environmental and agricultural data necessary to support the learning process and enable real-time decision-making to ensure sustainable agriculture. In the scholarly community, Professor Habib has made a substantial impact as editor-in-chief of "3" international journals, regional editor of "2" international journals, associate editor of "12" international journals, and an editorial board member of "20" international journals. His leadership roles have spanned numerous international conferences, where he has served as chair, program chair, co-chair, publication chair, and publicity chair. His influence is further underscored by his involvement every year in more than "50" annual international conferences as conference chair, program chair, and member of advisory and scientific program committees. Professor Habib's editorial contributions also include being the Chief Guest Editor for "18" special issues in international journals and organizing over "60" workshops and special sessions at international conferences, alongside organizing and implementing "3" industry-focused international workshops in Robotics and Automation. His collaborations span a global network, working with research groups and universities in Japan, South Korea, the US, Austria, the UK, Malaysia, and more. He has delivered over "60 invited talks at international conferences, universities, and the industry. Professor Habib's prolific writing includes the publication of "24" books, over "38" book chapters, more than "90" journal papers, and over "245" conference papers. His research interests are diverse and impactful, focusing on: 1. Nanorobotics systems for biomedical purposes. 2. Machine Learning in Cancer Detection, Diagnosis, and Analysis. 3. Human Adaptive and Friendly Mechatronics (HAFM). 4. EchoMechatronics and Bioinspired Design. 5. Service and Autonomous Robots, including Humanitarian Demining. 6. Robotics for the physically impaired using biological signals (EEG and EMG). 7. Electrical Autonomous Cars/Vehicles enhanced by Machine Learning techniques. 8. Pioneering in Industry 4.0/5.0, including IoT, Cyber-Physical Systems (CPSs), and WSNs. 9. Development of Intelligent Drone Systems: Structure, Control, Intelligence and Navigation. 10. Research in Telecooperation, Distributed Mixed Realities, and Collaborative Control. 11. Intelligent and Nonlinear Control Systems. Driven by a philosophy combining innovative scientific development with technological advancement for societal betterment, Professor Habib's approach to Robotics, Mechatronics, and Smart Systems is deeply rooted in creating solutions that enhance human lives and benefit social development.

***

Anita Devare has extensive experience in the Field of Applied Mathematics. Her interest lies in the computational mathematics and finding solutions to the engineering problems.

ManojDevare received BSc & MSc degree from North Maharashtra University and PhD in Computer Science from Bharati Vidyapeeth University, India. He is working as Associate Professor at Amity Institute of Information Technology, Amity University Mumbai. He has 17 years of teaching and research experience in Computer Science. His PhD thesis was based on the Congestion Controlling in Homogeneous and Heterogeneous Computer Networks. Dr. Manoj won the "Best Paper Award", in ICSCI 2008. He has served as Post Doctorate Fellow at Centre of Excellence on HPC, Uni. of Calabria, Italy under young researcher scheme. He has been developed Desktop Cloud system. His research deals with IoT, Cloud, Machine Learning and innovative applications of those technologies to scientific problems. He has 02 Patents pending; he edited 05 book chapters, 09 Int. Journal papers, 15 conference Papers. He has delivered 08 national and international keynote speeches. He is connected with review process of reputed international journal papers, edited chapters from the reputed journals like Elsevier's FGCS, Springer's review system, Symbiosis International Journal.

Dankan Gowda V is currently working as an Assistant Professor in the Department of Electronics and Communication Engineering at BMS Institute of Technology and Management in Bangalore. Previously, he worked as a Research Fellow at ADA DRDO and as a Software Engineer at Robert Bosch. With a total experience of 14 years, including teaching and industry, Dr. Gowda has made significant contributions to both academia and research. He has published over 170 research papers in renowned international journals and conferences. In recognition of his innovative work, Dr. Gowda has been granted six patents, including four from Indian authorities and two international patents. His research interests primarily lie in the fields of IoT and Signal Processing, where he has conducted workshops and handled industry projects. Dr. Gowda is passionate about teaching and strives to create an engaging and effective learning environment for his students. He consistently seeks opportunities to enhance his knowledge and stay updated with the latest advancements in his field.

Priya Jadhav, Assistant Professor with more than 12 years of extensive teaching and research experience in the field of engineering, I have developed a specialized knowledge base in Material Science, Manufacturing, and Robotic Process Automation. This is demonstrated by my authorship of several articles in high-impact journals, which attests to my expertise and commitment to advancing the field through rigorous academic research. My research interests include exploring the intersection of engineering and technology to drive innovation and create solutions to real-world problems. I am dedicated to staying current with the latest advancements in the field, and I continuously seek to integrate these developments into my teaching and research. Overall, my diverse background in teaching and research, combined with my passion for inspiring and challenging students, makes me a valuable asset to any academic institution or organization seeking to advance the field of engineering.

Sergey Jatsun, Scientific degree: Doctor of technical sciences Academic title: Professor In 1979 Jatsun Sergey has received PhD degree in Mishkole technical University in Hungary. In 1990 received degree Doctor science in a field of Dynamics and durability of machines devices and apparatus in Riga, Latvia. In 1990 he defended his doctoral dissertation: "Theory and Synthesis of Vibratory, Technological Processes and Vibromachines for the Processing of Dry Materials". He is author of more than 250 publications and has more than 100 patents on inventions. He is a member of the International Academy of Ecology and Safety and was awarded the token "High Achiever of Invention and Rationalization" and "USSR's Inventor". He has the silver and golden medal of exhibition of achievements of national economy of USSR and Russian. He awarded the title of Honored worker of science of the Russian Federation. Jatsun S.F. is a member of the editorial council of the journals: "Kursk State University News" and "Mechanism and Machine Theory" (Germany), "Izvestiya SWSU". He has prepared more than 40 PhD students.

Goran Klepac, Ph.D., works as a Director of Big data analytics in Croatian Telecom (T-com). Previously he worked as director of Stretegic development in in Raiffeisenbank Austria d.d., Croatia. In several universities in Croatia, he lectures subjects in domain of data mining, predictive analyitics, decision support system, banking risk, risk evaluation models, expert system, database marketing and business intelligence. As a team leader, he successfully finished many data mining projects in different domanins like retail, finance, insurance, hospitality, telecommunications, and productions. He is an author/coauthor of several books published in Croatian and English in domain of data mining.

Naoyuki Kubota, He graduated from Osaka Kyoiku University in 1992, received the M.E. degree from Hokkaido University in 1994, and the D.E. degree from Nagoya University, Japan in 1997. He joined Osaka Institute of Technology in 1997. He joined the Department of Human and Artificial Intelligence Systems, Fukui University as an associate professor in 2000. He joined the Department of Mechanical Engineering, Tokyo Metropolitan University in 2004. He was an associate professor from 2005 to 2012, and has been a professor since 2012 at the Department of System Design in Tokyo Metropolitan University, Japan.

N.M.G. Kumar is currently working as a professor at the Dept. of EEE, School of Engineering, Mohan Babu University (Erstwhile Sree Vidyanikethan Engineering College), Tirupati, AP, India, and obtaining his B.E. in Electrical and Electronics Engineering from Bangalore University at SMVITS, Bangalore. Obtained M.Tech (PSOC) and Ph.D. at S.V.U. College of Engineering, S.V. University, Tirupati. He has 25 years of experience in teaching and research. He published 25 international journals (50 national and international conferences), four textbooks, 6 book chapters, and 6 patents. He is an IEEE member and a member of IAENG. His areas of interest are power system analysis, planning, security, state estimation, FACTS devices' application in transmission and distribution systems, power system reliability studies, real-time application of power systems, and non-linear controllers' applications to power systems.

Alessandro Massaro (ING/INF/01, FIS/01, FIS/03) carried out scientific research at the Polytechnic University of Marche, at CNR, and at Italian Institute of Technology (IIT) as Team Leader by activating laboratories for nanocomposite sensors for industrial robotics. He is in MIUR register as scientific expert in competitive Industrial Research and social development, and he was and scientific director of MIUR Research Institute Dyrecta Lab Srl. Member of the International Scientific Committee of Measurers IMEKO, IEEE Senior member, and IEEE TC34 Technical Committee Secretary, received an award from the National Council of Engineers as Best Engineer of Italy 2018. He was Research and Development Manager of LUM Enterprise in 2022 and 2023. He is in the World Top 2% Scientist list of Stanford University. Actually he is Associate Professor in Electronics at Department of Engineering of LUM University of Casamassima (BA, Italy).

Fusaomi Nagata, He received the B.E. degree from the Department of Electronic Engineering at Kyushu Institute of Technology in 1985, and the D.E. degree from the Faculty of Engineering Systems and Technology at Saga University in 1999. He was a research engineer with Kyushu Matsushita Electric Co. from 1985 to 1988, and a special researcher with Fukuoka Industrial Technology Center from 1988 to 2006. He is currently Professor at the Department of Mechanical Engineering, Faculty of Engineering, Tokyo University of Science, Yamaguchi, Japan.

Takenori Obo, He received the Ph.D. degree from the Department of Human Mechatronics Systems, Graduate School of System Design, Tokyo Metropolitan University, in 2014. In April 2014, he became a specially appointed Assistant Professor with the Human Mechatronics Systems Course, Faculty of System Design, Tokyo Metropolitan University. In October 2015, he was specially appointed as an Assistant Professor with the Intelligent Mechanical Systems Course, Faculty of System Design, Tokyo Metropolitan University. Since April 2017, he has been an Assistant Professor with the Department of Computer Applications, Faculty of Engineering, Tokyo Polytechnic University. In April 2022, he became an Assistant Professor with the Department of Mechanical Systems Engineering, Faculty of System Design, Tokyo Metropolitan University. He is currently with the Department of Mechanical Systems Engineering, Faculty of System Design, Tokyo Metropolitan University. He is a member of the Society of Instrument and Control Engineers, the Japan Society of Intelligent Information Fuzzy, the Robotics Society of Japan, and the Institute of Systems, Control and Information Engineers.

S.V. Ramanan is an Associate Professor at Department of Electronics and Communication you at PPG Institute of Technology

Jamil Safarov is a Professor at Azerbaijan Technological University and Director of Ganja Aluminium Complex.

Chyan Zheng Siow, He received the master's degree in computer science from the University of Malaya, Kuala Lumpur, Malaysia, in 2019, and the Ph.D. degree from Tokyo Metropolitan University, Tokyo, Japan, in 2024. He is currently a Research Assistant with the Department of Systems Design, Tokyo Metropolitan University. His current research interests include topological intelligence, fuzzy reasoning, robot navigation, lifelong machine learning, and multimodal learning. He is particularly active in time series forecasting and received the Japan FAN Outstanding Paper Award in 2023. In recognition of his research, he was honored with the Excellent Graduate School Research Award by the Japan Society of Automotive Engineers (JSAE), in 2025.

C.V. Suresh Babu is a pioneer in content development. A true entrepreneur, he founded Anniyappa Publications, a company that is highly active in publishing books related to Computer Science and Management. Dr. C.V. Suresh Babu has also ventured into SB Institute, a center for knowledge transfer. He holds a Ph.D. in Engineering Education from the National Institute of Technical Teachers Training & Research in Chennai, along with seven master's degrees in various disciplines such as Engineering, Computer Applications, Management, Commerce, Economics, Psychology, Law, and Education. Additionally, he has UGC-NET/SET qualifications in the fields of Computer Science, Management, Commerce, and Education. Currently, Dr. C.V. Suresh Babu is a Professor in the Department of Information Technology at the School of Computing Science, Hindustan Institute of Technology and Science (Hindustan University) in Padur, Chennai, Tamil Nadu, India. For more information, you can visit his personal blog at https://sites.google .com/view/cvsureshbabu/.

Egor Tistsov is a postgraduate student of the Department of Mechanics, Mechatronics and Robotics of the South-West State University, Junior Researcher.

Keigo Watanabe is an associate professor at Graduate School of Engineering, Sanyo-Onoda City University.

Fumiya Yagi is a second year graduate student. His research interest includes FCN, U-Net and YOLO for defect detection.

Index

Symbols

A

B

C

D

E

F

G

H

I

T

U

V

Y

www.ingramcontent.com/pod-product-compliance
Lightning Source LLC
LaVergne TN
LVHW061547290926
849162LV00004B/934